CASCADIA

McGRAW-HILL BOOK COMPANY

NEW YORK SAN FRANCISCO ST. LOUIS
DÜSSELDORF JOHANNESBURG KUALA LUMPUR
LONDON MEXICO MONTREAL
NEW DELHI PANAMA RIO DE JANEIRO
SINGAPORE SYDNEY TORONTO

BATES McKEE

Associate Professor
Department of Geological Sciences
University of Washington

CASCADIA

The Geologic Evolution of the Pacific Northwest

CASCADIA

The Geologic Evolution of the Pacific Northwest

Printed in the United States of America.

Library of Congress catalog card number: 74-169022

1234567890 VHVH 798765432

This book was set in Zenith by York Graphic Services, Inc., and printed on permanent paper and bound by Von Hoffmann Press, Inc. The designer was Janet Bollow; the drawings were done by Basil Wood. The editors were Jack L. Farnsworth and Michael A. Ungersma. Charles A. Goehring supervised production.

To the late

J. HOOVER MACKIN

colleague and friend

PART I GEOLOGIC PRINCIPLES

contents

PART II GEOLOGIC HISTORY OF THE NORTHWEST

list of illustrations

cornerstones of geology

the evidence—rocks and the landscape

the imperfect art of estimating geologic time

why the earth deforms

cordilleran landscapes

the columbia and rocky mountains

the north cascades

central and western british columbia

the san juan islands

the klamath mountains of oregon

the coast range province

the cascade range of oregon and southern washington

the cascade volcanoes

the blue mountains province

southeastern oregon—a volcanic highland

snake river country

the columbia plateau

the puget-willamette lowland

rocks as crystal balls

appendixes

preface

■ This book provides an introduction to the evolution of the Northwest landscape. The author of a book on the geology of the Northwest has two very significant factors working in his favor. One is the exceptional nature of the subject itself, for this is not only a beautiful region but also one that has had a particularly interesting geologic history. This factor explains in part the second favorable factor—the very high interest of most residents of or visitors to the Northwest in the scene around them. An appreciation of the geology of any region adds considerably to the pleasure of even a very brief visit.

The text is aimed at the student or interested layman who does not necessarily have any great knowledge of geology or the Northwest, though professional geologists may also find the book a useful overview. Technical terms are used where appropriate, but they are explained in the text and defined in the Glossary at the end of the book.

Appendix A provides useful information on rock classification and a summary of geologic time terms. The other appendixes suggest field trips for those who wish to explore the Northwest by car, by boat, or on foot.

The reader may be surprised by the seeming uncertainties and debatable questions raised in the text, but the nature of geologic interpretation is such that the evidence available does not often lead to a single, simple explanation. On the contrary, professional geologists may be disappointed by the sometimes simplistic discussions of very complex problems. Hopefully the treatment herein strikes a reasonable balance between meaningful generalities and rigorous scientific honesty.

"Cascadia" seemed to be an appropriate title for several reasons. For one thing, although the Northwest includes a very much larger region than that encompassed by the Cascade Range, those mountains are one of the area's most notable features. For another, the great crustal unrest that has characterized the geologically recent history of western North America has been known for many years as the *Cascadian Orogeny.* (*Orogeny* is synonymous with "mountain building.") Finally, "Cascadia" was the name used early in this century to denote a mythical landmass that supposedly lay in the northeastern Pacific Ocean, immediately beyond the present edge of the continent. The erosion of this landmass was thought to be responsible for great volumes of sediment deposited hundreds of millions of years ago in seas that covered what is now Washington, Oregon, and British Columbia. Recent investigations of ocean basins have cast serious doubts on the possible existence of such ancient oceanic landmasses, but the reader of Chapter 4 will note that some of the newer theories of earth deformation are no less imaginative than the concept of Cascadia.

The author of any regional geology book owes a tremendous debt to his colleagues and his predecessors in the profession, for his personal familiarity with such a large area must necessarily be limited. In the Additional Reading and References, I have tried to include for each chapter the significant publications (especially those of a somewhat general nature) for each region, with the hope that geologists will accept this as proper credit for work other than my own. I would also like to thank my colleagues and the several private concerns and state and federal agencies that have supplied materials for many of the illustrations in the book. The Department of Geological Sciences, University of Washington, and the Geology Department, University of New England, Armidale, New South Wales, Australia, assisted me in numerous ways, and this assistance is gratefully acknowledged.

Finally I would like to thank my wife, Pamela, who has participated ably and cheerfully in virtually every aspect of this endeavor.

Mount Saint Helens, a young volcano in the Cascade Range of southern Washington. View is from the west. (Photograph courtesy of Jan Fardell.)

□ Geology seeks to interpret the earth around us as the end result of processes that are operating today and have operated in the past. The ultimate aim is to interpret the history of the earth. The historical record exists in the rocks and the landscape. Understanding this record demands both a time scale for reference and a sound knowledge of physical and chemical processes that operate within the earth and at its surface. For this reason, geologic studies should begin with a look at processes and the many important environments of the earth's surface—river valleys, deserts, mountains, glaciers, shorelines, and the depths of the oceans. The sediments, rocks, structures, and landforms developed in each environment relate to the processes operative there. Likewise, processes that act within the earth can be deduced from the study of rocks once buried far beneath the surface that have been exposed by uplift and erosion and also from earthquakes and other surface manifestations of subsurface events.

The first four chapters of this book establish the physical and temporal framework essential to geologic interpretation. The comparative brevity of the treatment is in recognition of the desire of most readers to get on with the subject at hand—the evolution of the Northwest scene. Many of the important principles are elucidated in the second part of the book, where they are discussed in the context of particular regions. Although this brief treatment of the principles of physical geology is not a proper substitute for one of the many excellent books on the subject, it should provide a background sufficient to enable the geologic initiate to progress satisfactorily through the second part of this book.

I GEOLOGIC PRINCIPLES

1 cornerstones of geology

■ Geologic interpretation is based on common sense. Some of the basic principles are so obvious that one wonders why they were ever formalized as "laws." Such emphasis suggests great, mysterious concepts arrived at by brilliant deduction after years of painstaking observation and experimentation. The answer is largely historical. What is accepted today as obvious may have sparked lively debate among our ancestors—a fact familiar to anyone with some knowledge of the history of science. Furthermore, scientists are notorious classifiers. Putting an important-sounding term on a simple concept or object adds scientific respectability. Unfortunately, it also promotes scientific jargon. This jargon may ease communication between specialists, but naturally it also creates a communication gap between scientist and layman. This chapter will look at a few geologic truths, all quite reasonable, quite elementary, and quite essential to geologic interpretation. You will feel that you knew most of it anyway. That is correct. It proves that you are prepared to read the history of the earth around you, fortified with your own powers of observation and reasoning.

PRINCIPLES OF STRATIGRAPHY

Stratigraphy is the study of *strata*—layers of rock or sediment that form at the earth's surface. This includes a great variety of materials, for it embraces all of the so-called sedimentary rocks (Chapter 2), plus all volcanic lava flows and ashfall deposits.

Consider the deposition of sediment in a lake. It accumulates in layers (*beds*) on the lake floor. Each bed is younger than the one under it and older than the one above. The principle is termed *superposition.* It is the most fundamental concept in stratigraphy. All sequences of strata increase in age downward, if the beds were formed at the earth's surface and if the entire sequence has not been turned upside down by earth deformation. The floor of a house must be laid before the carpets can be installed. Just so the "floor" beneath a bed of sediment or lava must predate its covering.

Other principles of stratigraphy might be deduced from these lake beds. For one thing, the bedding (stratification, layering) is horizontal or nearly so when the layers first form, simply because loose sediment cannot rest on steeply inclined surfaces. If you think about loose sediment in different environments, you are hard pressed to

1-1 *Lake beds along the north shore of the Columbia River east of Grand Coulee Dam. The sediment accumulated behind a natural dam formed by a glacier approximately 15,000 years ago. By the superposition principle, the beds toward the bottom are older than those above. (Photograph courtesy of John Whetten.)*

recall places where the angle of a slope of loose material is greater than 30° or 35°—a phenomenon known as the angle of repose of loose material in air. The significance of the near-horizontality of strata will become evident when we look at the architecture of the Northwest mountain ranges, for there we find many places where the stratification is nearly vertical. We conclude that folding or some other deformation of the layers occurred *after* they were deposited. Much of our interpretation of the deformational history of the earth is based upon the distortion or disruption of strata.

One other fundamental principle follows directly from superposition—the *Law of Faunal Succession.* If you accept the concept of organic evolution (and if you don't, you should stop reading right here and give this book as a gift before it gets shopworn), you can imagine that the rock layers near the bottom of a thick succession of strata might contain *fossils* (evidence of former life) that are somewhat different from those near the top. Certain beds contain fossils that are distinctive for the age in which the strata accumulated. By collecting and comparing fossils at different levels, you document evolutionary changes (that is, the succession) that took place in the animals (that is, the fauna) of the past. Of course, the same principles apply to fossil plants, but no one has proposed a "Law of Floral Succession." If strata of a certain age contain certain characteristic fossils, then it follows that beds that contain the same fossils are of approximately the same age; and, once the succession has been worked out, rocks can be dated by the fossils that they contain. There are other ways of determining the relative or absolute ages of rocks, but the use of fossils is one of the most traditional and remains one of the most useful.

1-2 *A dacite lava intrusion (dike) cutting glacial lake sediments on the flank of Mount Edziza, northern British Columbia. By the intrusion principle, the dike postdates the lake beds. (Photograph courtesy of J. G. Souther, Geological Survey of Canada.)*

PRINCIPLES OF INTRUSION AND DEFORMATION

We restrict the application of the superposition principle to strata that accumulate at the surface, applying another common-sense principle to subsurface events. Not all molten rock reaches the surface as lava—some of it "intrudes" other rock

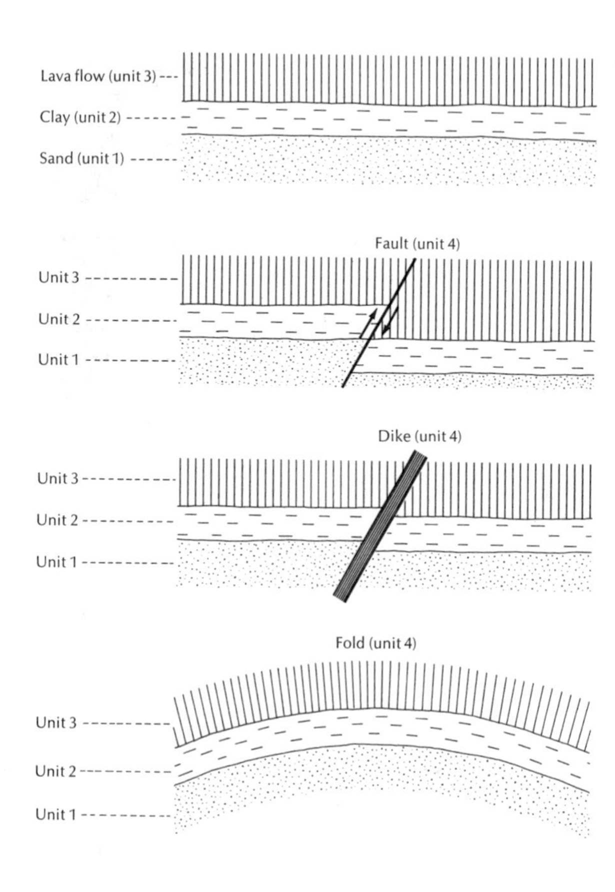

1-3 *Some basic principles of relative dating. By superposition, the sand (unit 1) is older than the clay (unit 2), which is older than the lava bed (unit 3). Folding, faulting, and intrusion are younger than all three units.*

underground. The rock melt, or *magma,* which is generated at depth, must of course intrude through overlying rock in order to reach the surface. We might refer then, to a principle of *intrusion,* which states that an intrusion is younger than the rock that is intruded (that is, the host or country rock). Again, not a very profound concept. However, the same reasoning applies to *faults* (fractures along which displacement has occurred) and to *folds* (bends in strata or other types of rock layers). In each case, deformation must postdate the formation of the rocks involved.

INTERPRETING THE LANDSCAPE

We have said that the ultimate aim of geology is to decipher the past history of the earth. The evidence is, of course, the rocks and the landforms that we have today, but on what basis can we deduce the past from the present? Suppose we look at a familiar feature like Mount Rainier. Almost everyone knows that this mountain is a volcano, but no one alive today has seen it erupt. The geologists say that it is a volcano, but most observers would probably surmise that Rainier is volcanic without the benefit of expert opinion. Why? Because of the mountain's shape. It looks like a volcano—like Mount Fujiyama, Mount Kilimanjaro, Mount Vesuvius, Mount Etna, and many other famous volcanic cones. The similarity does not end with the shape but includes the variety of rock and soil materials that make up Mount Rainier and these other volcanoes. To unravel the geologic history of Mount Rainier would involve a two-stage study. First, the mountain would be mapped in detail, and all the features of the rock materials and the landscape noted, using perhaps such principles as superposition to establish the sequence of events. But in order to understand the origin of certain rock types and landscape features, you would have to look elsewhere—specifically at other active volcanoes. There you would see the eruptive processes that produce the

1-4 *Mount Rainier from the east. Although deeply eroded by glaciation, the mountain retains the shape characteristic of a volcano. (Photograph courtesy of Washington State Department of Commerce and Economic Development.)*

different kinds of volcanic rock. The characteristics of each would be tied directly to the processes that formed it. The interpretive aspect is removed; it is all direct observation. Knowledge gained by this second stage, the study of volcanic processes, can then be applied to a mountain like Rainier in which the processes are no longer in action, or to any other volcano for that matter. This explains why geologists are so involved in the study of geologic processes—in rivers, landslides, glaciers, wind action, waves and currents in the oceans, or any of the other agents that have been important in the past history of the earth.

UNIFORMITARIANISM

The term *uniformitarianism* was coined early in the nineteenth century to encapsule the notion that the present is the key to understanding the past. By the middle of the century, it was the center of a great debate. The word itself is a bit unwieldy and sounds like a religious sect; furthermore, the controversy arose because of the religious fervor with which some of the adherents to the uniformitarian principle applied it. By studying processes we gain some idea of their rate; but how strictly can this information be applied to the past? If 1 inch of sand is deposited on the bottom of a particular river bed during a particular year, does this mean that sand is deposited in rivers at the rate of 1 inch per year? Of course not, for there are many factors that would influence the rate of sedimentation. Presumably, water has always moved downhill and has always been capable of eroding, transporting, and depositing sand. However, we cannot safely project a particular rate into the past, and this was one of the controversial issues in this great debate. The Uniformitarianists were gradualists. By noting the very slow rates of geologic processes and the complexity of the geologic record, they were impressed by the enormous amount of time implied by the total history. They had no means of directly measuring geologic time, but they believed that the evidence suggested hundreds of thousands or even millions of years. This belief was, and is, totally unacceptable to the Fundamentalists. They hold to a strict interpretation of the Bible, which they believe indicates a total age for the earth of approximately six thousand years. Such a short time period demands the creation of the geologic record by a series of short-lived cataclysmic events, and the opponents of the Uniformitarianists were accordingly dubbed "Catastrophists." The controversy generated more heat than light and encompassed the debate on evolution. Each side became more firmly entrenched in their beliefs, naturally going to great lengths to make their points convincing.

How do geologists view this subject today? Like most great debates, the final resting place for the pendulum of opinion is somewhere between the extremes. Geologists still accept a slow rate of development for most geologic features. However, they recognize that many events are in fact catastrophic, not only in terms of their effects on humanity but also in their immediate geologic effect. Major earthquakes and volcanic eruptions are familiar examples of short-lived events that produce major geologic change. Although these occurrences are common for the earth as a whole, for the resident of the affected area it may be a once-in-a-lifetime catastrophe. As a less spectacular example, geologists now realize that the amount of geologic work accomplished by a river during a few days of flooding may far exceed the results of years of normal flow. This must be understood in applying data from modern rivers to deposits of the past. Thus geologists are still Uniformitarianists at heart—they still believe in the value of studying modern processes to understand the ancient history of the earth—but they recognize the catastrophic nature of certain events and feel a natural caution about the strict application of modern rates to all past events. The Northwest has seen some geologic events that are difficult to imagine because nowhere on earth is anything quite comparable occurring today.

THE BALANCE OF NATURE

Perhaps the most important concept in geology—in fact, in all of science—is the idea of *equilibrium.* In its simplest terms, equilibrium implies a balance, but it is normally a dynamic and not a static balance. In other words, the apparent static nature of a particular situation really represents a balance of two or more forces that are capable of affecting a change. Many examples come to mind, but a few will suffice. You can dissolve a certain amount of sugar in a glass of ice tea; but you reach a point where no amount of stirring will bring about additional solution, and the extra sugar sinks to the bottom. This may appear to be a static situation, but at any moment a certain amount of sugar is, in fact, being dissolved. However, an equal amount is being precipitated, and so the equilibrium is really dynamic—solution and precipitation going on simultaneously at equal rates. The balance is very delicate. A slight change upsets the equilibrium and promotes that reaction that will restore equilibrium. Thus adding more water causes more sugar to dissolve until the solution is again saturated. Heating the solution causes the same effect. As long as there is excess sugar available, however, equilibrium will be restored. This example may seem remote, but geologists concerned with the deposition of chemical sedimentary rocks, such as limestone and gypsum in some ancient marine basin, are concerned with just such processes as heating and dilution and their effects on chemical equilibrium.

Concepts of organic evolution depend on equilibrium principles. A supposed change or adaptation of an organism to a change in the environment reflects the dynamic nature of the balance between life and its surroundings. Biologists speak of a "balance of nature" and are much concerned with the effects of Man on the natural fauna and flora of his environment. When we create disequilibrium (which we must because any balance is so precarious), certain changes that take place tend toward a new equilibrium. Balance and imbalance are of vital concern also to economists. Many of the "laws" of economics are simply predictions of what reactions will take place as an economy, upset in some manner, strives to regain equilibrium.

Finally, a geologic example. If we look at a river, we see how delicate the balance or equilibrium really is. (Perhaps you are tired of looking at rivers; but they are familiar to all, and geologically they are the most important agent at work on the earth's surface.) A river is capable of *eroding* (picking up), transporting, and depositing sediment, and at any one time a particular river is doing all these things. Yet geologists speak of mature rivers—rivers that are in equilibrium and in which the diverse processes are in reasonable balance over an extended period of time. Net erosion occurs during flood periods and gradual deposition, or sedimentation, at quieter times, but over the years these are fairly evenly balanced in a mature river. A river of a certain size flowing at a particular rate possesses a calculable amount of energy (energy of motion, or kinetic energy). Much of this energy is dissipated in friction and turbulence. What remains is available for geologic work—erosion and sediment transport. Thus as the energy is increased, the river is capable of doing more work.

1-5 *Cross-sectional view of a dam and reservoir, showing sedimentation in the reservoir and erosion below the dam.*

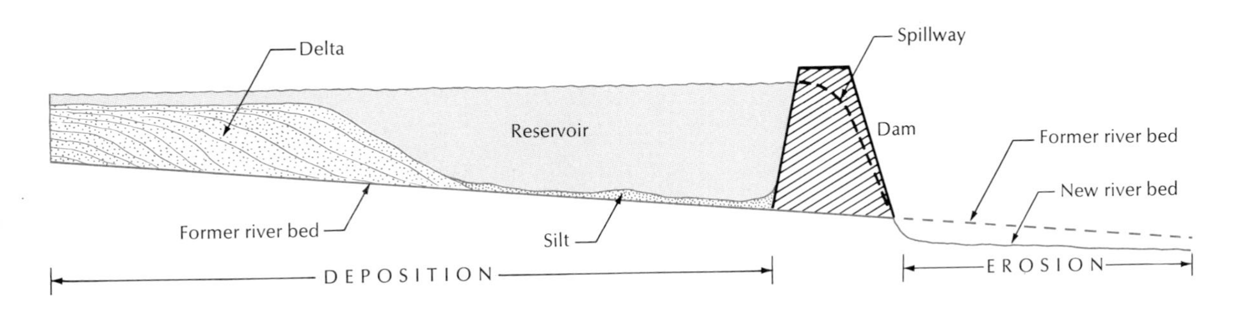

The river erodes, thus picking up and transporting more material, until the equilibrium is restored. Conversely, if the energy is decreased, deposition of sediment occurs. When Man builds a dam across a river the balance is upset. The results are soon very apparent. The velocity of the river is markedly reduced where it enters the newly formed lake. Sedimentation occurs at the head of the lake, and a delta forms. (The silting of reservoirs is a matter of serious concern in areas where the rivers normally carry a large load of sediment.) The water leaving the reservoir is normally relatively free of sediment, especially where it exits through turbines. (As sand and silt don't do much for this sort of equipment, engineers must design dams to keep sediment out of the turbines.) Immediately downstream from the dam, the now-clear water probably possesses the same amount of energy as the sediment-laden water did before dam construction. However, this energy is not being expended in sediment transport, and the river now possesses excess energy. The result is erosion—the river may rapidly cut a deeper channel below the dam. This was brought home by the construction of Hoover (Boulder) Dam forty years ago, because the downcutting below the dam "hung up" an irrigation canal system above the new equilibrium level of the Colorado River.

The point of all this is the delicate balance of nature, both at the present time and in the past. Change one condition, and all sorts of other changes will occur—changes that will tend to restore the system to an equilibrium condition. Many of these changes are predictable and can be verified by modern observations and experiments. In addition, by understanding the variables and how they interact, we can predict what happened in the past in response to known changes. Consider the rise of the Cascade Range. Scientists possess much information on the influence of mountain ranges on the climate of a region. In this case, the Cascades act as a barrier to prevailing onshore winds. The moist maritime air cools as it rises to pass over the range. This cooling causes condensation and precipitation, so that the western flank of the Cascades is an area of high annual fall of rain and snow. The air descending the eastern flank to the interior plateau regions is generally in an undersaturated condition. The contrast is quite striking—60 inches or more of precipitation per year at Snoqualmie Pass, 8 inches at Ellensburg. (This *rain shadow* effect is even more pronounced in the Olympic Mountains, where as much as 200 inches of rain and snow fall on the west side in contrast to 10 inches per year at Sequim, immediately northeast of the range.)

What has this to do with geologic interpretation? The geologic processes that operate in a region are influenced quite markedly by climatic conditions. Sediments that accumulate under arid conditions have some characteristics different from those of moister regions. Fossils are particularly useful, for the climate has a pronounced effect on the fauna and flora of a region. By careful study of the stratigraphy of eastern Washington and eastern Oregon, geologists have been able to date more precisely the actual rise of the Cascades, for this event produced gradual climatic changes, which are faithfully recorded in the geologic record. Many similar examples could be cited to illustrate how an understanding of the many variables that influence present-day processes leads to important conclusions about the past history of the earth.

2 the evidence: rocks and the landscape

■ It is not my intention to present herein any sort of complete account of the formation of rocks and the evolution of the landscape, for the reader may find very adequate and systematic discussions of these subjects elsewhere. Consequently, there follows a brief overview of rocks and processes for the benefit of those who have no familiarity with physical geology. The more advanced readers are invited to skip on.

THE ORIGIN AND CLASSIFICATION OF ROCKS

Geologists define three types of rocks—sedimentary, igneous, and metamorphic (see Appendix A). All *sedimentary* rocks originate at the earth's surface, either as solid particle material (*sediment*) or as chemical precipitates from water (such as limestone, gypsum, and salt). The solid particles are classified according to decreasing size into the following types—gravel, sand, silt, and clay. Initially, these deposits are not *lithified* (that is, made into rock), but if they are buried under younger beds they may become cemented or compacted into rock. If so, the corresponding rock terms are *conglomerate, sandstone, siltstone,* and *claystone.* The term *shale* is applied to either siltstone or claystone that has well-developed cleavage. We might back up for a minute and define a *rock* as a solid substance that consists of one or more minerals and a *mineral* as a naturally occurring substance with a well-defined composition, internal *crystal structure* (that is, arrangement of constituent atoms), and more or less distinct physical properties.

Igneous rocks form from a natural rock melt or *magma.* When magma reaches the earth's surface it is termed *lava,* and rocks that form from lava are *volcanic rocks.* The other type of igneous rock results from the cooling and crystallizing of magma below the ground. It is known as *intrusive igneous rock.* Granite is perhaps the best-known representative of this group. (Why is it that all writers of Westerns pick on either sandstone or granite for their scenarios? Perhaps it does sound better than a "nepheline-bearing leucotrondjhemite.") Igneous rocks are classified according to two parameters—grain size and mineral composition. Coarse-grained rocks have large crystals and generally form deep underground as intrusions. The large size of the

2-1 *Sandstone and siltstone beds in Hells Canyon on the Idaho side of the Snake River. The bedding planes and much of the original texture of the sediment are well preserved, even though the strata were deposited 200 million years ago. The many irregular cracks and fractures in the outcrop formed as the result of stresses applied after the strata were lithified. They are known as joints.* (*Photograph by Bates McKee.*)

grains reflects the slow rate of cooling that occurs at depth within the earth, since the long time available for crystallization allows large crystals to grow. Conversely, volcanic rocks crystallize relatively rapidly and have fine-grained or even glassy texture. Thus, in a sense, the classification based on texture is also genetic, for coarse-grained rocks generally crystallize in deep intrusions while fine-grained rocks form at or near the surface.

Metamorphic rocks are the most difficult type to understand. All of them form deep within the earth where the processes of metamorphism cannot be observed directly. Conditions at depth are different, of course, than they are at the surface—most notably higher temperatures and pressures. Rocks that may be relatively stable at or near the surface are not in equilibrium with the prevailing conditions deep underground, and reactions occur to restore the equilibrium. Most commonly this results in the formation of new minerals, for example the recrystallization of clay to mica. ("Clay" refers both to a size of sediment and to a mineral group.) Sometimes the only reaction that occurs during metamorphism is the enlargement of preexisting grains, as in the metamorphic recrystallization of limestone to marble. Perhaps the concept of minerals crystallizing at depth is not too difficult, but remember that the minerals are not grow-

2-2 *Wrinkled top of a congealed lava flow, Craters of the Moon National Monument, Idaho. Rippled pattern indicates that the flow moved from left to right. Such features are often retained in the stratigraphic record and are important in interpreting ancient volcanic rocks. (Photograph by Bates McKee.)*

2-3 *Metamorphic gneiss in the Coast Mountains 10 miles west of Kemano, British Columbia. The interpretation of such complex rocks that have deformed and recrystallized deep underground is difficult because the processes cannot be observed directly. (Photograph courtesy of J. A. Roddick, Geological Survey of Canada.)*

ing in a melt. (If they were, the resulting rock would be, by definition, igneous.) Thus *metamorphism* involves crystallization or recrystallization of minerals in a solid state, without extensive melting. The major implication of this is that the process must be a very gradual one. Once again, the concept of equilibrium is critical. Metamorphic reactions are driven by the disequilibrium between rocks and their environments—normally higher temperatures and higher pressures than those of the condition under which the original rocks formed. (In almost all instances, this deeper emplacement of rocks within the earth results from burial under younger strata.) If the various metamorphic reactions bring the rock to equilibrium with these new conditions, the character (mineralogy and texture) of the rock then reflects the conditions of metamorphism. Many millions of years later, when the rock may be exposed once again at the surface as a result of uplift and erosion, we are able to estimate the previous metamorphic environment.

PROCESSES THAT LEVEL THE LAND

How permanent are the "everlasting" hills? Viewed in the perspective of our life span, they seem unchanging, except, of course, for occasional large rock avalanches in the mountains. The evidence for gradual change is quite apparent to the trained eye, and various refined observational techniques have allowed some measurement of the rates of change. The rates seem very slow—inches per year or even inches per century for some important processes. Such barely perceptible rates might seem trivial, but given a time scale of millions and even billions of years' duration (Chapter 3), the time is there for dramatic change. The surface of the land is in many regions very mountainous. Geologists have estimated, however, that it could be reduced to a relatively low-relief surface in a matter of a few tens of millions of years at the present rates of erosion.

The processes that level the land are readily observable. Consider a high peak in the Cascade Range, such as the volcanic cone Mount Adams (Figure 2-4). From a distance, many of the Cascade volcanoes appear relatively smooth-sided and little

2-4 *Mount Adams, a volcano in the Cascade Range, southern Washington. View toward the west, with Mount Rainier in middle distance right and Olympic Mountains in far distance. See text for discussion. (Copyrighted photograph courtesy of Delano Photographics.)*

modified by erosion. But look more closely. The evidence of vigorous wearing away of the cone is quite dramatic. Canyons are carved in its flank, revealing the outward-dipping beds of lava and volcanic debris that originally formed a smooth surface. The larger canyons, high on the mountain, are occupied by glaciers. The glaciers are active, moving at a rate of a few feet per year. They do important geologic work. Certainly they pluck out and grind up the underlying rock and transport this debris downslope. But note how much rock debris mantles the surface of some of the glaciers. Where does it come from? It either falls or washes down onto the ice from the canyon sides above, then is carried along this slowly moving glacial conveyor belt. The glacier in the left foreground is relatively free of surface rock, in comparison with those to the right. The difference reflects the steeper, taller, and more actively eroding canyon walls rising above the latter glaciers.

If any doubt remains about the significance of glacial erosion in the modification of cones such as Mount Adams, contrast the north (right) and south (left) flanks. The glaciers are most numerous and largest on the northeast and north sides of the mountain, which are protected from the warm midday and afternoon sun. It is no coincidence that the glaciated part of the cone has the deep canyons, in marked contrast to the smoother profile of the south flank. The glaciers are not positioned by the canyons; they produce them. Moreover, Mount Adams is hardly unique, for all of the larger Cascade cones have a concentration of glaciers on their eastern and northern flanks.

Glaciers are important geologic agents on high mountains, but what about the large proportion of the earth's surface free from ice? Clearly, glacial action cannot be invoked there, at least at the present time. What other processes are significant? Consider the canyon of the Imnaha River in the northeastern corner of Oregon (Figure 2-5). There is no evidence that glaciers ever occupied this impressive canyon (although approximately 15,000 years ago they did occur locally on the highest peaks of the Blue Mountains in the distance). The most obvious geologic agent eroding this area is the Imnaha River and its numerous small tributary streams. But that is not a very impressive flow of water. Just how effective is running water in a moderately arid region like this? Look for yourself. The layers of rock exposed in the canyon walls are congealed lava flows of the formation known as the Columbia River Basalt. These flows erupted approximately 15 million years ago and covered much of eastern Washington, eastern Oregon, and western Idaho. Individual flows covered thousands of square miles, and the flows seemingly spread almost like water. In canyons such as the Imnaha, each lava flow can be traced continuously around the canyon walls, so we know that they once covered the area now occupied by the canyon. (The same conclusion would be reached by noting the flat plateau surface, underlain by the lava flows, in the middle distance. Clearly this surface has been dissected by canyon erosion—as if irregular pieces had been cut from a layer cake.) The tremendous volume of missing rock must have been carried off by the Imnaha River. The river had at most 15 million years to accomplish this. Even assuming a moister climate and hence a larger river in the past, this remains a most impressive testimonial to the efficiency of running water as a geologic agent. Fifteen million years is only $\frac{1}{3}$ of 1 percent of the probable age of the earth.

Is the Imnaha River the only geologic agent at work here? Certainly the river is responsible for transporting almost all the sediment out of the canyon; but it has not, on its own, covered every part of the canyon. The river is supplied with sediment by its many tributaries, but not all of the rock that arrives is carried by running water. Much of it—in fact most of it—falls, rolls, slides, or creeps into the Imnaha and its tributaries from the canyon sides. The controlling factor is gravity. The overall process of downslope movement of material by gravity is known as *mass-wasting*. The rate of movement may vary from the imperceptible creep of soil to the free-fall velocity of rock cascading from a cliff. Many separate mechanisms are recognised, but the detailed classification of mass-wasting processes need not concern us here. Suffice it to say that mass-wasting is a major geologic agent in any hilly or mountainous region.

2-5 *Imnaha River Canyon, northeastern Oregon. View toward the southwest, with the Blue Mountains in the distance. The canyon is eroded in the Columbia Plateau, which was built up by numerous outpourings of fluid basaltic lava. See text for discussion. (Copyrighted photograph courtesy of Delano Photographics.)*

But we have overlooked one vital step in the evolution of a deep canyon carved in a plateau. How do we transform solid rock into sediment available for transport? The breakdown of rock exposed to the atmosphere to form loose material is known as *weathering*. There are two distinct but related types of weathering. One involves the mechanical breakup of the rock, without significant chemical modification. This is referred to as *disintegration*. The second type involves various chemical changes brought about by the reaction between minerals and the atmosphere, surface water, or organisms. This is known as *decomposition*. Disintegration and decomposition work together to transform rock into loose debris or soil. The rate of transformation of the rock depends on very many factors (such as rock type, climate, slope angle, and types of organisms present), but no rock is immune. Weathering processes are the great providers of fragmental material at the earth's surface.

A closer look at the walls of the Imnaha Canyon (Figure 2-6) is instructive. Parts of individual basaltic lava flows crop out in the hillside as distinct

2-6 *Basaltic cliffs within the Imnaha River Canyon. The grassy slopes between the rock ledges are underlain by loose rock debris, slowly being moved by mass-wasting processes to the valley bottom. (Copyrighted photograph courtesy of Delano Photographics.)*

layers, separated by grassy stretches underlain by loose basaltic rubble. The loose rock is en route downhill, creeping perhaps to the lip of the next ledge, then falling to a lower slope. Paths of more actively sliding rock appear as dark stripes down the grassy slopes. The ledges are slowly eroding back, as frost action and other weathering processes dislodge the rock and make it available for mass-wasting transport. The breakup of basalt is not as formidable a task as one might suppose. Note that the rock ledges contain numerous cracks (*joints*). Many of these were developed during the cooling stages of the lava flow and are characteristic of basaltic layers. The prevalence of such joints in the Columbia River Basalt is more apparent in Figure 2-7. Blocks that have fallen off from above form an apron of rock along the base of the cliff known as *talus*. (The rounded basaltic pebbles in the trench walls in the foreground are flood-plain gravels deposited by the Palouse River.) Anyone who believes that canyon walls, such as those pictured here, are geologically stable environments, is invited to climb one.

To review, we have traced the transformation of solid rock to sediment in a stream or on a glacier through various processes of weathering and mass-wasting. We have used lava rock in our examples, but the general picture would have been the same had we started with granitic rock in the North Cascades or sedimentary strata from the Blue Mountains. All the movement of material has been downward, the ultimate source of energy being gravity. Running water and glacial action are two of the major geologic agents that are leveling the mountains of the Northwest, but of course there are others. Wind action is an effective force, especially where there exists an abundant supply of loose, fine-grained material. Today the most obvious regions in which to examine the transport of material by wind action are the areas of sand dunes along the coast (Chapter 11). In the recent past, parts of the Columbia Plateau east of the Cascades were also covered by drifting dunes (Figure 2-8).

2-7 *A bluff of Columbia River Basalt near the mouth of the Palouse River, southeastern Washington. See text for discussion. Excavation unearthed artifacts and fossil remains belonging to Marmes Man, who occupied the rock shelter as long ago as 10,000 years. (Photograph courtesy of Washington State Department of Commerce and Economic Development.)*

Coastal regions are other sites of vigorous geologic activity. The principal energy is provided by wave action and by currents generated by both wind and waves. Wave erosion produces some of the sediment along a coastline, but often much of it is river sediment, which, after being delivered to the oceans, is redistributed along the shoreline by currents.

Groundwater (subsurface water that saturates the cracks and cavities in rock) plays an important but unglamorous role in landscape evolution. Not all the products of rock weathering are removed as sediment; much is transported as material dissolved in underground or surface waters. Surprisingly large quantities of materials in solution reach the ocean from the land. Most is carried by rivers, whose dissolved load is supplied largely by groundwater springs and seeps. In regions underlain by soluble rock (such as limestone), the movement of groundwater can dissolve large underground caves and caverns. This process is relatively unimportant in the Northwest, where most of the underground flow is through pores and along joints and bedding planes in relatively insoluble strata. A significant proportion of the dissolved matter in groundwater is derived from processes of chemical weathering at the earth's surface.

2-8 *Sand dunes in The Potholes area south of Moses Lake, Washington. The sand was derived largely from the floodplain of the Columbia River and carried to the east (left) by the prevailing wind. The dunes are no longer actively migrating, having been stabilized by vegetation and surface water. (Photograph courtesy of David A. Rahm.)*

THE FATE OF SEDIMENT

Sediment moves downslope. The transporting agent can be any of a variety of media, as we have seen. The journey is likely to be interrupted by numerous pauses along the route. Equilibrium principles apply, so that the overall transport process is in reality a complex combination of erosion and deposition, constantly adjusting to changing energy conditions.

Early in this chapter, we discussed the erosive and transporting ability of glacial ice, but ultimately a glacier too must deposit its load of sediment. Consider Figure 2-9, a roadside exposure on the Columbia Plateau a few miles west of Grand Coulee Dam. The hammer rests on a hard rock surface that was smoothed and shaped by the passage of glacial ice approximately 15,000 years ago. Again, the rock here is a lava flow of the Columbia River Basalt. The ability of ice to smooth strata as hard and irregularly fractured as basalt is proof of the erosive power of a glacier. When ice melts away, its load must then be dropped. There it is, resting on the polished and striated surface over which the ice passed. Undoubtedly, the pebbles will move on eventually, for this part of the plateau is being eroded by streams tributary to the nearby Columbia River. It may take many thousands of years, but the fate of those stones is already sealed—they are destined to become part of the bed load of the Columbia River. Just as certainly, they will eventually come to the mouth of the river. (Actually, with all of the dams on the Columbia this is no longer a certainty. Small silt- and clay-sized particles abraded from the pebbles will make it through, but the pebble-sized material will have to wait for the dams to disappear, or at least for the reservoirs to fill completely with sediment.)

One might imagine that the edge of the Pacific Ocean would be the final resting place for Columbia River sediment, either in the estuary or along the coast of Washington and Oregon. Some is trapped here; but there is still a slope beyond the shoreline, and some sediment reaches the deep sea floor. (Sedimentation off the Oregon and Washington coastline is discussed in Chapter 5.) Given enough time, the many geologic agents at work could not only wear down the continents

2-9 *Glacial grooves and striations along Highway 174 west of Grand Coulee Dam, Washington. The basaltic bedrock was shaped by ice movement from right to left, parallel to the handle of the hammer. Overlying gravel was deposited by meltwater during the waning stages of glaciation. (Photograph by Bates McKee.)*

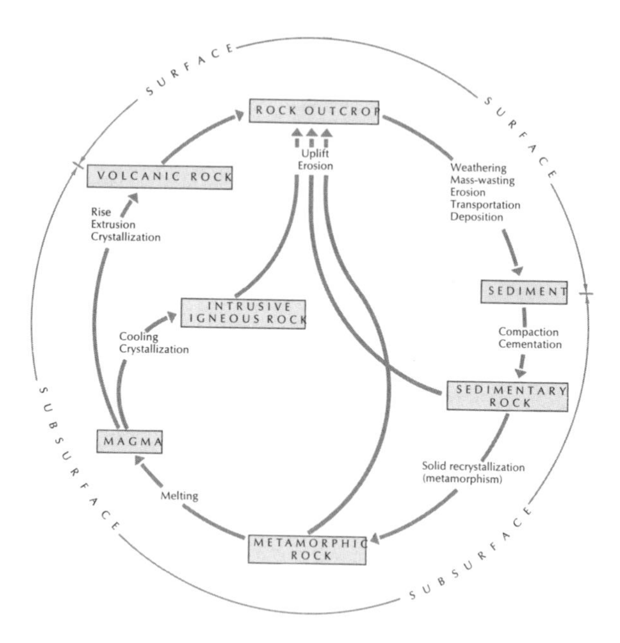

2-10 *The rock cycle, illustrating the relationships between geologic materials and processes.*

but also could partly fill in the oceans. We say "given enough time" these things would happen, but surely millions or even billions of years is theoretically long enough. Why do we still have mountains and ocean deeps?

DEFORMATION OF THE EARTH'S SURFACE

We retain mountains and ocean deeps because the earth's surface will not stay put. The surface is a compromise between all the erosive agents that tend to level it and all the internal forces that tend to deform it. The former are more easily studied and more readily understood. The geologic record demonstrates that mountains and oceanic trenches have probably always been present. The wearing down or filling in of one is balanced by (perhaps even causes) the formation of another. The surface of the earth rises, falls, bends, and breaks in response to stresses from within; and the smoothing effects of surface processes must constantly do battle with newly created surface relief.

Demonstrating, even measuring, the deformation of the earth's surface cannot answer our question about the causes of internal stresses. For the moment, accept the instability of the earth's crust, and the fact that rocks which form or sediments which accumulate at the surface can be buried and can sink to considerable depth before uplift and erosion expose them again to the air. The causes of earth deformation will be discussed in Chapter 4.

THE ROCK CYCLE

This chapter has introduced many of the processes whereby rock at the surface is broken down and transported, perhaps for great distances, before finding a permanent (or at least semipermanent) home. Geology is concerned, of course, with the creation as well as the destruction of rock. Sedimentary rock is made from sediment by compaction under the weight of younger sediment and by the precipitation of minerals (especially quartz and/or calcite) as cement between the grains. The process is a slow one. Normally, only sediments that remain buried for millions of years are converted into something solid enough to be considered as rock. What happens to sedimentary rock? Several possibilities exist. If the earth's surface is raised and erosion strips off the cover, weathering will convert the rock back to sediment. Conversely, it may be buried even deeper by the deposition of new strata above. If so, the temperature and the pressure may rise to the point where certain of the minerals in the sedimentary rock are no longer stable; and they will alter, very slowly, to more stable minerals. This would be the first stage of metamorphism—in the case of a marine shale, the first casualties would probably be the clay minerals. These would slowly recrystallize to form minute grains of mica. This recrystallization would cement the rock more firmly, turning it into a *slate.* With time, the mica grains would enlarge, and perhaps other new minerals would form, producing a *phyllite* or a *schist.* (See Appendix A.) Once again, uplift and erosion can interrupt these transformations at any stage. But what happens if the temperature and pressure of metamorphic rocks continue to increase and they are not returned to the surface? They start to melt. The melting is not uniform—the minerals with the lowest melting points melt first, followed slowly by the higher-temperature minerals. The end result is the generation of a magma (that is, a naturally occurring rock melt.) This magma may ultimately cool underground as an igneous intrusion, or it may rise to the surface and erupt as a lava to build a new volcano. Thus, in a sense, we have come full circle. We started with an exposed volcanic rock, and many millions of years later we have once again a volcanic rock. The entire sequence of events is embodied in the concept known as the rock cycle (Figure 2-10). The theory is useful, but in practice few materials ever travel the full circle. Most buried strata are returned to the surface by uplift and erosion from depths insufficient to produce melting. Most lava comes from the melting of rock that has not seen the stars since the time of the earth's beginning.

3 the imperfect art of estimating geologic time

■ The question of the age of the earth has been one of the most controversial issues of recent centuries, and quite naturally geologists have been central in the debate. We have mentioned how the uniformitarian approach of nineteenth-century geologists made them regard the earth as being very old indeed. Just how old they could not say, but they felt certain that the earth's age was some hundreds of thousands or even millions of years. As it has turned out, these pioneer thinkers were themselves conservative in their casual estimates; the now-accepted age of 4 to 5 billion years would have astonished them almost as much as their time scale offended the Catastrophists.

A number like 5 billion has little meaning, even in this age of gigantic numbers. Perhaps a translation to distance helps to understand the enormity of geologic time. Imagine some primeval (but immortal) creature capable of moving 1 inch per year. If he had been around at the start and was still moving today, he would have traveled a distance greater than three times the circumference of the earth at the equator. The concept of the vastness of geologic time is, perhaps, the most important contribution that the geologic profession has made to human culture.

The twentieth century has seen the development of techniques that allow us to use the natural radioactivity of certain rocks to determine the approximate time of their formation. Rocks nearly 4 billion years old have been dated by these means. One might imagine that direct methods of measuring time would make obsolete all of the previous means of estimating age, but these new "absolute" measurements are used more as a supplement to traditional methods than as a substitute. Geologists put more faith in the principles of superposition and faunal succession than they do in numbers that come out of a machine. If the laboratory results contradict the field evidence, the geologist assumes that there is something wrong with the machine date. To put it another way, "good" dates are those that agree with the field data.

THE DEVELOPMENT OF THE GEOLOGIC TIME SCALE

Like all histories, geology needs a time scale. The geologic time scale (Appendix A) was not formalized all at once. It evolved mainly during the late eighteenth century and the nineteenth century, with periods and epochs added as new evidence was unearthed. Most of the naming was done in Europe, although American geologists substituted the Mississippian and Pennsylvanian Periods for what the Europeans still call the Early and Late Carboniferous. In some instances, the names of the periods reflect localities where strata of that age are well exposed. Other names reflect certain characteristics of the beds themselves. The relative ages of the different strata were determined by superposition, with *correlation* (that is, age equivalence) from one region to another based primarily on similarity of fossils and similar physical characteristics of the strata. Major periods of deformation (involving folding, fracturing and metamorphism of rocks, intrusions, uplift, and extensive erosion) were recognized. They produced gaps in the faunal record and marked differences in the physical characteristics between the strata that predated the deformation and those beds deposited subsequently. These major events were termed "revolutions." Since it was presumed that these periods of deformation were of worldwide extent, they were used to separate major intervals of time known as *eras*. Subdivision of eras were known as *periods*, and these in turn were divided into *epochs*. The modifiers, "early," "middle," and "late," were applied in a time sense (for example, "during Early Permian time," or "Late Cretaceous dinosaurs"), while the adjectives, "lower," "middle," and "upper," were used in a rock position sense (for example, "Lower Cambrian sandstone" or "Middle Jurassic limestone"). Until the twentieth century, the time scale was entirely qualitative, not quantitative. It showed the relative age of strata but not the absolute age of anything. There was no way of discerning how old things were if they predated human history.

PAST ESTIMATES OF GEOLOGIC TIME

The Englishman Sir Charles Lyell, the foremost geologist of the first half of the nineteenth century, was instrumental in establishing the geologic time scale. His textbook, published in 1839, contained a long section that emphasized stratigraphic principles and their implications for interpreting earth history. He was a strong advocate of the uniformitarian approach. When he contemplated a section of stratified rock, such as that seen in Figure 3-1, it suggested to him a long period of sediment deposition. Many of the sections he studied contained considerable thicknesses of marine sedimentary rocks of a type that today accumulate very slowly. Furthermore, he recognized numerous breaks or periods of erosion (*unconformities*) in the stratigraphic record. These were indicated by changes in the types of strata deposited and in the fossils that they contained. He put a number on geologic

3-1 *Sedimentary strata of the John Day Formation at the Thomas Condon-John Day Fossil Beds State Park, central Oregon. The layers consist largely of reworked volcanic ash. Many centuries were required for the accumulation of so much ash—a conclusion born out by sedimentary changes in the fossil fauna. Hill in left background is capped by congealed lava of the Columbia River Basalt. (Photograph courtesy of Oregon State Highway Division.)*

time, estimating that the stratigraphic record demanded at least 200 million years for its evolution. Several decades later, Charles Darwin, considering the faunal succession preserved in the rock record, concurred with Lyell's calculation. In Victorian England, this figure was quite shocking. To many, it demonstrated the foolishness of the entire uniformitarian approach, especially to those not prepared to consider the evidence and follow the reasoning of men like Lyell and Darwin. Their estimate was, of course, little more than a guess, but soon other scientists were to attempt more formal calculations.

John Joly, an Irish physicist, attacked the problem by calculating the amount of salt in the oceans. This was not a very difficult estimate to make, since the approximate volume of the oceans was known and the salt content of seawater is relatively constant. He assumed that the original waters in the oceans had been fresh and that the salt has been added primarily by waters on land dissolving rock materials and carrying the salt to the sea in rivers. He then calculated the average salt content of major rivers and estimated the volume of river water entering the oceans, hence the total amount of salt added per year. By simply dividing the total salt in the oceans by the annual increment increase, he arrived at an age for the primeval ocean of approximately 100 million years. The method gives so "young" an age for a number of reasons. The most important are the removal of salt from the seas by a variety of processes and a probable much lower rate of influx of salt to the seas in ancient times.

Lord Kelvin, another eminent physicist, made calculations based on the rate of heat flow from the earth's interior. Even in his day, this rate could be measured with some success on land. (In recent years, more accurate determinations have been made using heat probes placed on the sea floor.) He assumed that the earth had formed as a molten or even gaseous sphere. Since its formation, the earth has been cooling by radiating heat to the atmosphere. If all the heat is original, the age of the earth can be approximated by calculating how long it would take for a molten sphere of the dimensions of the earth to cool to its present condition. The result was 80 million years. Lord Kelvin himself did not quite believe this figure, and he stated a more probable age of 20 million years.

RADIOMETRIC AGE DETERMINATIONS

What do we know that Lord Kelvin did not? We know that certain elements in the earth are *radioactive;* that is, that they are unstable and tend to disintegrate, forming new elements or lighter varieties (*isotopes*) of the original element. Various forms of energy are produced by these natural disintegration processes, including heat. Consequently, the discovery of radioactivity near the turn of the century cast doubt on Lord Kelvin's assumption that surface heat flow represents loss of original heat only. It also opened up a lively debate as to whether the earth is, in fact, cooling at all, with many scientists believing that this planet may have formed from cold cosmic dust. On this subject, I pass to more knowledgeable authorities, but at least it is clear that the natural radioactivity of rocks precludes any simple calculations of the age of the earth from heat-flow measurements. This loss was quickly overshadowed by a much greater gain, for as scientists studied radioactive processes they realized that radioactive disintegration provided, for the first time, a means of directly measuring a rock's age.

Direct or radiometric age determinations involve rather sophisticated techniques and some complicating factors that are beyond the scope of this book. Since these determinations have proven so important to our interpretations of Northwest geology, we might look briefly at the principles behind the method.

Imagine that a certain element, which we shall designate *X*, is incorporated in the crystal structure of a mineral that crystallizes in a granite intrusion. *X* is unstable, and in time it disintegrates and forms the stable element *Y*. Thus the older the granite, the less of the original *X* it contains and the more *Y*, the daughter product. Each type of radioactive element has a characteristic decay rate, which is measurable and which scientists believe has remained constant throughout geologic time. Thus the rate of decay of X would be the same regardless of the type of mineral or rock that served as the host. The decay rate is represented by the *half-life,* the time required for one-half of the element to disintegrate. If the half-life of element X is 1 million years and 1 cubic foot of our granite contained 1 ounce of X at the time of its crystallization, 1 million years later only $\frac{1}{2}$ ounce of X would remain, the rest having transformed into Y. After 2 million years only $\frac{1}{4}$ ounce would be left (one-half of the remaining $\frac{1}{2}$ ounce), and after 3 million years, $\frac{1}{8}$ ounce would be left. In other words, during the passage of a certain period of time, equivalent to the half-life of the element, one-half of the remainder disappears. This is accompanied by a corresponding increase in the daughter product, Y. (See Figure 3-2.)

How is this knowledge used to date rocks? What must be known are the half-life of the element in question and the precise proportions of the parent element to the daughter product. Thus in our simple example, if analysis showed that the proportion of X to Y was 1:32, an age for the granite of 5 million years would be indicated. Of course, the real world is a little more complicated than our example, but these few basic principles apply in all types of radiometric age determinations.

Would not direct measurements like these do away with the dating of rocks by fossils? Why do we bother with a traditional time scale any more—isn't it better to say that a limestone is 550 million years old than to say that it is of Cambrian age? As I suggested earlier, the geologist has more faith in the fossil evidence than in a machine date, and this reflects some of the uncertainties of radiometric determinations and the interpretation of results. Furthermore, even if a measured age is reasonably accurate, the determination is time consuming and costly (several hundred dollars per sample), and many rocks do not contain radioactive elements in sufficient quantity to permit direct dating. To be more specific, the elements with radioactive isotopes of use in radiometric dating are uranium, thorium, rubidium, potassium, and carbon. None of these is ideal. Uranium, thorium, and rubidium are not present in most rocks in sufficient concentration to permit accurate dating. Potassium is common, but the principal daughter product of its disintegration, argon, is a gas; and, because argon can escape by leakage from some minerals, the potassium-argon method sometimes gives spurious results. Radiocarbon dating is useful only for approximately the past 50,000 years. The

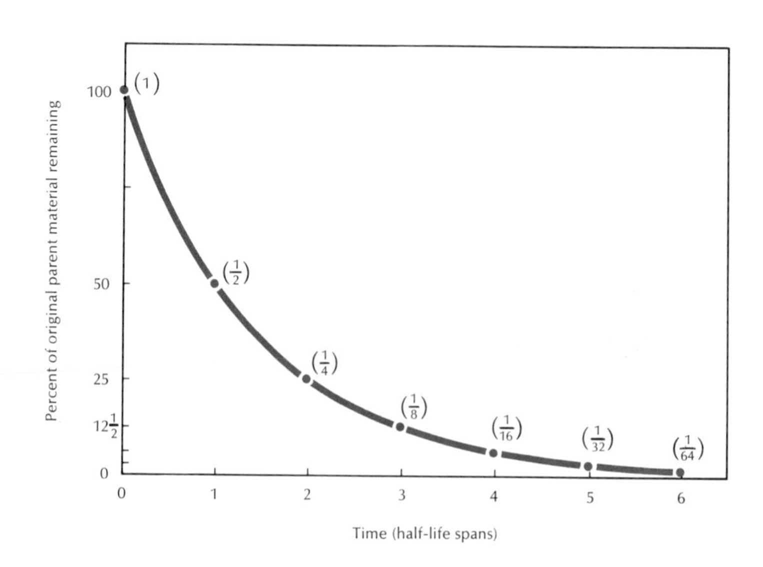

3-2 *The radioactive decay process. Vertical axis shows percentage of parent element remaining, horizontal axis represents half-life time. This curve is characteristic of all radioactive elements—only the actual time varies, for each element has its own half-life.*

problem here stems from the short half-life of radiocarbon and the initial low content of the radioactive isotope of carbon relative to common carbon. In samples older than 50,000 years, the amount of radiocarbon remaining is too small to measure. The other methods have the reverse of this problem—the half-lives of the radioactive elements are so long (billions of years) that for samples younger than a million years not enough daughter product has formed to provide an accurate ratio.

AN ILLUSTRATION OF DATING PROBLEMS

Consider the rock shown in Figure 3-3. The original material was deposited probably as mud in the ocean. The sediment was derived from erosion of the land and, in all likelihood, also from volcanic ash that settled to the sea floor. The mud was changed slowly to shale as it was buried and compacted under younger sediment. When this happened is not known, except that it preceded the Jurassic Period. Deeper burial resulted in metamorphic recrystallization, and the shale was transformed to phyllite (the Darrington Phyllite of the North Cascades of Washington; see Chapter 7). The phyllite was strongly deformed by folding and fracturing during the Cretaceous Period and also probably during the prior metamorphism. Even then its troubles were not over, for this particular outcrop lies close to a granitic intrusion of Early Tertiary age. The heat from this nearby body of magma caused additional metamorphic recrystalliza-

3-3 *An outcrop of metamorphic rock in the North Cascades, Washington. The rock was deposited originally as silt and clay in the ocean. Subsequently it was deformed, metamorphosed several times, and injected by granitic intrusion. A radiometric age determination of such a complex rock would be difficult to interpret. (Photograph courtesy of Peter Misch.)*

TABLE 3-1 Some radioactive elements used in dating

PARENT ELEMENT	DAUGHTER ELEMENT	HALF-LIFE, YEARS
Carbon 14*	Nitrogen 14	5.7 thousand
Potassium 40	Argon 40	1.33 billion
Rubidium 87	Strontium 87	50 billion
Thorium 232	Lead 208	13.9 billion
Uranium 235	Lead 207	710 million
Uranium 238	Lead 206	4.5 billion

*Number after element refers to the atomic weight of the element's radioactive isotope.

tion. Mineralized fluids from the intrusion (and probably also some sweated out from the phyllite itself) crystallized as the many light-colored veins seen in the photograph. Finally uplift and erosion of the Cascade Range exposed this outcrop.

Now, what would be learned from a radiometric age determination of this rock? Could you learn, for instance, the time at which the sediment was deposited on the sea floor? To do this you would want to date the original mud, represented by the dark layers in the outcrop. Undoubtedly this contains potassium, and a logical choice of dating methods might be the potassium-argon technique. Recall though that argon, the daughter product of the disintegration process, is a gas. If any of the potassium-bearing minerals recrystallized after the deposition of the mud, at least some of the argon would have escaped. But the mud did recrystallize at least twice—once in the metamorphic transformation of shale to phyllite, and later during the emplacement and cooling of the nearby granitic magma. A radiometric age determination might tell you very little. If all the argon had been lost during the second recrystallization, your date would be that of the intrusion, not the mud. (The effect of total argon loss would be to restart the radioactive clock.) If some but not all of the argon escaped during the metamorphic events, the age determined would be less than that of the mud, but greater than that of the granitic rock. And if none of the argon had escaped (which, looking at the outcrop, is exceedingly unlikely), would the date obtained necessarily be that of the deposition of the mud? No. The potassium-bearing minerals were derived, at least in part, from the erosion of older rocks exposed on land. This means that they already had undergone some radioactive decay and contained some argon prior to their deposition as mud. In other words, sediment deposition does not necessarily restart the radiometric clock.

Summing up, the interpretation of radiometric age determinations is a complicated business. Dating igneous rocks is the simplest, for minerals crystallizing from a melt are not likely to contain argon, for instance, from rocks older than the magma. Dating sedimentary rocks is possible if great care is exercised to separate and analyze only minerals that formed at the time of deposition of the sediment. Age determinations of metamorphic rocks are most complex. Unless metamorphism caused a total recrystallization of the rock, the date obtained will be some compromise between the age of the metamorphism and the age of the parent material in the rock. Finally, in the case of potassium-argon measurements, argon can leak from minerals even without the help of metamorphic recrystallization. Consequently any potassium-argon date must be regarded as a minimum age—minimum because the argon is the daughter product.

By now the reader should appreciate the comments made early in this chapter about the caution most geologists employ in interpreting "precise" dates. If only someone could find a diagnostic fossil in that old oceanic mud of Figure 3-3!

4 why the earth deforms

■ The mountains we see today are for the most part young features produced by relatively recent deformation and uplift of the earth's surface. The origin of mountains has intrigued scientists for many centuries. The subject has generated as much speculation as any other in the entire field of geology. No comprehensive, unifying theory of earth deformation has been widely accepted by the professionals. Lively debates about this have filled the geologic literature. A revolution of sorts is currently under way, and many geologists are convinced that at last we are gaining a much clearer understanding of earth deformation.

We can best grasp the significance of this revolution by tracing its evolution. The task of reconciling our geologic data with the new concepts has barely begun, and the literature on the geology of the Northwest does not reflect our new understanding of earth deformation. Clearly, the geology has not changed in the past few years, but new vistas of interpretation are opening and must be explored. It will be many years before a book can be written that adequately explains the structural evolution of the Northwest. For now, we must accept a relatively high degree of uncertainty about the causes of earth deformation.

THE GEOSYNCLINE THEORY

The geosyncline theory, first proposed in the middle of the nineteenth century, was an early attempt to explain the development of mountains. That mountains were due to deformation had long been recognized. The strata of many ranges are sharply folded, faulted, and, in some cases, metamorphosed. Furthermore, the uplift implied in mountain building (*orogeny*) was documented by the discovery in some of the world's major mountain systems of sedimentary rocks containing marine fossils at elevations of thousands and even tens of thousands of feet above sea level. The geosyncline theory was proposed by James Hall, the New York State Geologist, as an explanation for the origin of the Appalachian Mountains—specifically the folded and faulted sedimentary part known as the Valley and Ridge Province. The stratigraphic record consisted of marine rocks ranging in age from Cambrian to Permian (thus encompassing the entire Paleozoic Era). It had a total thickness of approximately 8 miles. Hall realized that the Appalachians were not unique. Other major mountain chains consist largely of similar thick, marine, sedimentary sequences. These too had been deformed and uplifted after a long history of subsidence and accumulation in a large trough or basin. The term "geosynclinal" (subsequently changed to *geosyncline*) was applied to

4-1 *The Pickett Range in the North Cascades, Washington, an excellent example of a geologically young mountain range. (Photograph courtesy of Peter Misch.)*

the large trough and its sedimentary fill. The theory supposed that mountains were derived by the deformation and uplift of geosynclines.

The significance of this concept cannot be overemphasized. The geosyncline theory, expanded and modified, has dominated orogenic thought for the past hundred years. The most significent modification resulted from the realization that the Valley and Ridge Province constituted only a part of the original Appalachian Geosyncline. The crystalline metamorphic and igneous rocks to the east and northeast (including New England and the Maritime Province of eastern Canada) were also a part of the Paleozoic stratigraphic record. This two-part division of the geosyncline into a sedimentary, relatively unmetamorphosed segment lying close to shore (the continental interior) and a volcanic, highly metamorphosed and intruded offshore segment was found to be characteristic of other mountain systems. The sedimentary part was termed the *miogeosyncline* and the volcanic part the *eugeosyncline* (Figure 4-3). Clearly, deformation, metamorphism, volcanism, and igneous intrusion were much more important aspects of the mountain-building process than Hall had realized. This did nothing to diminish the significance of his concept.

The geosyncline theory should be deemphasized in the years to come as geologists realize it has outlived its usefulness. It had some problems from the start. First, it was not really an explanation of mountain building, and the various refinements could not make it one. It was a recognition of the sequence of events that typify the development of a major mountain chain. It was not an explanation for why these events occur; and, a century after

4-2 *The Rocky Mountains north of Prince George, British Columbia. Uplifted marine strata, such as seen here, led to the formulation of the geosyncline theory for the origin of mountains. (Photograph courtesy of James Crawford.)*

Hall, geologists continue to debate why the trough forms in the first place, why it subsides for so long, and why it ultimately deforms and rises to produce mountains. In other words, it is not a theory that explains mountain building.

A second major obstacle has been the failure of geologists to agree upon modern examples of a geosyncline, although many have been suggested. Part of the problem here is that the ultimate test of a geosyncline is whether in some future age these beds will be deformed and uplifted to form mountains. Proposed modern geosynclines include the continental shelf off the east coast of the United States and the gigantic sedimentary prism offshore along the Gulf Coast. Perhaps someday there will be the Mississippi Alps, but many geologists question the inevitability of this. The fact that we cannot identify modern geosynclines suggests that the theory has been too confining. Part of the problem is with the concept of a trough with two well-defined sides. By proposing continental shelves as possible modern geosynclines, geologists have recognized that the margins of continental plates may be the sites of future mountains. Thus to understand orogeny we must learn more about continental margins. The geologic record of past geosynclines must be reinterpreted in the light of modern continental margins. We must mentally remove the seaward side of the trough margin, or else think of it in terms of island arcs or maybe continental plates that have subsequently moved someplace else.

4-3 *The geosyncline concept. Section A represents the original concept of a shallow, subsiding trough adjacent to a relatively stable continental margin. The sediment that fills the geosyncline is derived in large measure from an offshore landmass, such as Appalachia or Cascadia. Section B shows a more modern interpretation. The geosyncline has within it a volcanic archipelago, which is an important source of sediment. Much of the geosyncline is in deep water.*

CONTINENTAL DRIFT—A NEW RESPECTABILITY

Few geologic theories have been as controversial and have polarized thought so effectively as the theory of continental drift. The concept that conti-

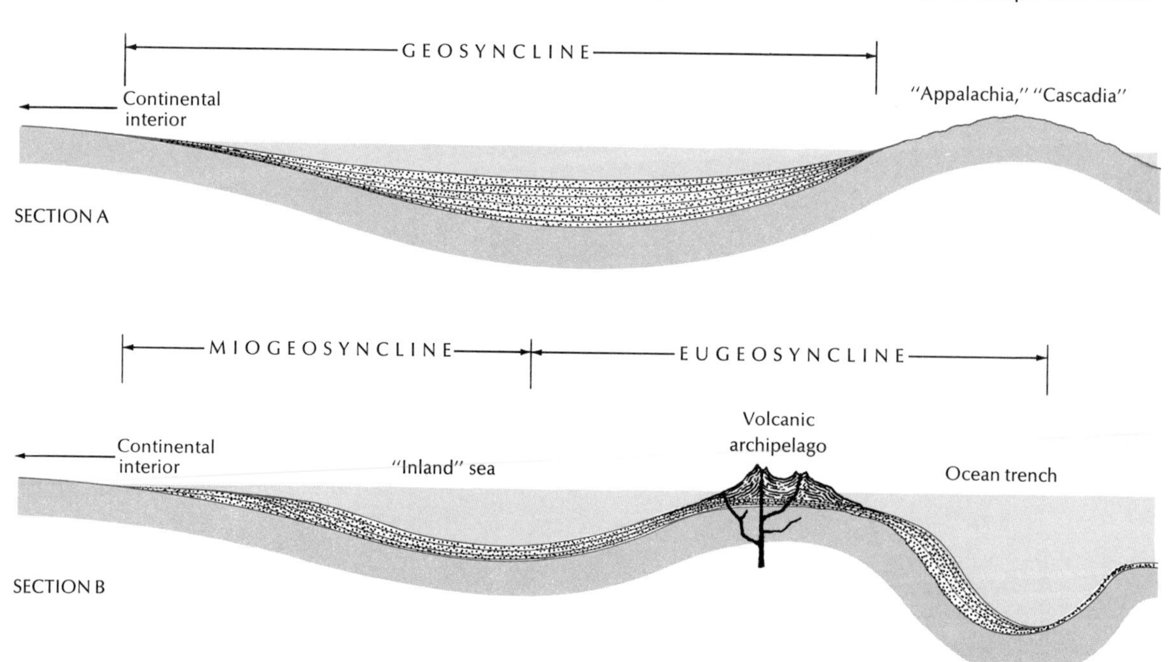

nents might shift position on the earth's surface is an old one, first suggested by the parallelism of many coastlines on opposite sides of oceans (Figure 4-4). This could be explained by the breakup of one or two original "supercontinents." The pieces then drifted apart and rotated to their present positions. The fit is best for the southern hemisphere. South America, Africa, India, Australia, and Antarctica, by suitable rotations, can all be fitted together reasonably well into one landmass, which has been referred to as Gondwanaland. The continents of the northern hemisphere have been reconstructed into a supercontinent called Laurasia, but the evidence has never been quite as straightforward. This explains in part why the theory of continental drift has had more support from geol-

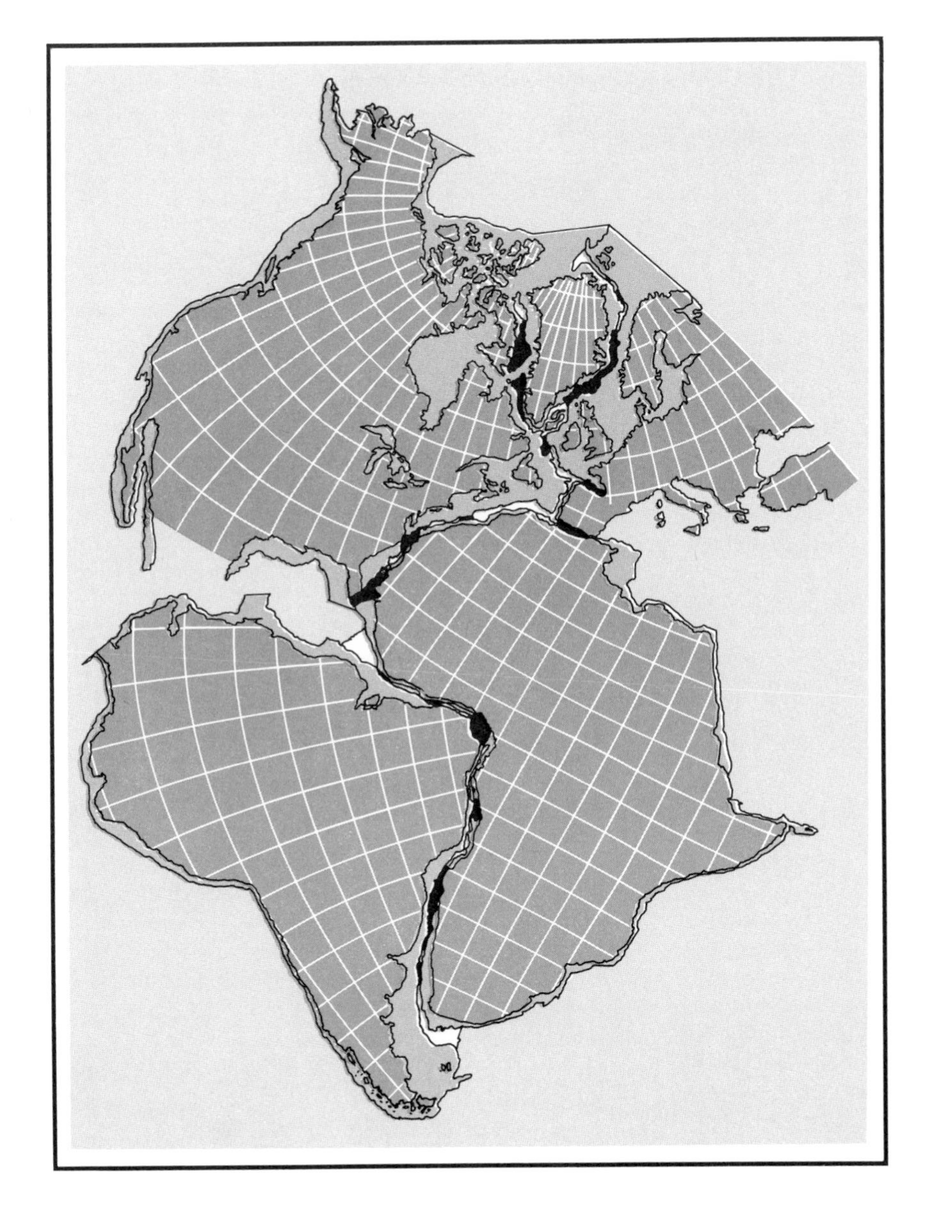

4-4 *The "fit" of continents on the two sides of the Atlantic Ocean has been recognized for many decades as strongly suggestive of continental drift. In this reconstruction, the black regions represent places where overlap of the continental shelves occurs; the white regions indicate a gap. (From Hurley, 1968, p. 52. Copyright © 1968 by Scientific American, Inc. All rights reserved.)*

ogists working in the southern hemisphere.

Basically, there is nothing too unreasonable about the supposition of drifting continents. Geologists have long recognized that the outer part or *crust* of the earth consists of rock that is lighter than that making up the earth's interior. In one sense, the crust "floats" at the surface. Furthermore, the crust under continents is thicker and lighter than oceanic crust, and hence it floats higher. This does not mean that the earth's interior beneath the crust is molten—only that it is plastic and can deform slowly over a long period of time.

The geologic evidence for continental separation or drift is exceedingly varied. It includes not only the remarkable fit but also the distribution of fossil floras and faunas; evidence from ancient glaciations; the apparent positions of the magnetic poles of the earth in the past (as deduced by measurements of magnetism in rocks from the different continents); and the matching of many rock formations and structures. Much new evidence in support of drift has been found in the past decade, but a considerable body has existed for a long time. Why then was the theory of continental drift so controversial and generally so scorned (especially in North America) for so long? There are many reasons. For one, many geologists believed firmly in the permanence of continents and ocean basins. To them, sliding pieces of the crust around seemed very exotic. Perhaps more important, the "drifters" were not able to explain adequately the mechanism whereby continental fragmentation and migration occurred. Each time they proposed a cause, some geophysicist would make a few calculations with his slide rule and demonstrate that the suggested force was wholly inadequate to do the job. For these and other reasons, most geologists regarded all the evidence for continental drift as interesting but certainly explainable in other ways. The theory remained relatively dormant, attracting few new adherents.

General acceptance has come in the past few years. In part, this is due to additional geologic evidence from the continents. The biggest factor has been the tremendous amount of geological and geophysical evidence obtained from the oceans. The continents are regarded now as rafts or plates that are carried along by currents that move slowly in the soft, plastic layer of the earth (the so-called *asthenosphere*) that immediately underlies the relatively rigid crust and immediate subcrust (referred to together as the *lithosphere*). This interpretation removes the mechanism problem that stumped the earlier drifters. The continents assume a passive role rather than that of plates plowing along independent of the asthenosphere. The mechanics of flow are still not well understood. However, the possibility, even probability, of slow movement of material within the earth is firmly established.

SEA-FLOOR SPREADING

Intensive investigation of the 70 percent of the earth's surface that lies underwater was initiated during the Second World War. It has been accelerating ever since. Prewar geologists presumed that oceanic areas were generally rather monotonous regions without much surface irregularity or sedimentation, except around volcanic islands and immediately adjacent to continental regions. Earth deformation was considered to have been concentrated largely in mountainous regions on land and locally along continental margins. Many geologists saw no reason to consider the possibility of worldwide patterns of deformation. But evidence from the oceans changed that.

For one thing, the largest mountain chains in the world, dwarfing the continental systems, were discovered in the oceans. As some of these (such as the Mid-Atlantic Ridge and the East Pacific Rise) were traced, they were found to continue on and on, until finally they all connected into a gigantic globe-encircling system at least 70,000 miles long! Suddenly a mountain chain like the Appalachians seemed like a pretty small part of a much grander pattern of earth deformation. Geologists were forced to take a new look at the problem on a much larger scale. The oceanic ridges received some intensive study, especially using various geophysical methods. They were found to possess many characteristics in common, such as:

1. The ridges are broad, irregular ranges that rise from the sea floor as much as 10,000 feet or more. Some are 1,000 miles or more wide. Their flanks are typically rough, with many steep slopes suggesting much faulted structure.
2. The rocks beneath ridges are mostly basalt lava. There is virtually no sediment cover near the crest, but sediment thickness gradually increases down the flanks of the ridge.
3. The ridge crest is cut in places by a steep-walled symmetrical canyon thousands of feet deep. Its morphology strongly suggests a down-dropped fault block, or *graben*.
4. The ridge crest and graben are offset along major transverse fracture zones. Offset is normally less than 100 miles but may be much longer. The fracture zones may be thousands of miles long, and are expressed by linear troughs and escarpments. They are referred to as *transform faults*.
5. Shallow earthquakes are generated near ridge crests and along that part of the transform fault that connects offset ridge crests.
6. Ridge crests are areas where abnormally large amounts of heat are flowing to the earth's surface. They are also sites of volcanic activity. This sometimes builds volcanic islands above sea level, such as the Azores and Iceland on the Mid-Atlantic Ridge.
7. The lower part of the lithosphere beneath ridge crests is unusually soft.
8. Anomalous "stripes" are found in the magnetism in oceanic crustal rock. These parallel the ridge crest and are interpreted as the remnant magnetic field of the earth frozen into the volcanic rock of the crust as it crystallized. The stripes with high magnetic intensity have a preserved field that parallels and reinforces the present magnetic field of the earth. Those with low intensity have a reversed magnetic field. (Scientists know from the study of remnant magnetism in volcanic rocks on land that the earth's magnetic field has been reversed at times in the past. See Figure 4-5.) The pattern of magnetic anomalies or stripes in the ocean crust is remarkably similar on the two sides of the ridge crest, that is, one flank is the mirror image of the other.

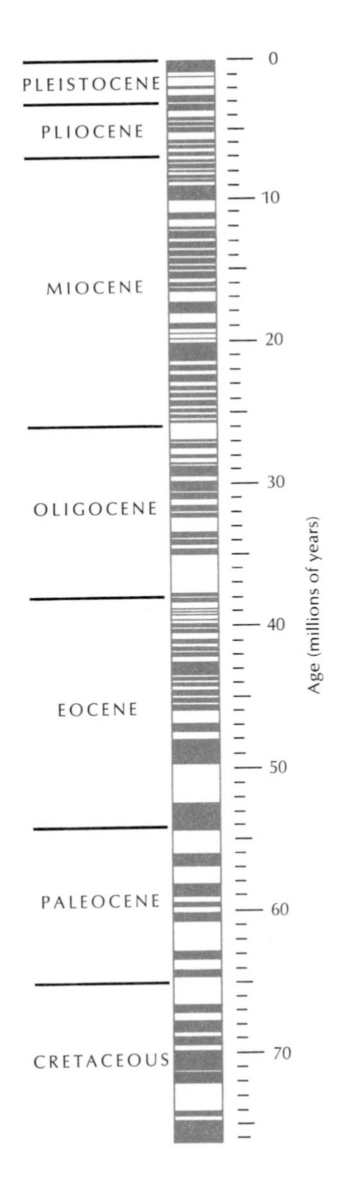

4-5 *A magnetic "time scale" for the past 76 million years. Colored bars indicate periods when the earth's magnetic field was "normal" (that is, like today), while white bars represent periods of reversed magnetism. Data for the past 4 million years came from measurement of dated lava flows—that rest from oceanic anomalies assuming a constant spreading rate. (From Bullard, 1969, p. 73. Copyright © 1969 by Scientific American, Inc. All rights reserved.)*

This and other evidence was marshaled a few years ago to support a concept that ocean basins are spreading away from ridges. Oceanic crust is created by igneous intrusion and extrusion at the ridge crest. The new crust moves down the flank of the ridge, opening up a new rift or graben, which is the site of renewed crustal generation. In a sense, the mechanism is comparable to two immense conveyor belts moving away from the ridge crest and carrying off newly created sea floor. The transform faults separate different conveyor systems. The offset pattern of ridge crests reflects perhaps an original irregularity in the position of spreading centers. The oceanic plates on opposite sides of the transform fault between ridge crest segments are moving in opposite directions (which produces the earthquakes) but beyond the end of the crests the plates are moving in the same direction (and hence are seismically dead; see Figure 4-6).

The implications of the theory of sea-floor spreading were enormous. For one thing, the width of the magnetic stripes was potentially a measure of the rate of crustal generation and spreading. If a positive stripe was 50 miles wide and had been generated at the crest during an epoch of normal magnetic field that had lasted 500,000 years, the suggested rate of spreading on one side is 50 miles in 500,000 years, or 1 mile per 10,000 years, or approximately 6⅓ inches per year. The time scale for magnetic reversals for the past 10 million years has been established by studying well-dated volcanic rocks on land. Using this, the rate of spreading from various ridge crests has been determined. The correlation between the width of magnetic reversal patterns at sea and the magnetic time scale is excellent, indicating a relatively constant spreading rate backward in time for particular ridges.

The concept of sea-floor spreading suggested a great many geologic and geophysical tests that could be made in the oceans to confirm the hypothesis. These are being run at present, and virtually all the results point toward the essential correctness of the theory.

The probability of sea-floor spreading requires a whole new look at the question of earth deformation. Continental drift gains immediate respectability, for if the Atlantic Ocean is in fact opening up at an appreciable rate, then South America and Africa should be moving apart at a rate equal to the spreading rate. And as a matter of fact, the suggested spreading rate deduced from the magnetic pattern is just sufficient, if projected backward in time to when the geologic evidence suggests continental separation (that is, the Mesozoic), to account for the amount of separa-

4-6 *An actively spreading oceanic ridge segment, offset along a transform fault. Earthquake activity should be concentrated along the transform fault only between the two ridge axis segments, for beyond these the sea-floor plates move at the same rate.*

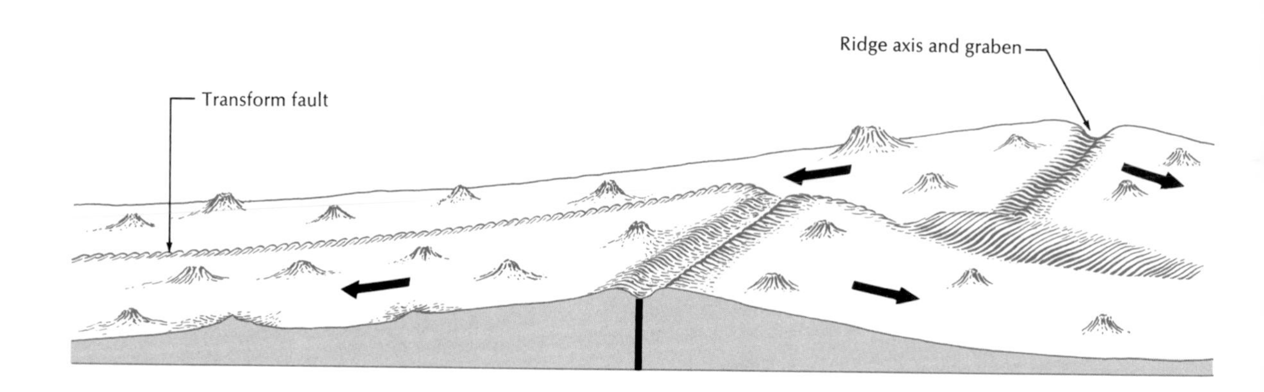

tion we find today. But if the oceans are all opening up, does not this require that the earth is expanding, for we cannot allow much shrinking of the continents? It does, unless oceanic crust is somehow being destroyed at ocean margins. Presumably the mechanism of crustal destruction would have to involve some process of sinking the crust back into the earth's interior, that is, of thrusting it under other pieces of crust. This process has been termed *subduction.*

Subduction does seem to occur. If we look at the distribution of world crustal deformation (as measured by earthquake and volcanic activity) the most unstable regions are the oceanic ridge crests, the Mediterranean–Near East–Himalayan zone, and the margin of the Pacific Ocean. The first is a region of crustal generation and spreading, the second appears to be a complex zone of continental collision (involving the impingement of the African and Indian plate remnants of Gondwanaland against the Eurasian remnant of Laurasia), and the third a region of crustal underthrusting or subduction. The Pacific rim, including as it does the Pacific Northwest, is of particular concern to us, and must be looked at more closely.

THE PACIFIC MARGIN

The continent-ocean border around the Pacific Ocean varies quite markedly from place to place, although earthquakes and volcanism are characteristic throughout most of the region. Clearly the "take-up" from sea-floor spreading can be accomplished in different ways, as illustrated in Figures 4-8 and 4-9. Along the western margin of central South America the oceanic plate seems to be underthrusting the continental plate. This produces (1) a zone of earthquakes that dips under the Andes and marks the position of the descending oceanic plate; (2) a deep ocean trench (the Peru-Chile Trench) at the continental margin marking the downbuckling of the oceanic plate; and (3) folding, faulting, and volcanism in the Andes. Across the Pacific, in the Tonga-Kermadec Trench, one oceanic plate appears to be underthrusting another. A volcanic island arc has formed above the descending plate. A similar situation prevails in the Aleutian Island chain. Japan appears to be a fragment of the Asian continent that has drifted relatively eastward and is currently being underthrust by the Pacific plate.

The most complex region appears to be (unfortunately) the western margin of the North American continent. The continent probably has overridden the East Pacific Rise. This rise enters the Gulf of California and apparently reappears as the Gorda and Juan de Fuca Rises off the Northwestern coast. The Gulf of California has opened up over

4-7 *A whimsical interpretation of sea-floor spreading and subduction along a continental margin. (Courtesy of Marguerite Stroh.)*

4-8 *Two possible types of crustal plate collision. In Section A, a generalized reconstruction of the western margin of South America, the oceanic crust underthrusts the continent. This results in the formation of a trench, a subcontinental zone of deep earthquakes, and surficial uplift, deformation, and volcanism. In Section B, typical of parts of the western Pacific, one ocean plate underthrusts another, producing many of the same features as in Section A.*

the past few million years, probably as a result of continued spreading of the rise under the edge of the continent. The famous San Andreas Fault of California is now interpreted as the transform fault that connects the offset segments (East Pacific and Gorda Rises) of the ocean ridge system.

The relationship between the oceanic and continental plates offshore from Oregon, Washington, British Columbia, and southeastern Alaska is particularly problematic. Magnetic patterns and seismic activity indicate that the Gorda and Juan de Fuca Rises are now axes of spreading, but underthrusting of the continent by the oceanic crust

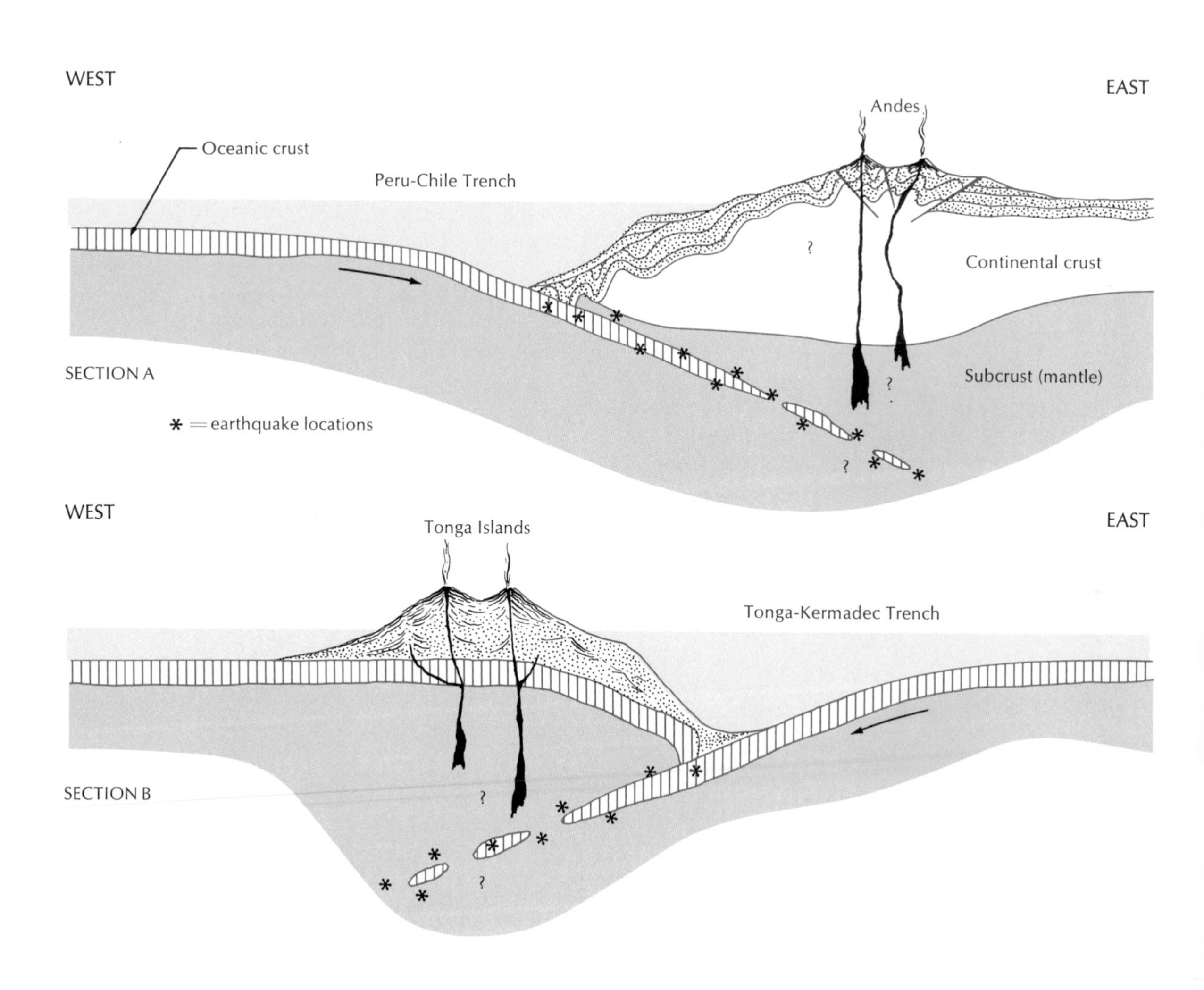

cannot be proved. The disappearance of the magnetic stripes under the continental edge (Figure 4-10) and the presence of dormant volcanoes in the Cascades and Coast Mountains of British Columbia suggest underthrusting. However, the deep trench, common elsewhere around the Pacific, is missing. Moreover, although the region is moderately active seismically, the characteristic zone of earthquakes descending under the continental margin cannot be identified. The unhappy truth is that, at present, the region we are most interested in is one of the world's most difficult ones to understand.

IMPLICATIONS OF THE NEW THEORIES

The new theories have been called "The New Global Tectonics" (*tectonics* is the study of the broader structural features of the earth and their causes), or simply "Plate Tectonics." Suddenly we are in a frantic period of reevaluation of the rock record and reinterpretation of the geologic history of areas mapped long ago. Where are ancient plate boundaries? Where are the subduction zones of the past? How can we reconcile the geosyncline theory with modern theories of plate tectonics? How can we reconstruct the paleogeography of a region until we know from where all of the pieces came and when they arrived at their present positions? Hardly a week seems to go by without

4-9 *Six major crustal plates of the earth and their relative motions, assuming that the African plate is stationary. Major trenches around the Pacific where crustal underthrusting is probably occurring are numbered. Spreading rises are indicated by the double solid lines with a medial dashed line. The numbers identify the following major trenches: (1) Acapulco-Guatemala, (2) Peru-Chile, (3) Kermadec-Tonga, (4) Mariana, (5) Japan, (6) Kuril-Kamchatka, and (7) Aleutian. (From Bullard, 1969, p. 74-75. Copyright © 1969 by Scientific American, Inc. All rights reserved.)*

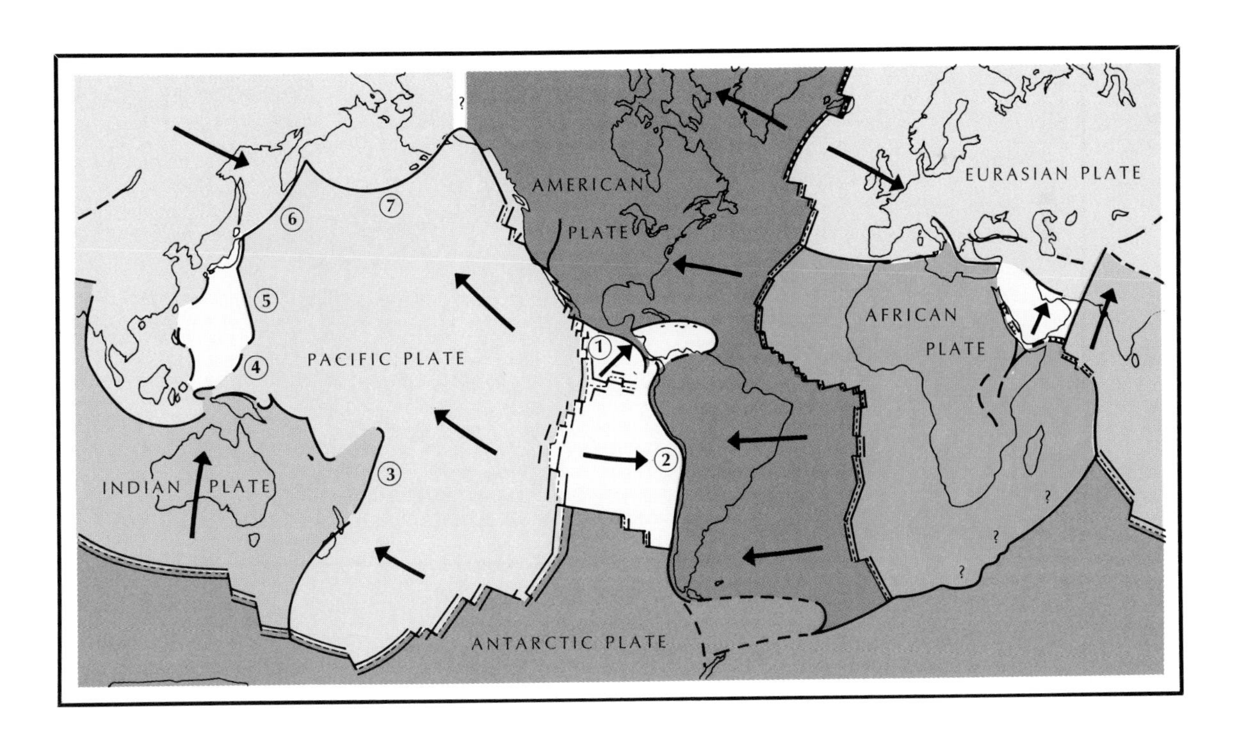

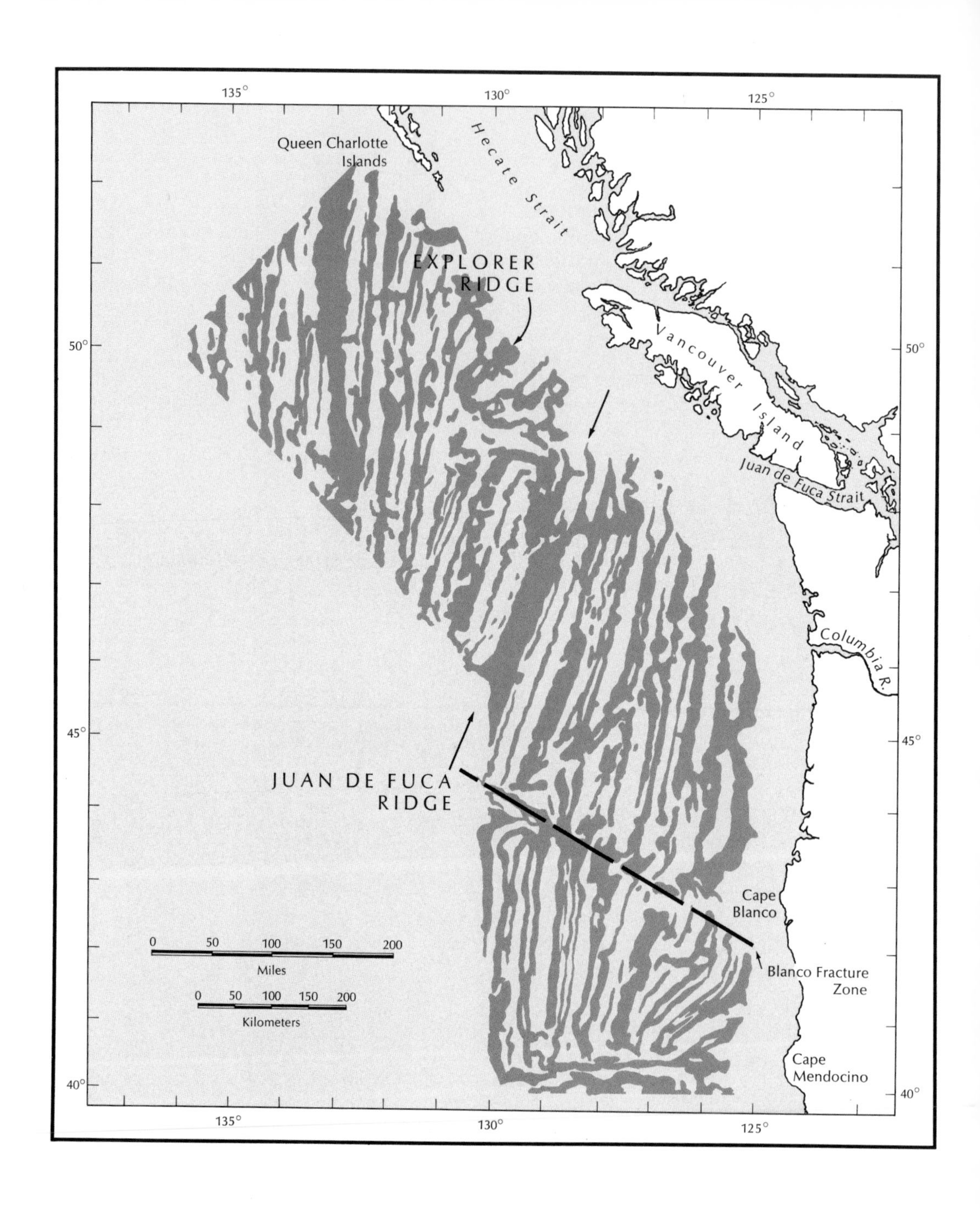

4-10 *Magnetic-field anomalies off the coast of the Northwest. Positive anomaly fields (normal remnant magnetization) are shown by the colored stripes. See also Figure 5-5. (After Raff and Mason, 1961, p. 1,268.)*

4-11 *The core of the Olympic Mountains from the west. Mount Olympus in the middle distance. Regions such as these must now be considered as possibly representing pieces of oceanic crust that underthrust the continental margin. See Chapter 11. (Copyrighted photograph courtesy of Delano Photographics.)*

someone, somewhere, organizing a symposium on plate tectonics at which eminent scientists propose new models to explain the geologic pattern of particular regions. Suddenly earth deformation and mountain building seem explainable in the light of present-day processes we can go out and study, resulting in a new reinforcement of the principle of uniformitarianism.

But the revolution is just beginning. The implications of these new tectonic theories are so numerous that satisfactory geologic reinterpretation will take many years and demand much additional supporting evidence. We now believe that oceanic crust is both created and destroyed, and that magnetic anomalies record the past movements of crustal plates. It appears that the oldest remnants of crust preserved in oceanic areas are no older than Mesozoic. Therefore, the record for the previous history (which includes 95 percent of geologic time) must be found in the continents, for the oceanic evidence has been destroyed.

The reader of ensuing chapters will note some reference to plate tectonic theory and its possible role in interpreting certain aspects of the geologic history of the Northwest. For the most part, these references are minor, and seemingly insufficient to warrant the emphasis of this chapter. However, future interpretations of the geology of the region will certainly be made using plate tectonic theory. A more important consideration is cultural. The scientific writer of the twenty-first century could easily select these new tectonic theories as the most significant contributions made by geologists. Anyone at all interested in geology should have some familiarity with these concepts. They are not simple, but then if they were we should have thought of them long ago.

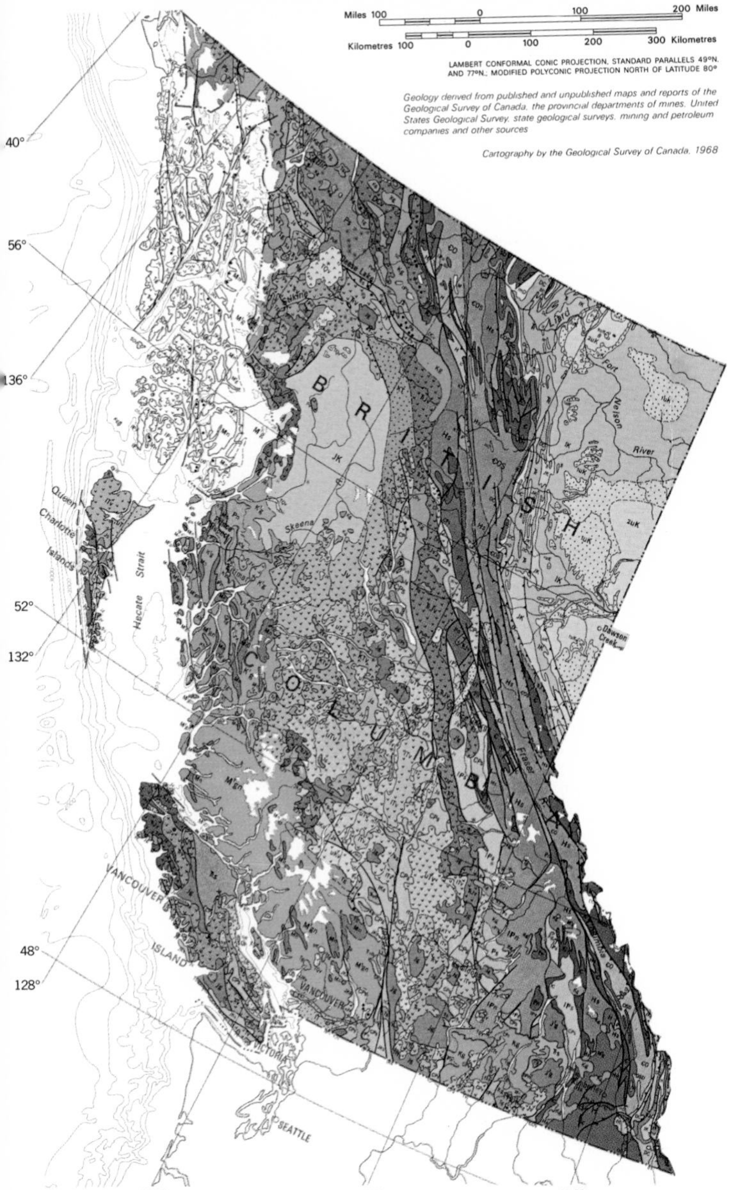

Miles 100 0 100 200 Miles
Kilometres 100 0 100 200 300 Kilometres
LAMBERT CONFORMAL CONIC PROJECTION, STANDARD PARALLELS 49°N. AND 77°N.; MODIFIED POLYCONIC PROJECTION NORTH OF LATITUDE 80°
Geology derived from published and unpublished maps and reports of the Geological Survey of Canada, the provincial departments of mines, United States Geological Survey, state geological surveys, mining and petroleum companies and other sources
Cartography by the Geological Survey of Canada, 1968
BRITISH COLUMBIA
Queen Charlotte Islands
Hecate Strait
VANCOUVER ISLAND
VANCOUVER
VICTORIA
SEATTLE
Skeena
Fort Nelson River
Fraser
Dawson Creek
40°
56°
136°
52°
132°
48°
128°
124°
120°
116°

LEGEND

Age of rocks is indicated by capital letters. Modifications are shown to the left by lower case letters (l - lower, m - middle, u - upper) and/or by numbers (1 being the oldest)

CENOZOIC

QUATERNARY
- Q — Q, Qv

TERTIARY
- T — uT, uTv, plT, miT
- T — oT
- T — T, Tv, lT, lTv, pT
- T — Tg, Ty, Tb

pl - *Pliocene*, mi - *Miocene*, o - *Oligocene*, p - *Paleocene*

LATE TERTIARY AND QUATERNARY
- TQ — TQ, TQv

CRETACEOUS AND TERTIARY (UPPER CRETACEOUS AND TERTIARY)
- KT — KT, KTv

MESOZOIC

CRETACEOUS

UPPER CRETACEOUS
- K — uK, uKv, 1 - 9uK; 1 - *Cenomanian*, 2 - *Cenomanian to Santonian*, 3, 4 - *early Campanian*, 5 - *mid-Campanian*, 6 - *late Campanian*, 7 - *Cenomanian to late Campanian*, 8 - *Campanian and Maestrichtian*, 9 - *Maestrichtian*

LOWER CRETACEOUS
- K — lK, lKs, lKv; *mainly Albian; may include Upper Cretaceous and Upper Jurassic*

CRETACEOUS
- K
- K — Kg, Ky

JURASSIC AND CRETACEOUS (LATE JURASSIC AND LOWER CRETACEOUS)
- JK — JK, JKs, JKv

TRIASSIC AND JURASSIC
- ŦJ — ŦJs, ŦJv

JURASSIC
- J — J, Js, Jv
- J — Jg, Jb

TRIASSIC
- Ŧ — Ŧ, Ŧs, Ŧv
- Ŧ — Ŧg, Ŧb
- Ŧ — Ŧub

MESOZOIC
- M — Mn
- M — Mg, Mgn
- Gabbro dyke

PALAEOZOIC

PERMIAN
- P — P, Ps, Pv
- P — Pub

CARBONIFEROUS

PENNSYLVANIAN
- P — P, Pv

MISSISSIPPIAN
- M — M, Ms, Mv
- C — Cg

CARBONIFEROUS AND PERMIAN
- CP — CP, CPs, CPv

DEVONIAN AND CARBONIFEROUS
- DC — *May include some Permian*
- DC — DCs, DCv
- DC — DCub

DEVONIAN
- D — D, Dv, uD, uDv; *may include some Middle Devonian*
- D — mD
- D — lD, lDs, lDv, lDn; *may include some Middle Devonian*
- D — Dg, Dy, Db

SILURIAN AND DEVONIAN
- SD — SD, SDs, SDv

SILURIAN
- S — S, Ss, Sv, lS, lmS, mS, uS

CAMBRIAN AND ORDOVICIAN
- ЄO — ЄO, ЄOs, ЄOv

PALAEOZOIC
- OSD — OSD, OSDs, OSDn; *Ordovician, Silurian, Lower and Middle Devonian*
- ЄOS — *Upper Cambrian, Ordovician, and Silurian*

ORDOVICIAN
- O — O, Os, Ov, uO, mO, lO
- O — On
- O — Og, Oy, Ob
- O — Oub

PALAEOZOIC AND? OLDER
- P
- P — Ps, lPs, *may include some older rocks*
- P — Pn, lPn, *may include some older rocks*

CAMBRIAN
- Є — Є, lЄ, mЄ, uЄ
- Є — Єs, lЄv, Єn

PROTEROZOIC

HADRYNIAN
- H
- H — Hs, Hv
- H — Hg, Hy
- H — Hb, Hd
- Gabbro dyke

HELIKIAN
- H
- H — Hs, Hv
- Gabbro dyke

PROTEROZOIC
- P — Pn

Note: Age designations of the sedimentary and volcanic rocks and their metamorphic equivalents are meant to be the time of their original deposition or extrusion as determined by palaeontological methods for Phanerozoic rocks and by structural, lithological and radiometric means for Precambrian rocks. Age designations of the igneous and plutonic rocks are meant to be the time interval of the orogeny during which they were intruded or emplaced.

For the igneous and plutonic rocks a prime (ʹ) is used to indicate an age designation that is older than the radiometric age obtained mainly by potassium-argon methods. These rocks probably form part of a crystalline basement that has been regionally metamorphosed, but it is not known whether they have been remobilized or remelted.

Lithology indicated by patterns or by lower case letters to the right of the age designations

Carbonate, shale, sandstone, conglomerate, coal, evaporites, iron-formation; cratonic cover and miogeosynclinal, exogeosynclinal and foredeep sequences; little metamorphosed or unmetamorphosed	no pattern
Slate, phyllite, schist, greywacke, quartzite, conglomerate, chert, iron-formation carbonate; eugeosynclinal sedimentary sequences; mainly metamorphosed, partly high grade	s
Volcanic flows and pyroclastic rocks; may include some intrusive rocks; unmetamorphosed or metamorphosed	v
Mainly basic and intermediate volcanic rocks	∴
Mainly acidic and mixed volcanic rocks	∴
Gneiss and schist derived from sedimentary rocks; may include volcanic rocks and granitic material	n
Gabbro, diorite	b
Diabase	d
Amphibolite; mainly metamorphosed basic extrusive and intrusive rocks	m
Granitic gneiss, migmatite; undifferentiated plutonic, sedimentary and volcanic rocks; includes some unmapped areas	gn
Granite and allied plutonic rocks	g
Syenite and alkalic rocks	y
Dunite, pyroxenite, serpentinite	ub

- Geological contact (mapped, unmapped or underwater)
- Fault (mapped, extrapolated or underwater)
- Fault (inferred from aeromagnetic maps)
- Normal fault (hachures on hanging wall)
- Thrust fault (teeth on hanging wall)
- Volcano, volcanic centre
- Ultramafic intrusion
- Bathymetric contours (depth in metres) — 500
- Glacier, ice-cap, ice-field

Geologic map of British Columbia, compiled by R. J. W. Douglas, 1966, 1967.

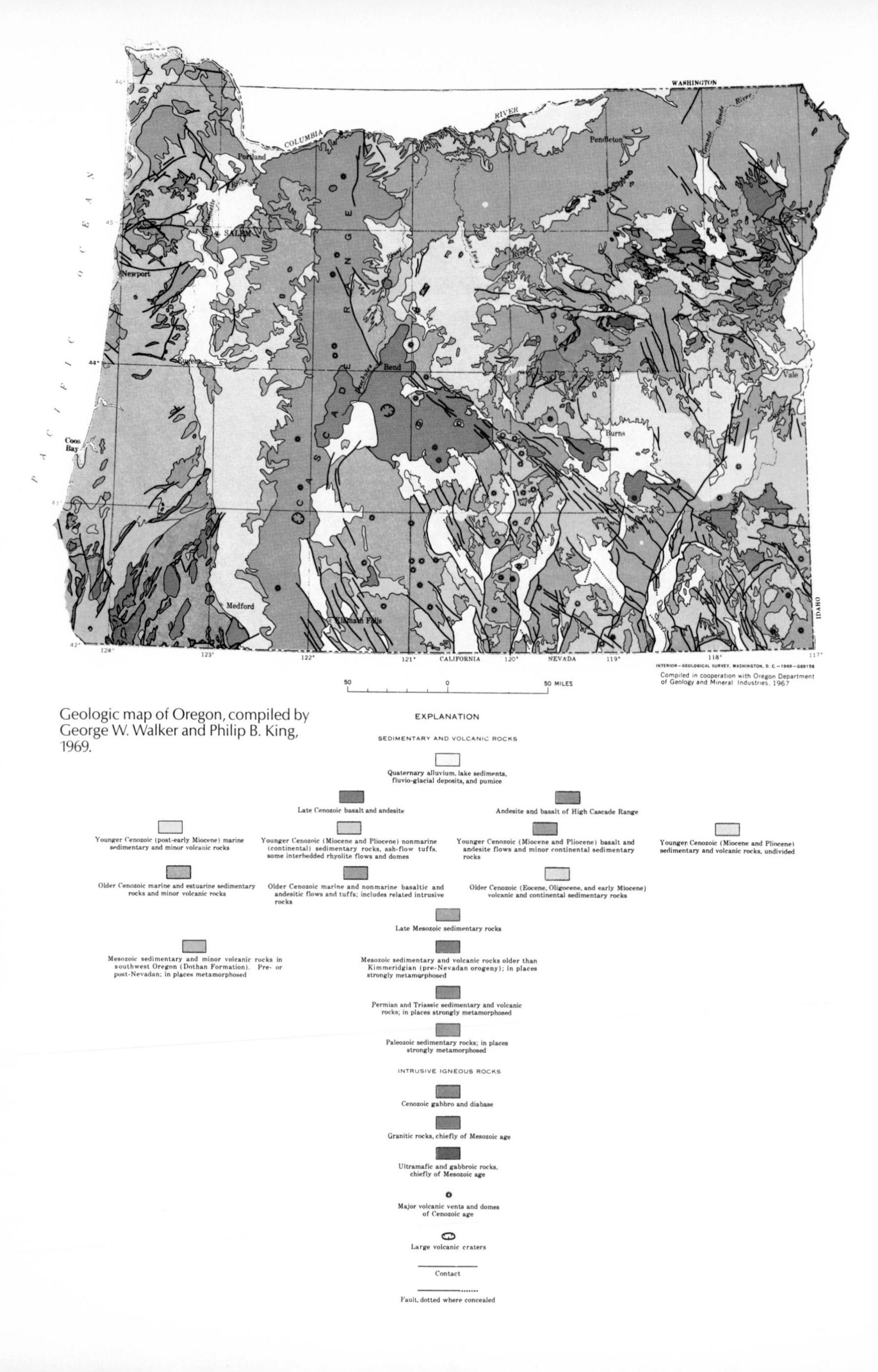

Geologic map of Oregon, compiled by George W. Walker and Philip B. King, 1969.

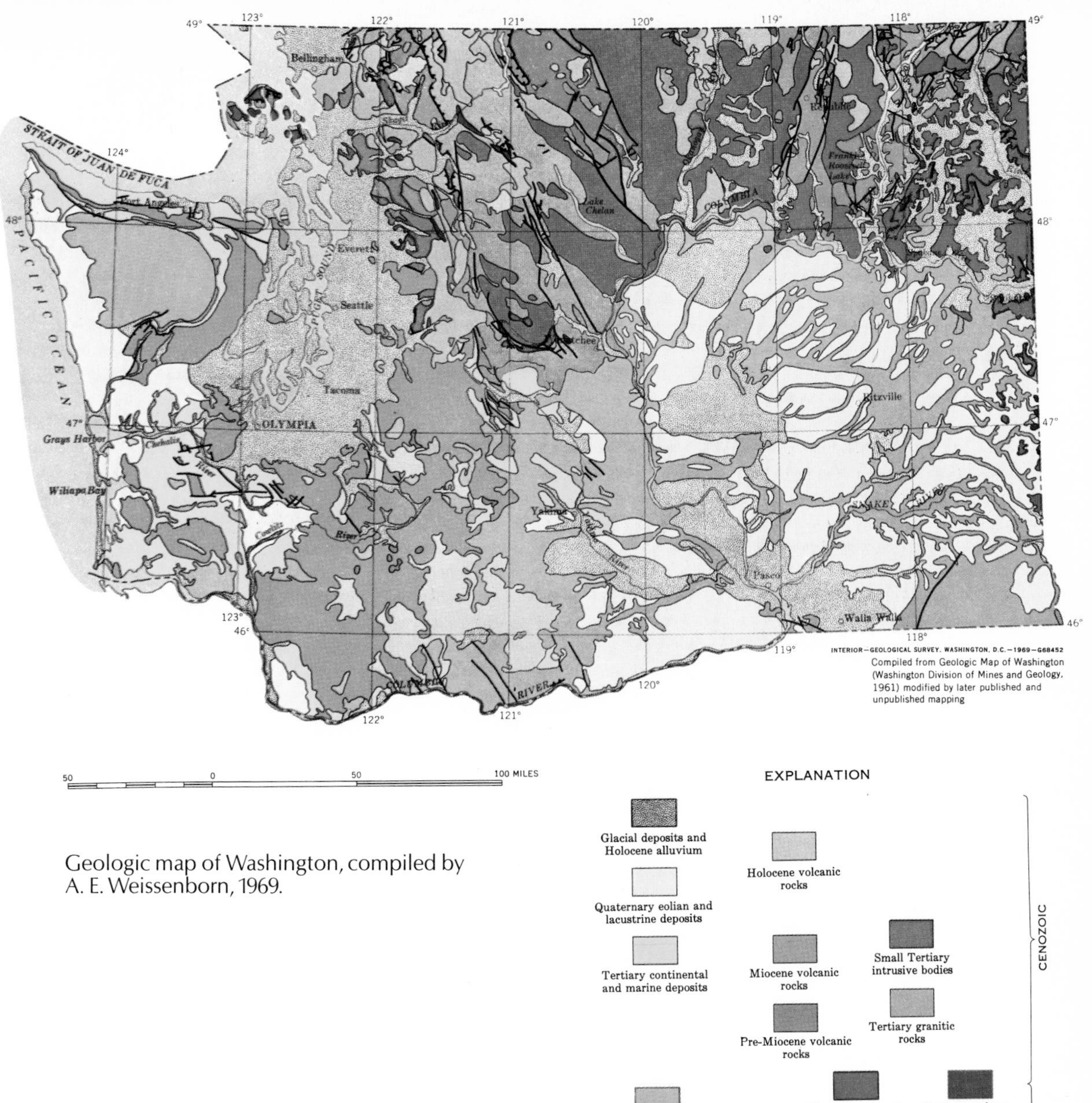

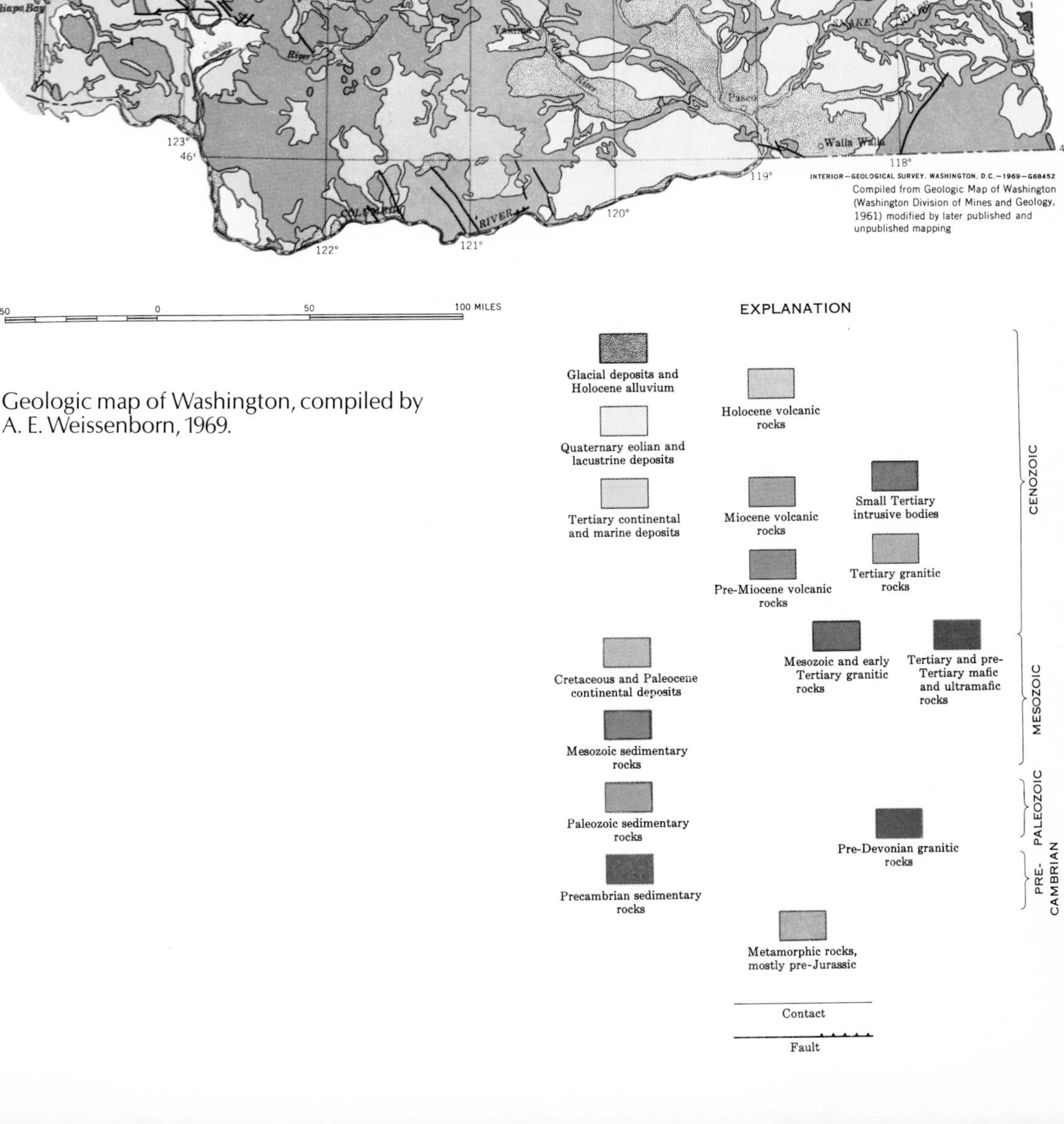

Geologic map of Washington, compiled by A. E. Weissenborn, 1969.

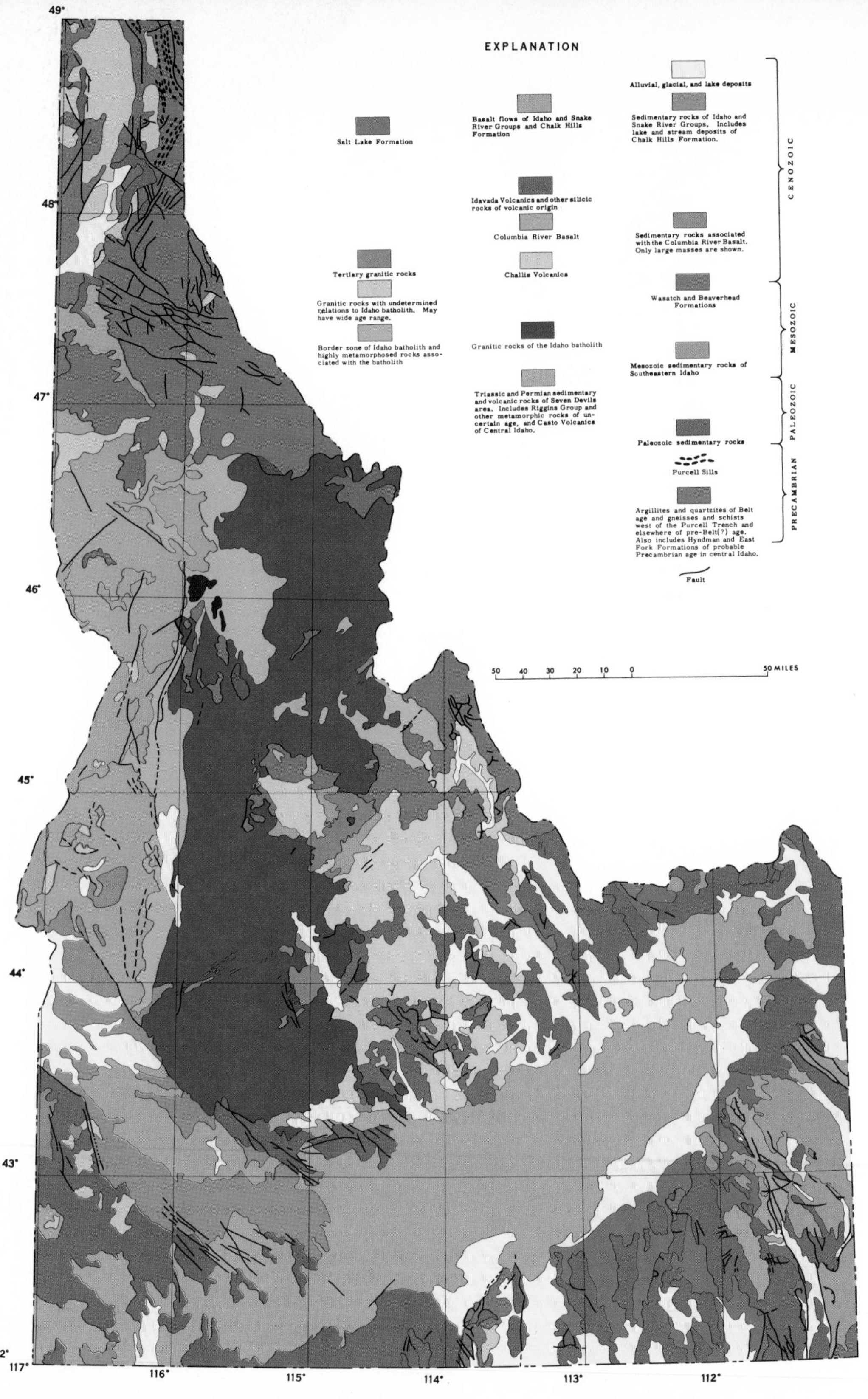

Geologic map of Idaho, geology by C. P. Ross, modified by A. E. Weissenborn.

□ The organization of the remainder of the book may appear curious, especially the order in which particular regions are discussed. The choice, which is neither random nor capricious, requires some explanation.

The provinces are defined largely on geographic or more precisely on physiographic criteria. Fortunately, the relationships between landforms and the underlying rocks and structures are sufficiently close to insure that in most cases a distinct physiographic province is also a distinct geologic province. However, no region can be treated entirely independent of surrounding regions, and obviously the history of the Northwest must be told as the sum of the various parts. Certain provinces tell us much about the ancient history of this part of the continent but relatively little about the events of the past hundred million years or so. Others, conversely, contain a valuable record of recent events, but the young strata effectively conceal the evidence of more ancient events. The antiquity of the record in each province has been an important factor in determining the placement of that region in the book. Early chapters in Part II discuss regions with significant strata of Precambrian or Paleozoic age, while subsequent chapters treat those with a better Mesozoic or Cenozoic record. This organization is not entirely consistent, for consistency would demand too much switching from one region to another. However, the compromise achieved does give some feeling for the geologic evolution of the entire Northwest.

The organization within each chapter is generally similar. The geography and overall geology of each region are outlined briefly, and then the geologic history is traced in somewhat more detail. Comparisons are made with previously mentioned events of a similar nature occurring in other regions, which serves to relate the history of each province to that of its neighbors. Exceptions to this pattern include Chapter 5, which introduces the entire region; Chapter 13, which treats the young volcanoes of the Cascades; and the last chapter, which attempts some predictions of future geologic events based upon trends discernible in the record of the recent past.

II GEOLOGIC HISTORY OF THE NORTHWEST

5
cordilleran landscapes

■ The highest mountains on the continent. The deepest canyons. The greatest concentration of active faults and earthquakes. The fastest ups and downs of the surface of the land. Virtually all the volcanoes; the glaciers. The most vigorous rivers. The fastest erosion rates. The greatest climatic variation. The largest deserts. The densest rain forests. The most devastating landslides. This is the Cordillera—a geologist's paradise.

The Cordillera includes all of the mountainous region of western North America. This is, perhaps, a third of the continent, but it is the wildest part, and seems larger than it really is. (I recall a conversation with a resident of the San Francisco Bay area, who told me he was going East for Christmas. Being originally from Connecticut, I inquired where in the East he was visiting. The reply was "Salt Lake City." This seemed preposterous, for before I moved West I regarded Salt Lake City as virtually a suburb of San Francisco. But then, viewed from New England, Chicago was "out West." I guess we all have some of the Texan in us, appreciating the dimensions of our own region and mentally shrinking distances elsewhere.) The major mountain ranges within the Cordillera include the Alaska and Brooks Ranges of Alaska; the Coast

Mountains of British Columbia; the Rocky Mountains, which extend from the Yukon almost to Mexico; the Cascade Range of Washington and Oregon; the Coast Ranges of the Pacific States; the Sierra Nevada in California; the Peninsular Ranges of southern California and Baja California; and the various Sierra Madres of Mexico. The Cordillera is bounded on the west by the Pacific Ocean. To the east lies a lowland—the Arctic Slope in Alaska, the Interior Plains of Canada and the United States, and the coastal plain of the Gulf of Mexico.

Washington, Oregon, and Idaho lie wholly within the Cordillera. British Columbia does too, except for its northeastern corner, which is east of the Rocky Mountain front. This is part of the Alberta Plateau, a division of the Interior Plains Region.

THE RESTLESS CRUST

The mountains of the Cordillera are young compared with most other parts of the continent. The rugged topography of the West reflects geologically recent deformation and uplift of the earth's crust. This does not imply that the rocks are all young or that there were not mountains in the Cordillera in more ancient eras. The general topography of the continent (and indeed of the world) provides a clear indication of regions that have been geologically unstable in the recent past. The Cordillera is an excellent example. Other signs of crustal deformation are earthquakes and volcanism. Here too the West provides an impressive record.

The previous chapter suggests a reason for this. The rim of the Pacific Ocean is the most extensive region of the earth where oceanic crust appears to be underthrusting continental margins. Present rates of sea-floor spreading and continental underthrusting are calculable to some extent, and earth satellite measurements in the near future offer great promise for direct determination of relative movements of all parts of the earth's surface. Projection of such data back into the geologic past will always involve certain assumptions and much interpretation. We are only beginning to agree on some of the ground rules for making such interpretations.

The geologic record contains much evidence that deformational processes acting today have operated in the past. However, the revolution in geologic thought, necessitated by the recognition of continental drift and sea-floor spreading, is too recent to have permitted much reinterpretation of geologic history. The most obvious shortcoming in most interpretations is a geographic one. Geologists have quite naturally assumed that the rocks we see today were formed close to their present position, and that major shifts of continents or pieces of continents have not occurred. But now it is a new ball game in which the rules allow for sliding pieces of the earth's crust around through geologic time, arriving at the present configuration (which is mobile) only yesterday. This puts a new premium on information that can be gleaned from the rocks themselves. Geologists are realizing that properties of a volcanic rock, for example, such as chemical and mineralogical details and the apparent relative position of the earth's magnetic poles "frozen" into the rock when it crystallized, are important clues to where the rock formed. These data set limits on interpretation. The general mobility of the earth's surface through time necessarily means that the relative positions of the pieces 1 million years ago is more certain than, say, 100 million years ago. Furthermore, geologists must work backward in time to understand relative plate positions, but traditional geologic interpretation (or at least discussion) works from the past to the present, evolving the record we see today from its earliest beginnings. This remains the most satisfactory method for relating history—geologic or otherwise. The reader is cautioned, however, that when we talk about 200-million-year-old volcanic eruptions on Vancouver Island, this does not imply that the region was then an island. Nor does it mean that the piece of the crust now known as Vancouver Island was even at its present latitude and longitude 200 million years ago. Today's landscape is a very young one.

CORDILLERAN TOPOGRAPHY

Most of the major mountain ranges of the Cordillera run north-south (Figures 5-1, 5-2, 5-3). Starting at the Pacific, we have a major range along the coast, which appropriately enough is referred to in Oregon and Washington as the *Coast Range*. At the south end, it merges with the *Klamath Mountains* of southern Oregon and northern California. At the north end, the Coast Range rises to its greatest elevation on the Olympic Peninsula of Washington in the *Olympic Mountains*. It is breached by rivers in only a few places—in southern Oregon, in the Columbia River region, at the south end of the Olympics (by the Chehalis River), and at the north end of the Olympic Peninsula where the mountains drop spectacularly into the Strait of Juan de Fuca. In British Columbia, the geologic and topographic continuations of the Coast Range are Vancouver Island and the Queen Charlotte Islands, which together define the *Insular Mountains Province*.

Inland from the coastal mountains lies a lowland. In Oregon, this is the valley of the Willamette River, a major tributary of the Columbia. In Washington, the interior lowland is Puget Sound and parts of the valleys of the Cowlitz and Centralia Rivers. The entire lowland segment from the Klamath Mountains to British Columbia is commonly referred to as the *Puget-Willamette Trough*. In British Columbia, the equivalent feature is the *Coastal Trough*, consisting of the Strait of Georgia, or Georgia Depression, and farther north, the Hecate Depression.

East of the lowland rises the *Cascade Range*, extending from Lassen Peak in northern California into southern British Columbia. The Cascades merge with the Coast Mountains of British Columbia, which continue north along the coast into southeast Alaska. These Coast Mountains are topographically and geologically unrelated to the Coast Range of Washington and Oregon.

The topography is more complex east of the Cascades and Coast Mountains. The pronounced north-south grain of the landscape partially disappears, and we find rather marked geologic differences from place to place. We have here the junction of several major physiographic provinces of western North America. Southeastern Oregon and southern Idaho belong to the *Basin and Range Province*—a vast region that includes Nevada and Western Utah and consists of mostly north-south trending minor ranges and intervening valleys. The *Blue Mountains* of central Oregon and the *Snake River Plain* of southern Idaho are situated north of the Basin and Range Province. The *Columbia Plateau* (sometimes referred to as the Columbia Basin) is located north of the Blue Mountains. It covers southeastern Washington and parts of northern Oregon and western Idaho. The many ranges of central and northern Idaho lie to the east. They are considered part of the *Northern Rocky Mountain Province*.

North of the Columbia Plateau is a situation even more complex, for there exists no well-defined eastern flank of the Cascades-Coast Mountains or western edge to the Rockies. Instead, mountains and high plateaus extend continuously from the Interior Plains of Montana and Alberta to the Pacific Ocean. The hills of northeastern Washington are referred to as the *Okanogan Highland*. They merge with the Cascades to the west along the Okanogan River. The vast upland region east of the Coast Mountains in southern and central British Columbia is known as the *Interior Plateau*. It is divided into various subprovinces, such as the Fraser Plateau, Thompson Plateau, Shuswap Highland, and Okanagan Highland. (Okanogan, like Kootenai [Chapter 6], is spelled differently on opposite sides of the border.) The Interior Plateau is bounded on the east by the Columbia Mountains Province, which includes the Monashee, Selkirk, Purcell, and Cariboo Mountains. The Rocky Mountains proper are separated from the Columbia Mountains by the Rocky Mountain Trench. South of the Canadian border, common usage extends the Rocky Mountain Province west to include the southern end of the Columbia Mountains and the Okanogan Highland.

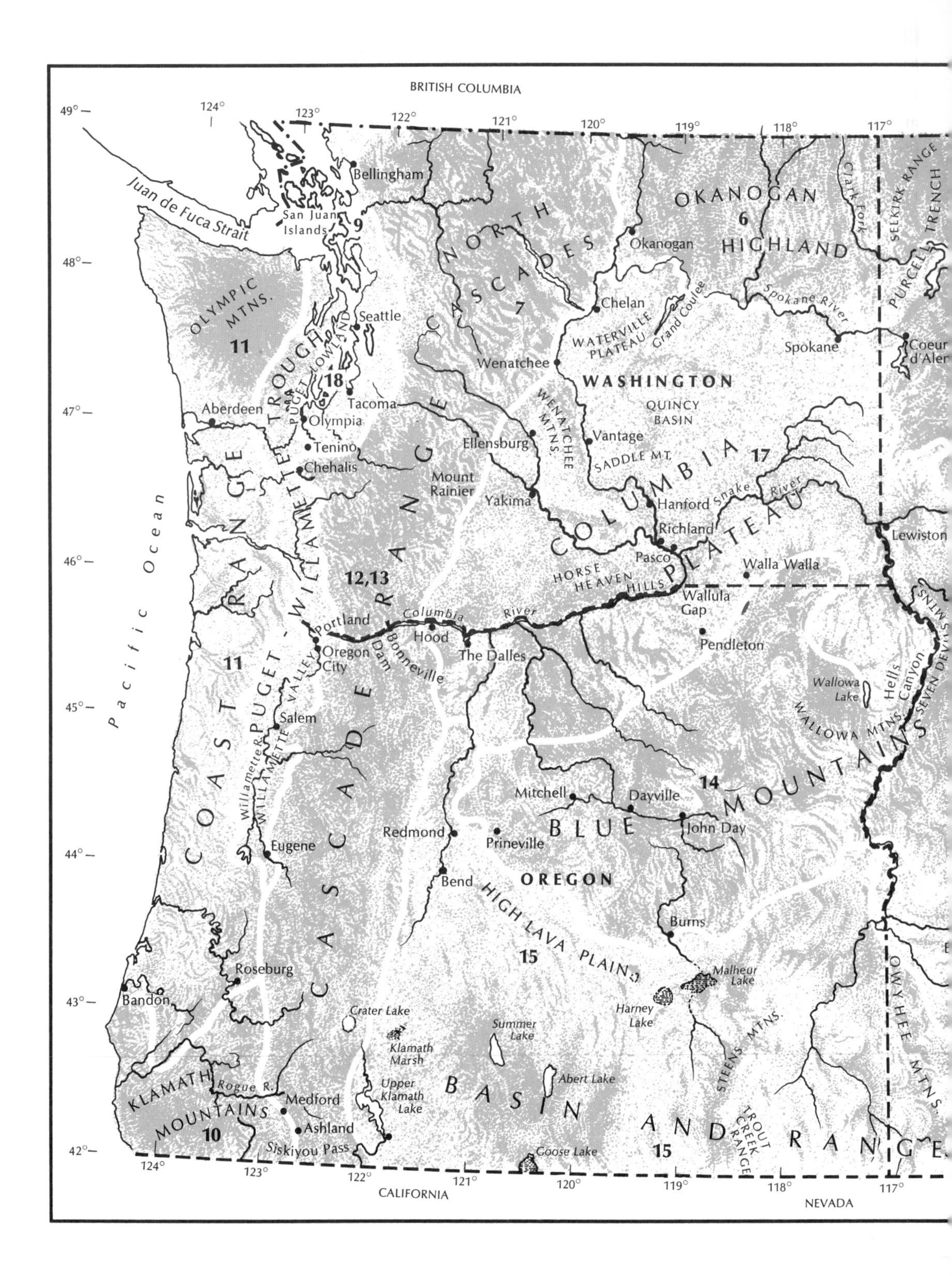

BRITISH COLUMBIA
Juan de Fuca Strait
San Juan Islands
Bellingham
OLYMPIC MTNS.
PUGET TROUGH
PUGET LOWLAND
Seattle
Tacoma
Olympia
Aberdeen
Tenino
Chehalis
NORTH CASCADES
OKANOGAN HIGHLAND
Okanogan
Chelan
Wenatchee
WATERVILLE PLATEAU
Grand Coulee
Spokane River
Spokane
Coeur d'Alene
SELKIRK RANGE
PURCELL TRENCH
Clark Fork
WASHINGTON
QUINCY BASIN
WENATCHEE MTNS.
Vantage
SADDLE MT.
Ellensburg
Mount Rainier
Yakima
Hanford
Richland
Pasco
Snake River
COLUMBIA PLATEAU
HORSE HEAVEN HILLS
Walla Walla
Wallula Gap
Lewiston
Pacific Ocean
COAST RANGE
CASCADE RANGE
PUGET-WILLAMETTE TROUGH
Portland
Columbia River
Hood
The Dalles
Bonneville Dam
Oregon City
WILLAMETTE VALLEY
Willamette R.
Salem
Pendleton
Wallowa Lake
WALLOWA MTNS.
Hells Canyon
SEVEN DEVILS MTNS.
BLUE MOUNTAINS
Mitchell
Dayville
John Day
Redmond
Prineville
Eugene
Bend
OREGON
HIGH LAVA PLAINS
Burns
Malheur Lake
Harney Lake
Roseburg
Bandon
Crater Lake
Klamath Marsh
Summer Lake
STEENS MTNS.
OWYHEE MTNS.
Upper Klamath Lake
Abert Lake
KLAMATH MOUNTAINS
Rogue R.
Medford
Ashland
Siskiyou Pass
Goose Lake
BASIN AND RANGE
TROUT CREEK RANGE
CALIFORNIA
NEVADA
6
7
9
10
11
12,13
14
15
17
18
49°
48°
47°
46°
45°
44°
43°
42°
124°
123°
122°
121°
120°
119°
118°
117°

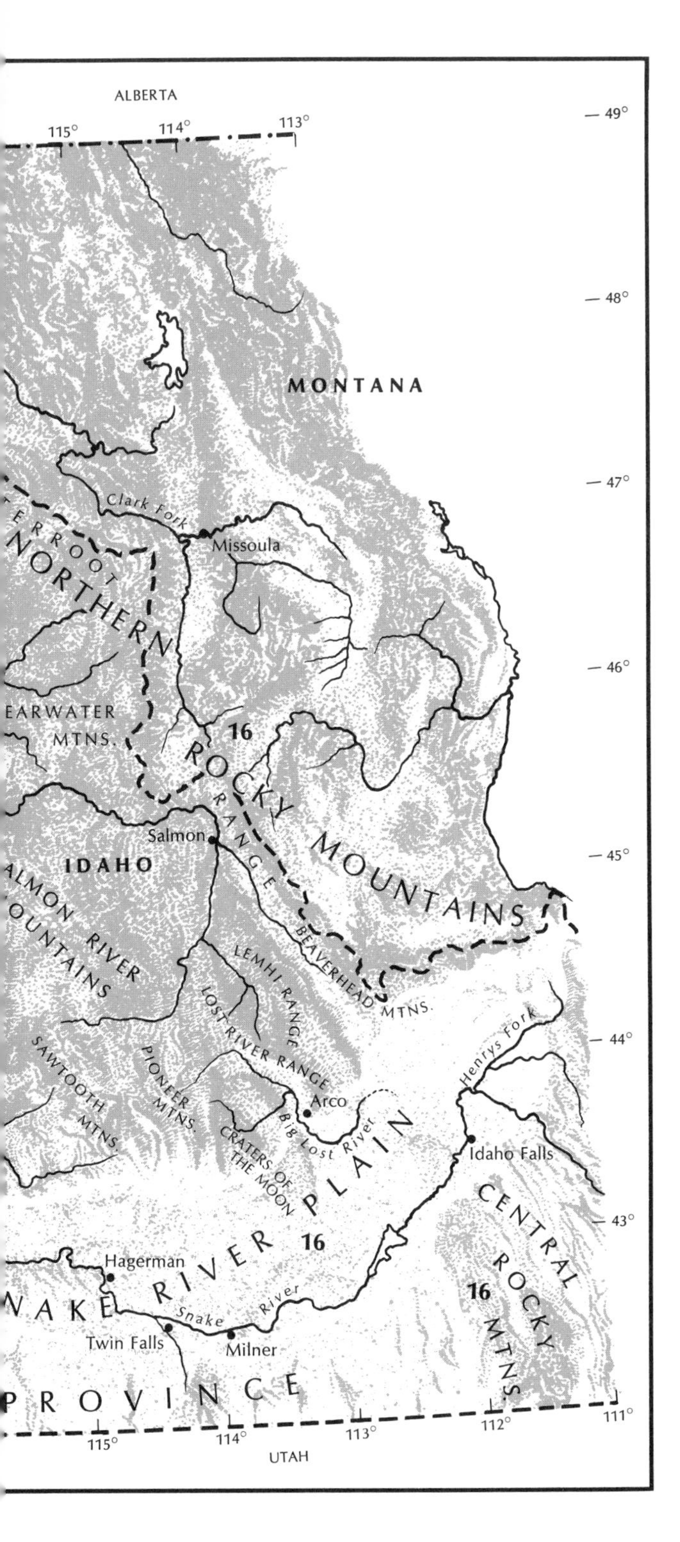

5-1 *Physiographic provinces of the northwestern United States. (Modified from Raisz, 1941.)*

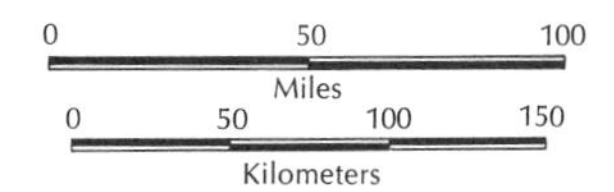

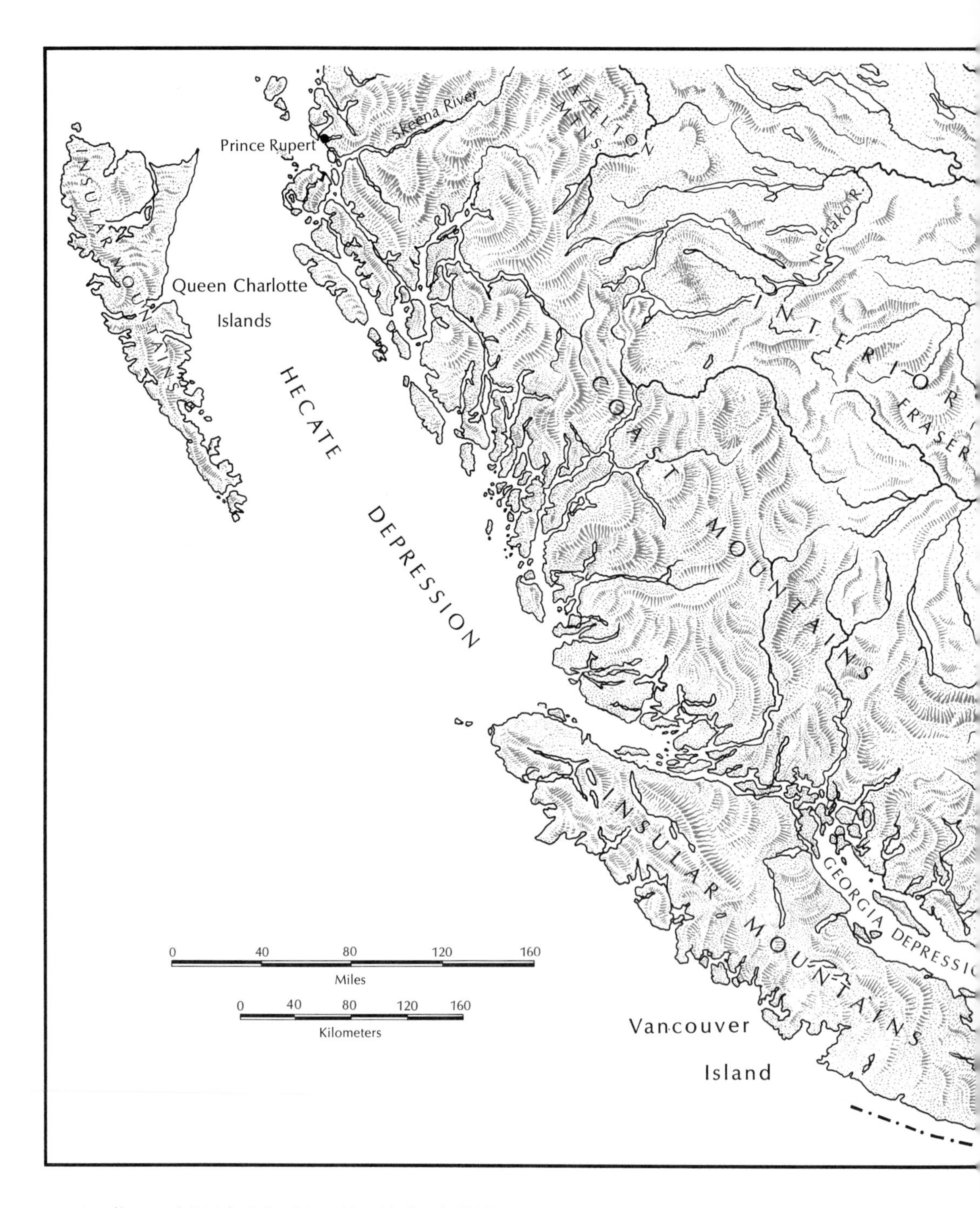

5-2 *Landforms of British Columbia. (After Holland, 1964.)*

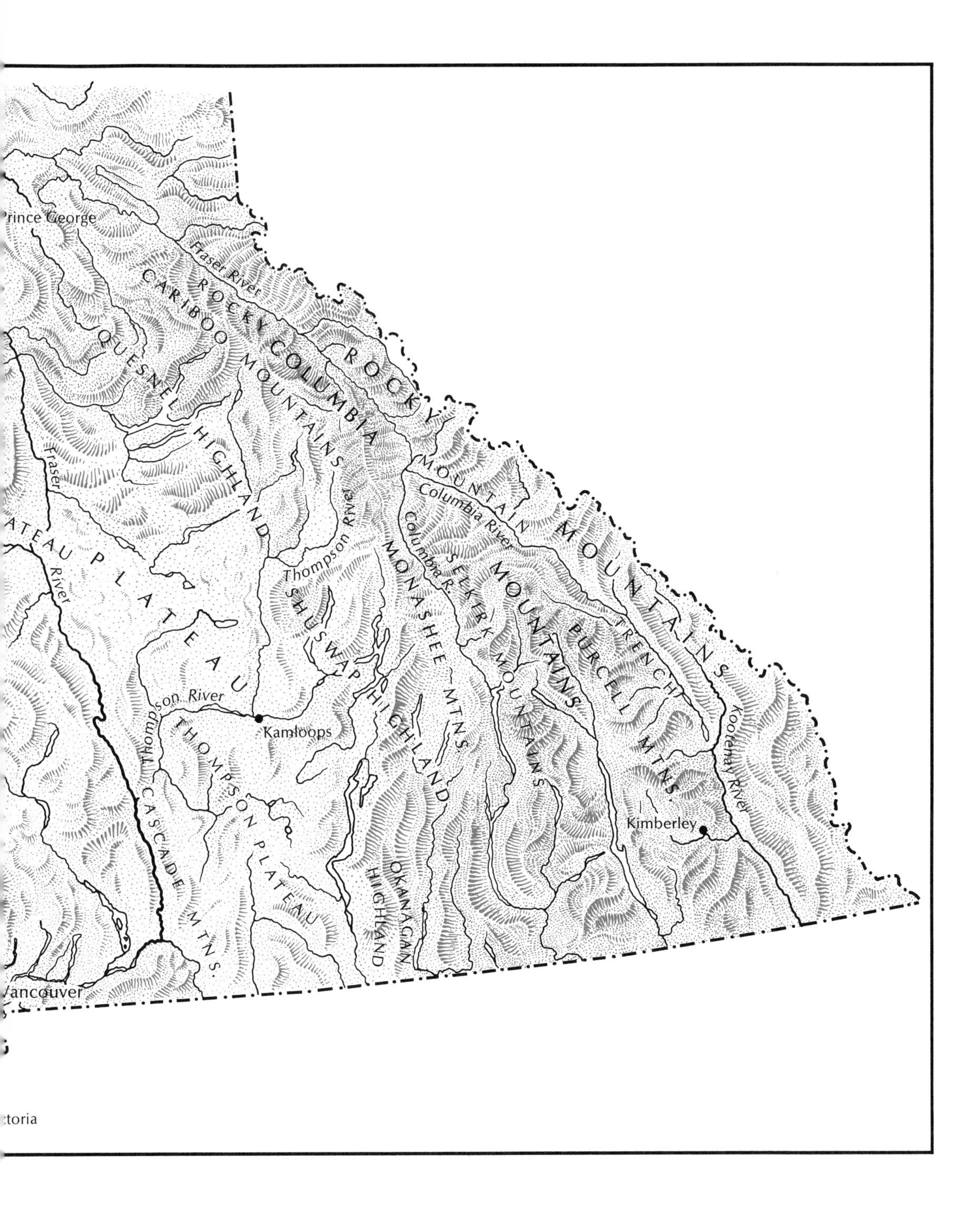

Prince George
Fraser River
ROCKY
CARIBOO
QUESNEL
COLUMBIA
MOUNTAINS
ROCKY
HIGHLAND
Fraser
ATEAU
PLATEAU
River
Thompson River
MOUNTAIN
Columbia River
Columbia R.
MONASHEE
SELKIRK
MOUNTAINS
MOUNTAINS
SHUSWAP
HIGHLAND
Thompson River
Kamloops
THOMPSON
Thompson River
CASCADE
MTNS.
THOMPSON PLATEAU
OKANAGAN
HIGHLAND
MTNS.
MOUNTAINS
PURCELL MTNS.
TRENCH
Kootenay River
Kimberley
Vancouver
toria

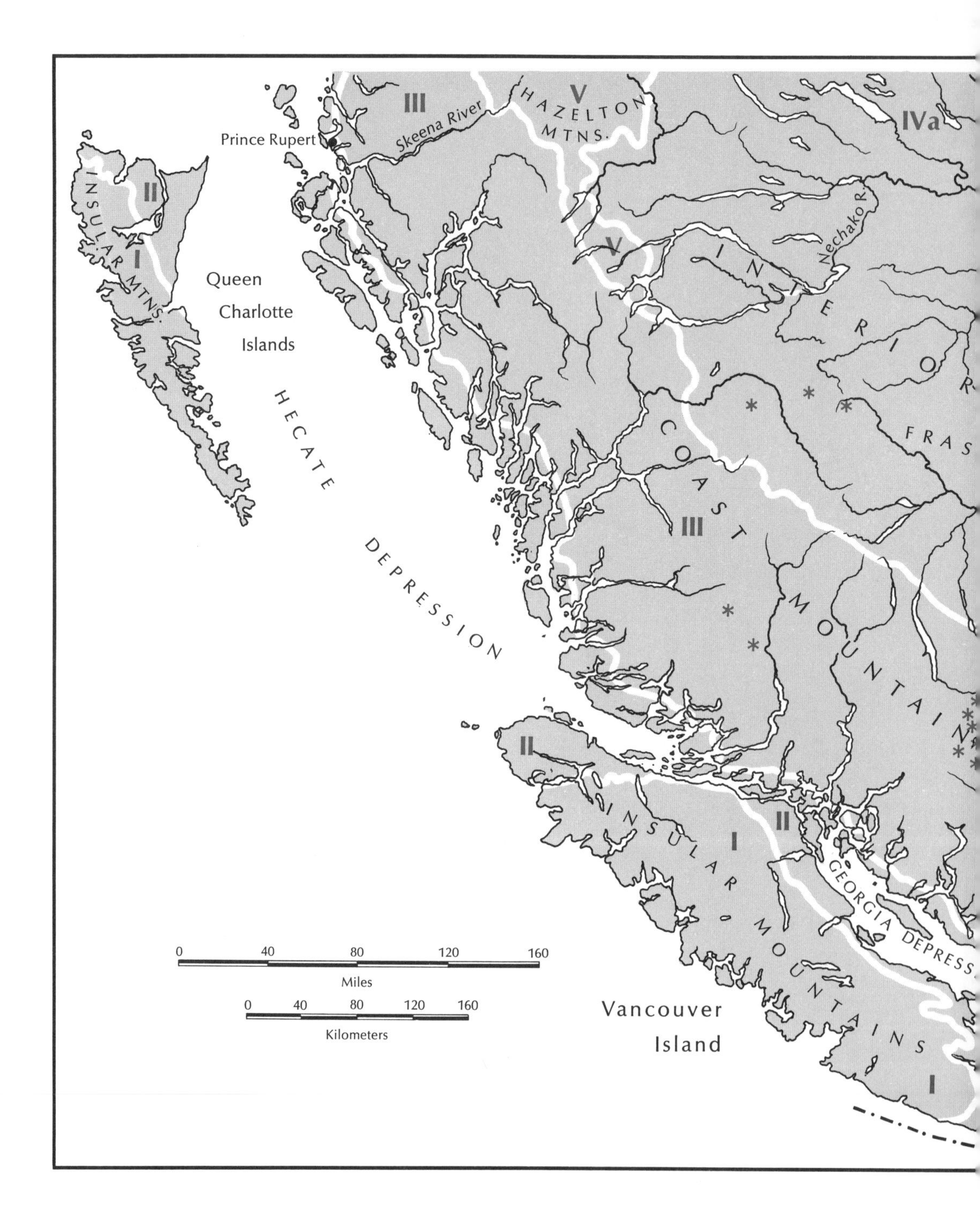

5-3 *Physiographic provinces of British Columbia. Major subdivisions are indicated by numbers as follows: (I) Outer mountain area; (II) Coastal trough; (III) Coast Mountains area; (IVa) Northern and southern plateau area; (IVb) Northern and southern mountain areas; (V) Central plateau and mountain area; (VI) Rocky Mountain Trench; and (VII) Rocky Mountain area. (After Holland, 1964.)*

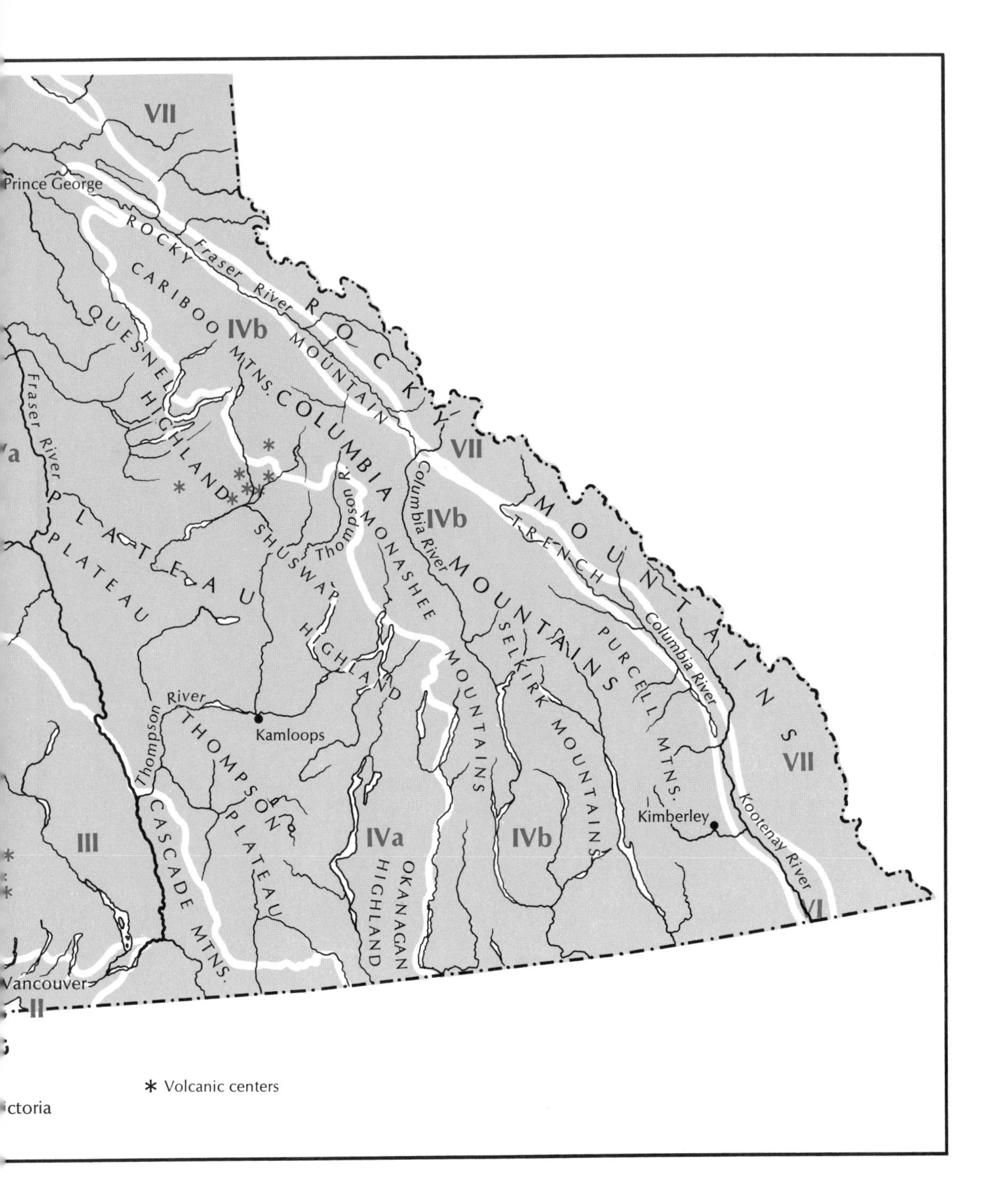

VII
Prince George
ROCKY
Fraser River
ROCKY MOUNTAIN
CARIBOO MTNS.
QUESNEL HIGHLAND
IVb
COLUMBIA MOUNTAINS
Fraser River
VII
Columbia River
Thompson R.
MONASHEE MOUNTAINS
IVb
TRENCH
ROCKY MOUNTAINS
PLATEAU
SHUSWAP HIGHLAND
SELKIRK MOUNTAINS
PURCELL MTNS.
Columbia River
Thompson River
Kamloops
THOMPSON PLATEAU
VII
Kimberley
Kootenay River
III
CASCADE MTNS.
IVa
OKANAGAN HIGHLAND
IVb
VI
Vancouver
II
* Volcanic centers

5-4 *Hills along the valley of the Okanogan River, north-central Washington. They consist of Paleozoic and Mesozoic sedimentary and igneous rocks, somewhat metamorphosed and smoothed by glaciation. (Photograph courtesy of Washington State Department of Commerce and Economic Development.)*

OFFSHORE LANDFORMS

The major physiographic features found off the coast of Oregon, Washington, and British Columbia are a continental shelf, a continental slope, a deep sea basin, and a "mid-ocean" ridge offset along several major fracture zones (Figure 5-5). The shelf slopes gradually seaward from the shoreline to a depth of about 600 feet. It varies in width from less than 1 mile off the Brooks Peninsula near the north end of Vancouver Island to about 50 miles opposite the Strait of Juan de Fuca. The average width of the shelf is about 25 miles. The continental slope passes from the shelf edge down to the margin of the deep sea floor. It has an average width of 30 miles and a slope of 3° to 4°. The sea floor between the continental slope and the mid-ocean ridge is known as the Cascadia Basin. Its average depth is about 10,000 feet, and the surface slopes very gently to the south. The maximum width of the basin is about 200 miles, measured off central Oregon. The ocean ridge is actually several separate ridge segments, offset along major fracture zones or transform faults (Chapter 4).. From south to north, the segments are the Gorda Ridge, the Juan de Fuca Ridge, and the Explorer Ridge. Each is a complex region of mountains, hills, and depressions. So are the fracture zones joining them. Some of the mountains rise to within several thousand feet of the ocean surface.

Recent studies of the continental shelf and slope off the Northwest coast and of the morphology and sediments of Cascadia Basin have produced most interesting results. Chapter 2 considered the fate of sediment. It implied that some could be transported to deep water, in fact some must be for there is not room to deposit it all in estuaries and on the continental shelf. Cascadia Basin, located between an actively eroding continent and an oceanic mountain belt, must be a gigantic sediment trap. The largest single source of sediment would be the Columbia River, but many lesser rivers plus the products of continuous wave erosion must each contribute material.

How does the sediment reach deep water? The positions of large fanlike deposits of sediment (Nitinat and Astoria Fans, Figure 5-5) suggest that much sediment is transported down adjacent submarine canyons cut into the continental shelf and slope. Note that the canyons, in turn, are located near sources of sediment, such as the Columbia River. A most important type of current moves down canyons. Known as a *turbidity current,* it is generated by sediment in suspension. A cubic foot of water with sediment in suspension weighs more than a cubic foot of clear water. Consequently, turbid water tends to move downslope, beneath clear water, for rates of settling of fine sediment are slow, and the turbulence of the flow keeps material in suspension. By this means, large quantities of fine-grained sand, silt, and clay can be transported great distances down gently sloping bottom surfaces. Where the gradient is steep, velocities of as much as 60 miles per hour develop in some turbidity currents. Velocity decreases and the current is spread somewhat as it issues from the mouth of a canyon onto the relatively flat sea floor. This is very similar to a mountain stream issuing from a canyon mouth onto a flat basin floor. Deposition of sediment occurs, producing an alluvial fan that slopes outward in all directions from the mouth of the canyon.

The rest of the Cascadia Basin is not featureless. Its surface is cut by well-defined channels. These are bounded by natural levees, not unlike those found along rivers, like the Mississippi, that are subject to periodic flooding. Most prominent are the Cascadia and Vancouver Channels. These join near the south end of Cascadia Basin and run west through a canyon in the Blanco Fracture Zone to reach the Tuffs Abyssal Plain. Occasional turbidity currents pass down these channels and carry fine sediment far out into the Pacific Ocean.

Recent detailed studies of the morphology and cores of the sediment in the Cascadia Basin have provided some interesting statistics. At present, most sediment deposited in or adjacent to Cascadia Channel comes from the Columbia River. It consists of silt and clay, indicating that the coarser

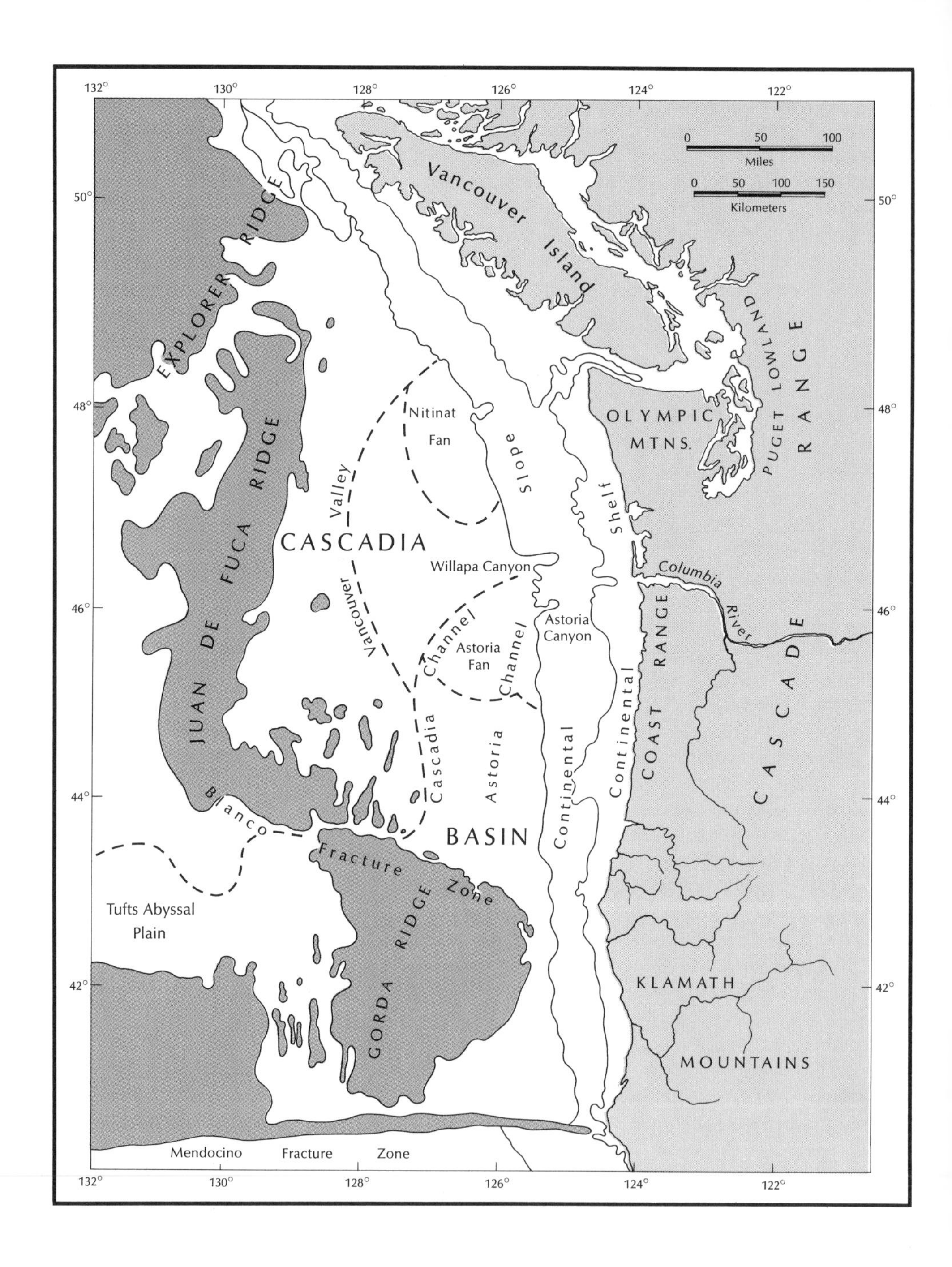

5-5 *Submarine physiography off the coast of Oregon, Washington, and southern British Columbia. Submarine mountains and hilly areas are shown in color. (After Griggs and Kulm, 1970.)*

sand-sized sediment remains behind in the estuary or is spread along the inner edge of the shelf by longshore currents. Cores from the basin floor contain some plant debris and shallow-water marine fossils, transported from the shelf. They also contain a layer of volcanic ash, known to have erupted from Crater Lake (Mount Mazama) approximately 6,600 years ago (Chapter 13). This ash, plus radiocarbon dates from the organic material present, permit calculations of sedimentation rates and of turbidity-current frequency for the recent past. Since deposition of the Mazama Ash, turbidity currents have swept down the channel about once every 500 to 600 years. They probably take less than two days to cross the entire basin, reaching a maximum velocity in the upper channel of perhaps 15 miles per hour. The largest one recorded had a total length of almost 400 miles and deposited sediment 10 miles to either side of the channel and as high as 300 feet above its floor.

Why are turbidity currents so infrequent when sediment is poured into the ocean continuously? Apparently, for most of the time, virtually all but the very finest grained clay is deposited on the shelf—the coarse sand nearshore and the silt and clay farther out. Some of the fine sediment accumulates near the head of the canyons, especially Willapa Canyon, for much of the sediment plume from the Columbia River is transported to the northwest. For centuries, it accumulates on the gentle slopes near the head of the canyon, and then something, an earthquake perhaps, triggers a slide. The motion puts material into suspension and the slide changes to a turbidity current. Within a few hours, a total of as much as 600 million cubic yards of sediment may be transported from the shelf edge to the floor of Cascadia Basin. Truly, this is a fine illustration of the short-term nonuniformity of some geologic processes. More sediment is transported to deep water by this one event than is accomplished during the centuries between turbidity currents. The average rate of sedimentation determined for the floor of the Cascadia Basin (several feet per thousand years along Cascadia Channel) does not really suggest the infrequency of sediment deposition.

THE GEOLOGIC FRAMEWORK

The prime concern of this volume is to unravel the geologic evolution of a relatively small part of the Cordillera of North America. However, a proper introduction to the subject requires a short summary of the history of the entire continent.

Central and eastern Canada (the Canadian Shield Province) contain the most complete record of the events of the Precambrian Era. Much of this vast region of relatively low topography was never covered by younger rock layers, and glacial scour and erosion in the Quaternary Period laid bare fresh exposures of these ancient strata. The rocks have been metamorphosed (recrystallized due to heat and pressure) to a greater or lesser degree. They consist of clearly identifiable volcanic and sedimentary strata, generally resting on crystalline schists and gneisses whose parentage can be interpreted only with some difficulty. Granitic igneous intrusions (also of Precambrian age) cut both types of rock. They were the source of much of the mineral wealth of this region. Unraveling the geology of this immense region of Precambrian rocks is difficult due to the near-absence of a fossil record and the metamorphic recrystallization that has obscured so much of the original detail. Radiometric age determinations have been of considerable help. The pattern that has emerged is basically one of periods of sediment accumulation and volcanic activity lasting several hundred million years, terminated by times of orogeny (mountain building), metamorphism, and granitic intrusions. Traditional geosynclinal patterns can be recognized. There is a slight suggestion of the oldest strata occurring in the center of the province with younger Precambrian strata marginal to the core.

Precambrian rocks crop out elsewhere on the continent. For the most part, they are of such limited exposure (due to burial beneath younger rocks) or so difficult to interpret and relate regionally that they have not yielded much data useful in paleogeographic interpretation.

The record for the Paleozoic Era is much better. By Cambrian time, the continent was surrounded largely by complex subsiding shelves or troughs in which great thicknesses of strata accumulated.

There was the Appalachian Geosyncline along the East Coast, the Ouachita Geosyncline through the Gulf Coast region, the Cordilleran Geosyncline from Central America north through Alaska, and the Franklin Geosyncline from northern Greenland through the Canadian Arctic. The strata of these geosynclines were ultimately deformed, uplifted, and added to the continent. This would seem to support the concept of continental growth by marginal geosynclinal evolution, especially since the continental interior remained relatively stable (and has subsequently), despite the great deformational activity around it. However, we can no longer regard the North American continent as a "closed system" divorced from other continents. The Atlantic Ocean apparently has expanded greatly since the Paleozoic Era. A reevaluation of the geologic record of the Appalachian Geosyncline strongly suggests that at times the North American and Eurasian continents were joined. Similar large movements have complicated the interpretation of the Cordillera, where small continental pieces may have been rafted in and added to the North American plate.

The geologic evolution of the two sides of the continent also differed markedly during the Mesozoic and Cenozoic Eras. The Appalachian Geosyncline was destroyed by several Paleozoic orogenies; since Late Triassic time, the East Coast has seen little deformation. The Cordilleran Geosyncline, also subjected to several Paleozoic deformations, persisted through the Mesozoic. The sea was excluded from most of the region only after major Late Mesozoic orogeny and uplift. Furthermore, deformation in the Cordillera has continued to the present. Thus, for the past several hundred million years, the East Coast has been relatively stable; and the Cordilleran region has been most unstable. This lends strong support to the theory that the North American plate has been moving relatively westward, with the Cordillera along the leading edge and the East Coast as the trailing edge. How the pioneers of the continental drift theory would enjoy reading the current journals!

THE NORTHWEST IN THE PRECAMBRIAN ERA

Our record is very meager. The only strata demonstrably as old as Precambrian are found in the Northern Rockies and the Columbia Mountains (Figures 5-6, 5-7). The strata are primarily sedimentary and monotonous. They consist of as much as 50,000 feet of sandstone, shale, and limestone, all slightly metamorphosed. The sediments were deposited in a shallow marine environment, probably quite similar to our present Mississippi Delta. They were deposited by rivers draining the Canadian Shield Province to the east. In the Columbia Mountains, some Upper Precambrian volcanic rocks were deposited on top of these marine sediments; but, on the whole, volcanism was not very pronounced.

To the west, we may have rocks as old as Precambrian, but this is not certain. The possible candidates are principally schists and gneisses at the bottom of the thick Paleozoic sequence of the North Cascades and the southern Coast Mountains of British Columbia. All that we know for sure is that these "basement rocks" are older than the Devonian Period. Considering the large size of the area we are treating, plus the fact that the Precambrian Era represents six-sevenths of geologic time, this is a very scanty record. The reason, of course, is burial. Most Precambrian strata lie under great piles of younger sedimentary and volcanic rocks, and if they ever reach the surface again, it will be only through large-scale uplift and erosional stripping of the younger cover.

THE CORDILLERAN GEOSYNCLINE

Paleozoic and Mesozoic rocks were deposited in the giant Cordilleran Geosyncline. The strata are generally marine sandstone, shale, and limestone

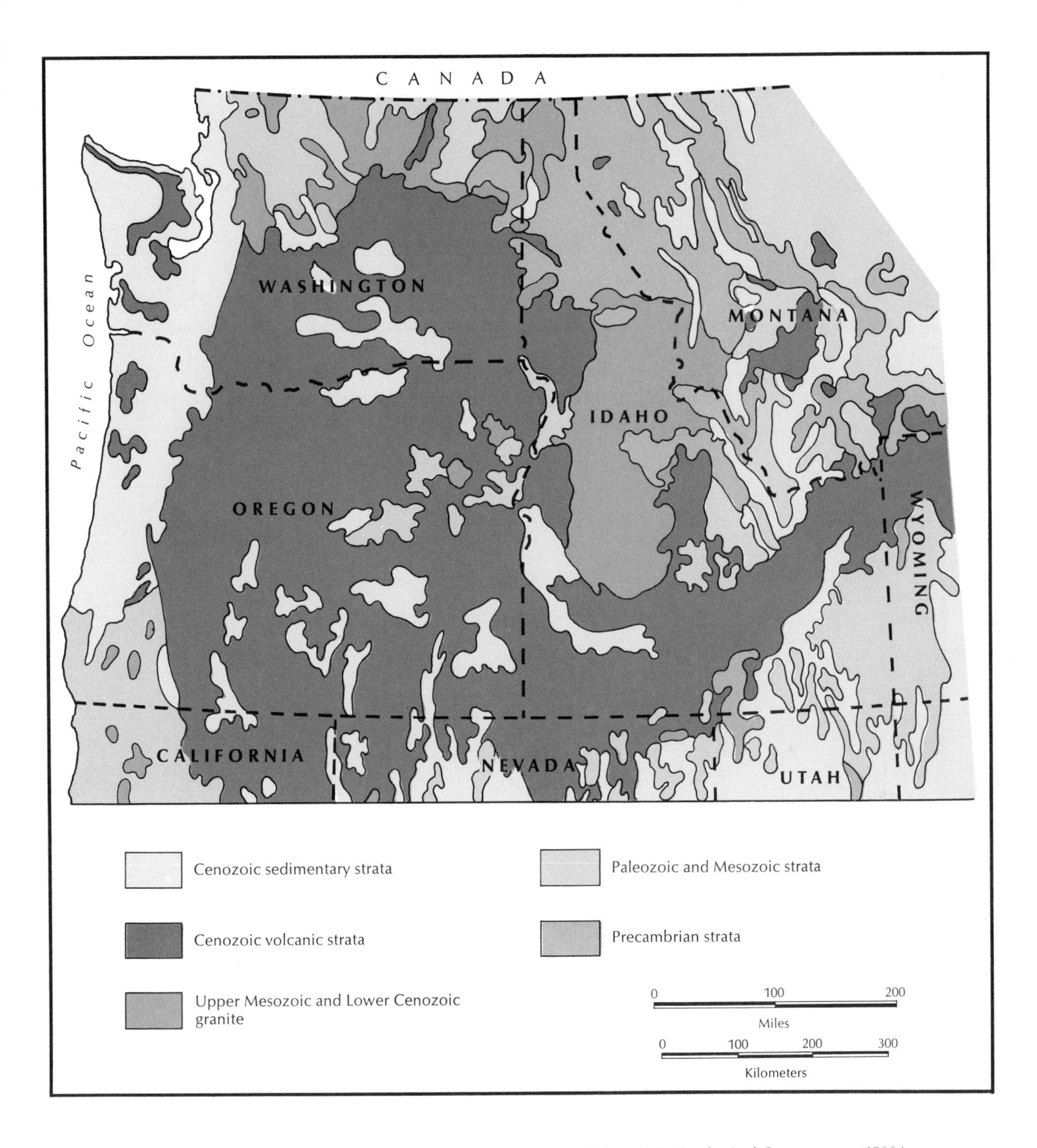

5-6 *Generalized geologic map of the northwestern United States. (After U.S. Geological Survey map, 1966.)*

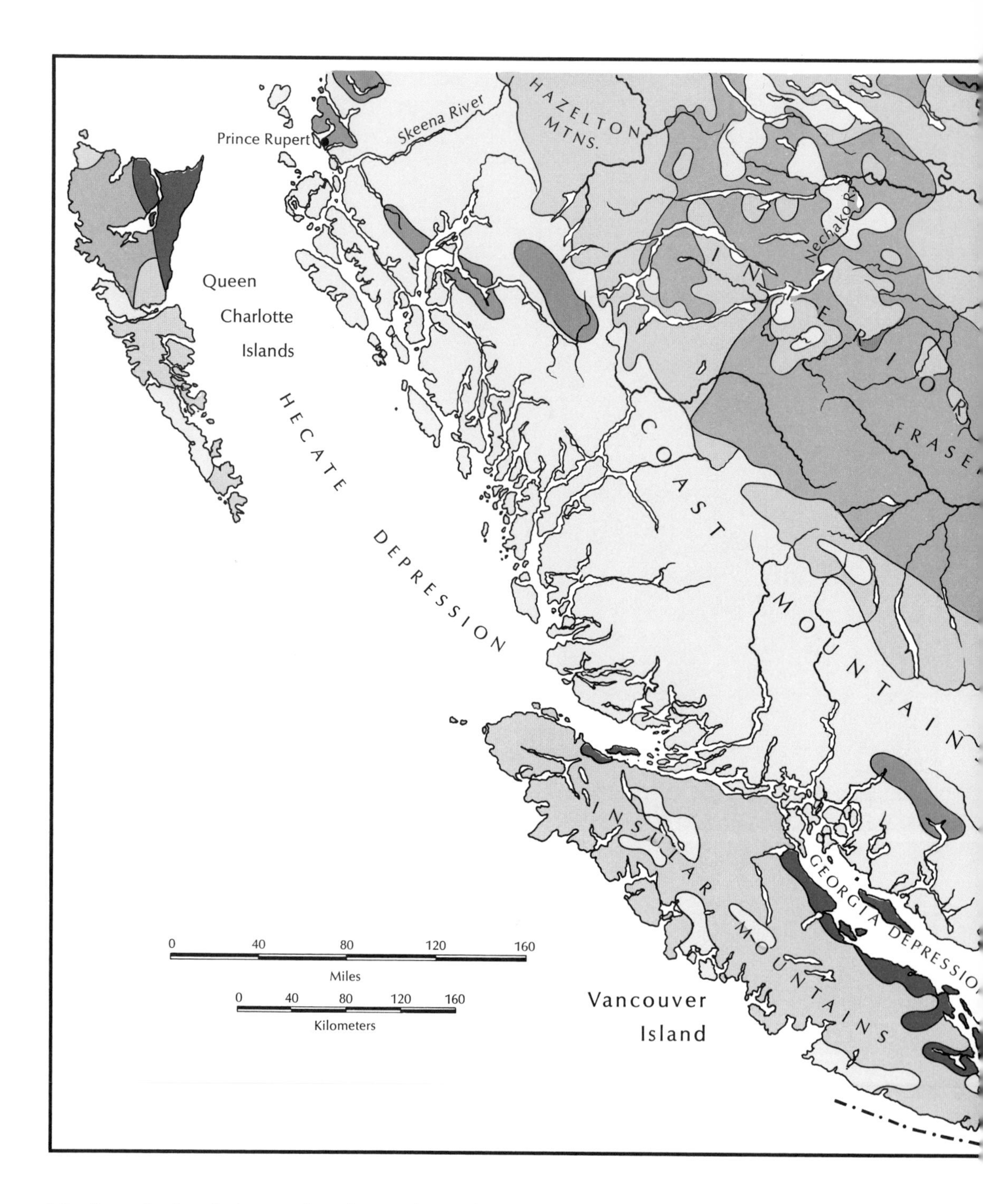

5-7 *Generalized geologic map of southern and central British Columbia. (After Holland, 1964.)*

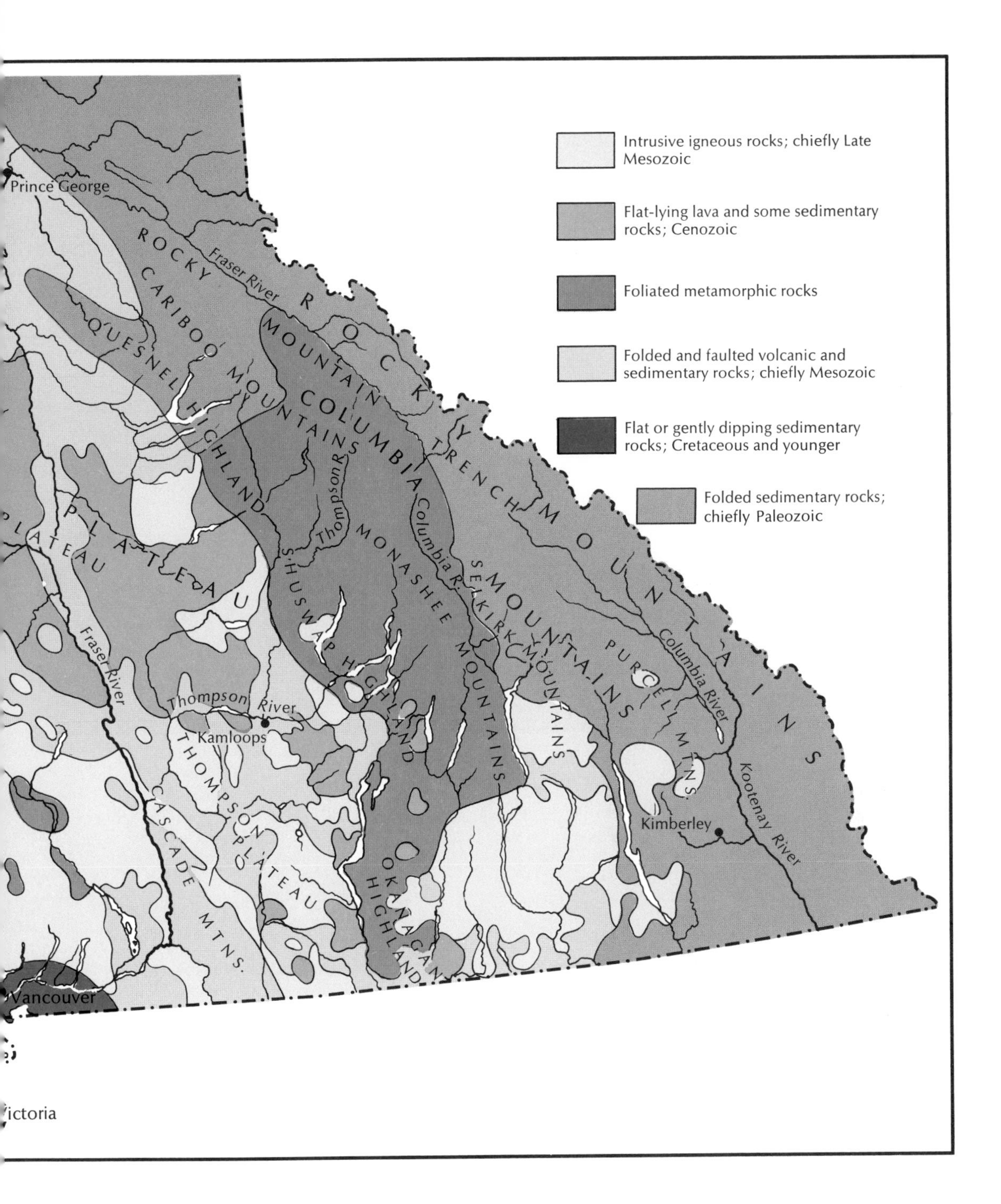

Intrusive igneous rocks; chiefly Late Mesozoic
Flat-lying lava and some sedimentary rocks; Cenozoic
Foliated metamorphic rocks
Folded and faulted volcanic and sedimentary rocks; chiefly Mesozoic
Flat or gently dipping sedimentary rocks; Cretaceous and younger
Folded sedimentary rocks; chiefly Paleozoic
Prince George
ROCKY MOUNTAIN TRENCH
ROCKY MOUNTAINS
Fraser River
CARIBOO MOUNTAINS
QUESNEL HIGHLAND
COLUMBIA
MONASHEE MOUNTAINS
SELKIRK MOUNTAINS
PURCELL MTNS.
Columbia R.
Columbia River
Thompson R.
SHUSWAP HIGHLAND
PLATEAU
Thompson River
Kamloops
THOMPSON PLATEAU
CASCADE MTNS.
OKANAGAN HIGHLAND
Kimberley
Kootenay River
Vancouver
Victoria

5-8 *The Rocky Mountain Trench southeast of Valemount, British Columbia. View southeast, with Rocky Mountains to the left and flank of Columbia Mountains to the right. (Photograph courtesy of K. V. Campbell.)*

in the Rocky Mountains, whereas much volcanic material is interbedded with marine sedimentary rocks to the west. Nowhere was sedimentation continuous. Paleozoic and Mesozoic strata crop out extensively in British Columbia and northern Washington, in central and eastern Idaho, in the Blue Mountains, and in the Klamath Mountains. These are regions of relatively recent uplift and erosion, so that younger Cenozoic rocks have been stripped away, providing "windows" through which we can see the older strata.

The time spanning the latter part of the Mesozoic and the beginning of the Cenozoic Eras was most traumatic for the Cordillera. The Cordilleran Geosyncline—never a very stable feature at best—was destroyed by large-scale folding and faulting, extensive metamorphism, the emplacement of gigantic granitic intrusions, and wholesale regional uplift. Before this time, most of the strata had been laid down beneath the sea, but thereafter the sea never again covered the continental interior. The Cenozoic deposits (which are locally very thick) are nonmarine, except near the Pacific shore. The Coast Range of Oregon and Washington was the site of a new geosyncline that formed early in the Cenozoic Era and that has been uplifted only recently.

THE CENOZOIC ERA—FIRE AND ICE

The Cenozoic record of the Northwest is perhaps the most interesting of all. Its aspect is volcanic and it includes one of the most remarkable collections of lavas and other volcanic strata found anywhere in the world. Some of the features are well known, such as the Cascade volcanoes and the great lava sheets of the Columbia Plateau and the Snake River Plain; but these lavas constitute only a small fraction of the total volcanic sequence. There are many sedimentary units interbedded with the lavas. They tell us much about the environment of deposition—especially those that contain fossils. As we might expect, the Cenozoic conglomerates, sandstones, siltstones, and claystones are composed to a large degree of fragments of volcanic material derived by erosion from nearby volcanic rocks. Many formations are rich in volcanic ash blown out of neighboring cones.

The past million years or so are of particular interest for another reason—glaciation. Glaciers exist on the highest peaks in the Northwest today, but these are only tiny remnants of once-vast ice sheets that covered most of Alaska, Canada, and the northern part of the "Lower 48." The glacial history of the Northwest is as interesting as that of any place in the world. Unreal as these glacial landscapes of a few thousands or tens of thousands of years ago seem today, they take on special meaning when we realize that there is no basis for assuming that we are actually out of the Ice Ages.

6 the columbia and rocky mountains

■ This is quiet country—a land of immense space and sky and seemingly endless forested mountain slopes. Geologically, it seems a quiet country as well, without the young volcanoes and frequent earthquakes that attest to the crustal unrest along the continental margin farther west. It was not always like this. One hundred million years ago, this eastern part of the Cordillera was wracked by violent earth movements. High mountains rose up out of the edge of the sea. Great slabs of rock a mile or more thick and many tens of miles long were thrust as much as 100 miles east toward the low interior of the continent. Solid rock was compressed so severely that the opposite sides of folds were flattened parallel to each other, and thousands of cubic miles of rock were completely recrystallized by metamorphism.

Deformations like this occurred throughout the entire Cordilleran region. The Rocky Mountains–Columbia Mountains area lay mostly to the east of the zone of most intense deformation. To unravel the Precambrian and Paleozoic history of the Cordillera, one starts in the east, for rocks this old farther west have been deformed so intensely that they tell us very little about ancient environments. Furthermore, the past 100 million years has seen continued uplift and erosion of the eastern part of the Cordillera. This has saved it from burial beneath thick sections of Cenozoic sediments and volcanic strata that conceal so much of the older record elsewhere.

GEOGRAPHY

The area in question lies north and east of the Columbia Plateau in Washington and Idaho, and east of the Interior Plateau of British Columbia. The topography of this region is complex, consisting of a great number of named ranges, basins, and rivers (Figures 5-1, 5-2, 5-3). The easternmost part of the Cordillera is known as the Rocky Mountain Province. In the United States, the various ranges of northwestern Montana and northern Idaho commonly are considered a part of the Rockies, although they lie west of the Rocky Mountain Trench and are continuous with the Columbia Mountains of southeastern British Columbia.

Perhaps the best way to introduce the topography of this region is to consider the drainage pattern. The continental divide is the crest of the Rockies. It extends from west-central Montana north along the British Columbia–Alberta border to latitude 54°N, then diagonally across the northeastern corner of British Columbia. West of the divide, all drainage is to the Pacific Ocean. Only two major rivers—the Columbia and the Fraser—breach the Cascades-Coast Mountains chain. Most of the other rivers of this vast region are tributary to one or the other of these giants. The drainage pattern is quite complex. Many major river segments run north-south, but they show some curious reversals. The Columbia River originates in the Rocky Mountain Trench southwest of Calgary and flows northwest for 150 miles. It then turns and flows south past Revelstoke and Trail into northeastern Washington. At the edge of the Columbia Plateau, it runs west and forms the southern border of the Okanogan Highland Province. It turns south again at the foot of the North Cascades at Pateros. The Fraser River also originates in the Rocky Mountain Trench. It flows northwest for almost 150 miles to Prince George, then south diagonally across the Interior Plateau, and finally west through the Coast Mountains–Cascades chain to the ocean. The most confused course of all is made by the Kootenay (Kootenai) River which heads in the Rockies only a few miles east of the headwaters of the Columbia River. It flows south along the Rocky Mountain Trench into northwestern Montana, then northwest across the top of the Idaho Panhandle and back into British Columbia. It then runs west out of Kootenay Lake and joins the south-flowing Columbia near Nelson.

The reasons for this complicated drainage pattern are several. They include recent deformation of the region, the tendency for rivers to follow more easily eroded strata, and changes brought about by Pleistocene glaciation. Many of the rivers run in deep glaciated valleys and form the borders between named subranges of the region. Thus the Kootenay and Duncan Rivers separate the Purcell and the Selkirk Ranges; the Columbia divides the Selkirks from the Monashee Mountains; the lower Fraser separates the Coast Mountains from the Cascades; and the lower Okanogan River divides the Okanogan Highland from the Cascades.

GENERAL STRUCTURE OF THE REGION

A useful method of introducing the distribution of rock types and structures in a region is by means of a geologic cross section (Figure 6-2). This is a vertical view into the earth's crust along a straight line. It shows in profile the configuration of the Earth's surface, as well as the rocks and structures, along this line. A typical cross section also portrays the geology to some depth below the surface. This is usually conjecture, although it may be based in part on information from wells, underground mine workings, subsurface geophysical data, or surface exposures away from the line of section. In general, the accuracy of a cross section

6-1 *The lower reaches of the Kettle River, northeastern Washington. The scene is typical of the Okanogan Highland. The strata exposed are of Paleozoic age. (Photograph courtesy of Washington State Department of Commerce and Economic Development.)*

will diminish with depth. A cross section is inherently less accurate than a map. Often, however, it is more useful in representing the geologic patterns (especially the structures) in a particular region. Such is the case with Figure 6-2, a very generalized cross section running east-west across the Rockies and Columbia Mountains in southeastern British Columbia.

The Interior Plains Province of Alberta is underlain by relatively flat-lying sedimentary strata of Paleozoic, Mesozoic, and Cenozoic ages resting on an eroded surface of Precambrian gneiss. The gneiss is a continuation of the crystalline rocks of the Canadian Shield Province to the east. The Paleozoic and Mesozoic strata were deposited in the eastern (shallow-water) part of the Cordilleran Geosyncline. The Cenozoic beds consist of nonmarine river and lake sediments. They were derived from mountains to the west that rose from the geosyncline as the Mesozoic Era drew to a close.

The Rockies consist of folded and fractured sedimentary strata of Upper Precambrian, Paleozoic, and Mesozoic age. The Paleozoic and Mesozoic formations correlate with those that underlie the Interior Plains. The Precambrian strata were deposited along the western margin of the continent before the beginning of the Cordilleran Geosyncline. They have a total thickness of almost 10 miles and consist largely of slightly metamorphosed sandstone, shale, and limestone. They are exposed best in the Purcell Mountains west of the Rocky Mountain Trench and in the Northern Rocky Mountain Province of northwestern Montana and northern Idaho. (State and national boundaries inevitably result in confusing geographic terminology. The Northern Rockies [U.S.] are south of the Southern Rockies [Canada]. Similar problems exist with Cascade and Coast Range nomenclature for Washington and Oregon.) Major faults in the Purcell Mountains are steeply dipping or vertical. In the Rockies, they dip westward at low to moderate angles, forming an impressive series of overlapping plates, all of which moved eastward near the end of the Mesozoic Era.

6-2 *Generalized geologic cross section through southeastern British Columbia.*

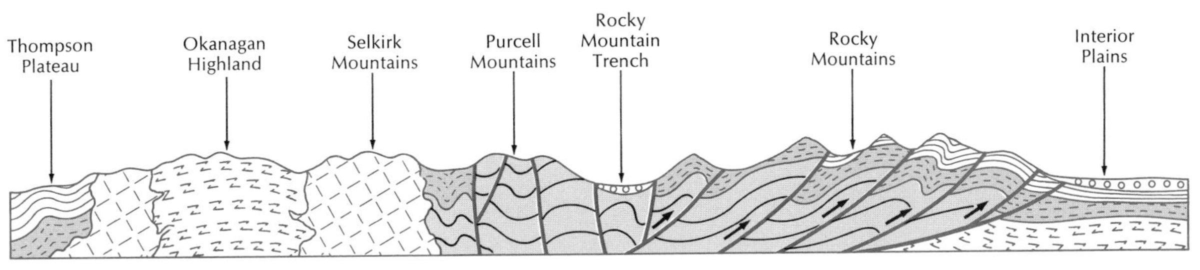

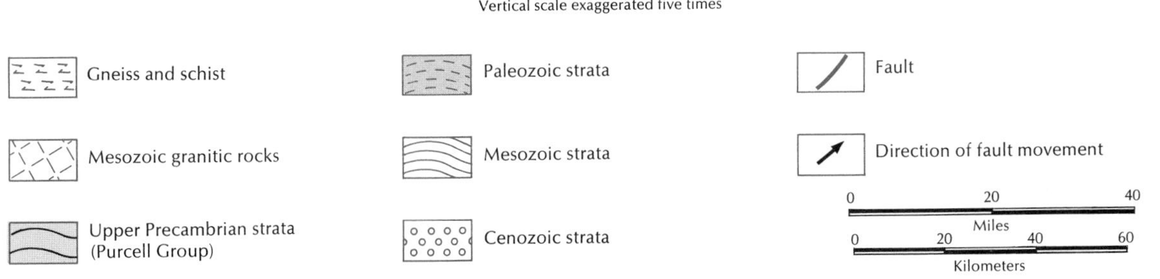

The geology west of the Purcell Mountains is more complicated—much more so than is suggested by the cross section. Furthermore, it changes quite markedly north and south away from the line of section, although the basic elements shown remain characteristic. These include the following:

1. Intensely folded and faulted marine sedimentary rocks in the Columbia Mountains. These range in age from Late Precambrian to Jurassic.
2. A complex mixture of schists and gneisses, forming a crystalline belt through the Okanogan Highland, Shuswap Highland, and Monashee Mountains. These metamorphic rocks were derived largely from the metamorphism of Paleozoic strata.
3. Large granitic igneous intrusions (batholiths), mostly of Jurassic and Cretaceous age.

6-3 *The Rocky Mountains north of Prince George, British Columbia. The strata are gently folded sedimentary rocks of Paleozoic age. The sharpness of the topography reflects recent alpine glaciation.* (*Photograph courtesy of James Crawford.*)

4. Relatively unmetamorphosed Paleozoic and Mesozoic strata, including much volcanic material, along the eastern edge of the Interior Plateau Province (Chapter 8). These formations have been folded and fractured extensively.

THE HISTORY IN BRIEF

For much of Late Precambrian, Paleozoic, and Mesozoic time, the Rocky and Columbia Mountains region lay somewhat to the west of the shoreline of the North American continent. Rivers draining the continent supplied sediment—much of it mud. Limestone beds formed in this shallow shelf environment during periods of less turbid water. Subsidence generally kept pace with sedimentation so that tens of thousands of feet of shallow-water marine strata accumulated during this long interval. Periods of minor uplift and erosion did occur and the record is by no means continuous.

To the west, the picture is less clear, and paleogeographic reconstruction is more difficult. The

rocks suggest deposition in deeper water. At times, a source of sediment lay west of the region—perhaps a volcanic archipelago or even a separate plate that was not originally a part of the North American continent. Throughout most of the eastern part of the Cordillera, the major episode of deformation was in the latter part of the Mesozoic Era and early part of the Cenozoic Era. The orogeny was a long-lived disturbance that began with metamorphism and the emplacement of large, granitic batholiths in the Columbia Mountains and Okanogan Highland. This was followed by immense large-scale movement by thrust faulting of the marine strata eastward toward the continental interior. The orogeny marked the end of the Cordilleran Geosyncline. The seas were expelled from the western interior of the continent in the Cretaceous and never returned.

The Cenozoic history of the eastern Cordillera has been anticlimactic. Local thick sequences of nonmarine sediments accumulated in troughs between eroding mountain ranges, much as they do today. By the middle of the era, the topography was probably one of relatively low relief. Late Cenozoic uplift, renewed vigorous erosion, and Pleistocene glaciation have combined to produce the present mountainous terrain.

PRECAMBRIAN STRATA

Northeastern Washington, northern Idaho, southeastern British Columbia, and western Montana have mountain ranges that consist entirely of rocks of Late Precambrian age. One can travel for hundreds of miles and see no other rocks. These strata are known as the Belt Supergroup south of the international boundary and as the Purcell Group in Canada (Table 6-1). They consist of well-stratified beds of limestone, shale, and quartz sandstone and have a total thickness of as much as 50,000 feet. The rocks appear to have formed principally in shallow water, in an environment that compares perhaps to the Gulf Coast region today. The sediment was derived by erosion of the continental interior. It was deposited by west-flowing rivers in gigantic, slowly subsiding deltas and shallow banks at the edge of the sea.

The only well-identified fossils in the Belt-Purcell rocks are algal masses (mostly genus *Collenia*) in the limestone. These are of little use in age determination. Sometime after sedimentation ceased, the beds locally were folded, faulted, and intruded by granitic rocks that have been dated by radiometric methods as approximately 800 million years old.

Throughout the region, these rocks are slightly

TABLE 6-1 Precambrian—early Paleozoic stratigraphy of northeastern Washington

AGE	PEND OREILLE CO., WASH.	COLUMBIA MOUNTAINS, B.C.
Silurian-Mississippian(?)	Marine shale and limestone	Lardeau Group
Ordovician	Ledbetter Slate	
Cambrian	Metaline Limestone Maitlen Phyllite Gypsy Quartzite	
Late Precambrian	Monk Phyllite Leola Volcanics (marine greenstone) Shedroof Conglomerate	Windermere Group
Precambrian	Belt Supergroup	Purcell Group

▬▬ Major unconformity

metamorphosed. The quartz sandstones have been transformed into white, pink, and green quartzite; the shales into slate or shiny phyllite; and the limestone into white, sugary marble. Some of the best and most famous exposures of Belt strata are in the mountains of Glacier National Park and Waterton Lakes National Park in northwestern Montana and southwestern Alberta. The Purcell Group is well displayed in its type area, the Purcell Range of southeastern British Columbia. Here it contains minor lava flows and volcanic ash beds.

Sedimentation resumed near the end of the Precambrian Era, producing the Windermere Group. It is unlike the Belt-Purcell in that it contains considerable thicknesses of volcanic material. Windermere rocks are found principally in the Okanogan Highland and the western part of the Columbia Mountains. The lowest formation is a course cobble and boulder bed, almost a mile thick in the southern Purcell Range. Above this are submarine volcanic rocks, overlain by shales, sandstones, and, to the north, limestones. Thus the overall composition of the upper part of the Windermere section is not unlike that of the Belt-Purcell sequence. Likewise, the environment of deposition was similar, with sediment derived from the east and deposited in a shallow, coastal trough.

BEGINNING OF THE CORDILLERAN GEOSYNCLINE

The Geologic Time Scale (see Appendix A), with its formal divisions, implies a certain hierarchy of significance to the time boundaries. In practice, some stratigraphic sections refuse to cooperate. In southeastern British Columbia and northeastern Washington, the Precambrian-Paleozoic boundary cannot even be identified. Windermere strata grade upward into fossiliferous Cambrian beds with no indication of a break in sedimentation. The oldest Cambrian formation is typically a white quartzite sandstone, passing upward into a thick, monotonous shale unit. This is overlain by a rather pure limestone. The names of these units vary from place to place, but the sequence is pretty much the same throughout this part of the Cordillera. Some excellent marine fossils, including the extinct trilobites, are preserved. The Cambrian section totals 15,000 feet in the Okanogan Highland of northeastern Washington.

Conditions did not change much for the next several hundred million years. Marine shales and shallow-water limestones were deposited during the Ordovician, Silurian, and first part of the Devonian Periods. At most localities, deposition was not continuous, for local uplift and erosion produced minor unconformities in the record. Ordovician shales in many parts of the region contain fossil graptolites—small, extinct invertebrates whose carbonized remains somewhat resemble pencil marks. These are important "index" fossils around the world for the Ordovician Period. Despite their relatively simple structure, graptolites evolved rapidly into a variety of recognizable forms. Consequently, they provide rather close dating of the enclosing strata. Graptolites occur almost exclusively in dark shales and are believed to have accumulated in near-shore mudbanks rich in seaweed. The group became extinct later in the Paleozoic Era.

THE AMAZING BURGESS SHALE

High on the flanks of Mount Stephen in the Rocky Mountains of British Columbia is exposed one of the most famous formations in the world. The unit is a dark shale, the Burgess Shale, similar in many ways to shales found elsewhere in the Paleozoic section of the geosyncline. It is of Middle Cambrian age. What makes the Burgess Shale unique is its preservation of a wide variety of fossil organisms, many of which are seldom found in the fossil record. The fossils include ten genera of worms, some still encased in their burrows or tubes. Conditions in the enclosing black mud were such that seemingly every detail of their morphology was impressed in the sediment or in the carbon film formed from the residues of their body substances. The Burgess Shale also contains other genera of wormlike organisms, three extinct subclasses of crustaceans, plus many other *Arthropoda,* including exquisitely detailed trilobites. No other formation in the world's geologic record has yielded comparable fossils, although the Jurassic Solenhofen Limestone of Bavaria has come close.

The fauna of the Burgess Shale and other Cambrian formations points up a problem that has bothered geologists for many years. The Cambrian Period was considered once to have been the beginning of life on earth. Cambrian fossils consequently should represent primitive life forms—the simple beginnings from which evolved the complex organic variety of subsequent periods. However, the fossil fauna of Cambrian strata is not simple, at least the various phylla represented show marked diversity of form. Moreover, when details are preserved, as in the Burgess Shale, the highly evolved, intricate anatomy of individual species is apparent. Geologists now regard the Cambrian Period as the beginning of preservation of a fossil record for organisms whose evolution must extend far back into the Precambrian. (Fossil algae, for instance, are now known to occur in rocks as old as 2 billion years.) The explanation for the first preservation of a decent fossil record in the Cambrian is not well established. The theories that have been proposed are too numerous to mention here. The question remains one of the more fascinating ones about the history of the earth yet to be answered.

LATE PALEOZOIC—EARLY MESOZOIC PATTERNS

The Cordilleran Geosyncline was probably never a simple submarine trough, accumulating sediment everywhere at the same time. Locally there were many pulses to uplift and erosion. The geosyncline may have been split in two by uplift in the Mississippian Period. This is suggested by the absence of marine strata younger than Mississippian from the eastern part of the Columbia Mountains and the Idaho Panhandle. Marine sedimentation did continue almost until the end of the Jurassic Period both east and west of this supposed uplift. The strata were markedly different on the two sides. To the east were deposited beds of well-washed sandstone, shale, and limestone—essentially a "clean" assemblage free from volcanic debris. Rocks to the west (Chapters 7 and 8) consist of submarine lava and "dirty" sediments rich in volcanic debris.

An accurate reconstruction of just what the margin of the North American continent looked like at this time is impossible. Although the Rocky Mountain region can be tied safely to the continent, the area to the west is too complex and still too unknown geologically for simple analysis. Almost certainly, the geography did not consist of just two parallel troughs separated by an uplifting landmass. Not only here but throughout the entire Cordillera, the original relationship between the eastern nonvolcanic basin (miogeosyncline) and the western volcanic basin (eugeosyncline) has been obscured by subsequent folding, faulting, metamorphism, and batholithic intrusion. Perhaps much of this deformation was caused by large-scale movement of separate crustal plates.

OROGENY IN THE MESOZOIC ERA

All the elements characteristic of the destructional phase of a geosyncline are found in the Late Mesozoic record of the Rocky Mountains-Columbia Mountains region. These include folds and faults ranging from small structures clearly visible in single outcrops to gigantic folds involving whole mountain ranges and faults tens or even hundreds of miles long. Also typical are igneous stocks and batholiths, formed by the crystallization underground of large volumes of molten rock created somehow in the orogenic process. Vast tracts of metamorphosed rock are equally characteristic. The origin of these features is seldom obvious; the interpretation of such terrain is seldom simple.

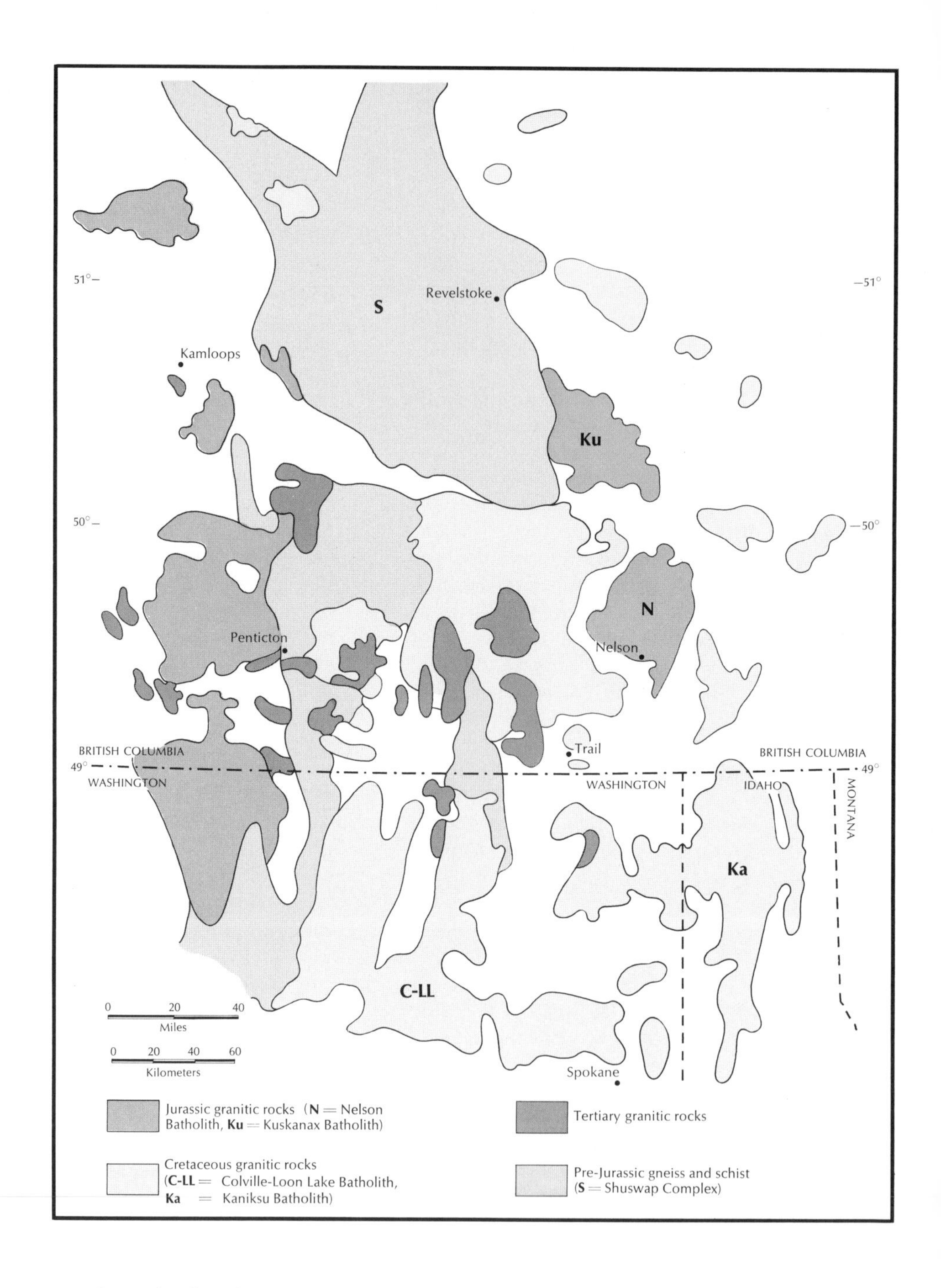

6-4 *Generalized geologic map showing the distribution of crystalline rocks in northeastern Washington, northern Idaho, and southeastern British Columbia. (After CIM Sp. Vol. 8, 1966, Fig. 10-1.)*

By the time the great deformation of that Late Mesozoic Era was over, the major rock units, except for the local cover of Tetriary strata, were in their present relative positions. The major events of the orogeny were metamorphism, emplacement of granitic intrusions, and thrust-fault displacements of many miles. Each process is distinct, but each is also characteristic of the climactic stage of orogeny in the history of most of the world's major mountain chains. The "typical" mountain chain has a core of highly deformed metamorphic rock (sometimes called a crystalline core). This is surrounded by less metamorphosed or even unmetamorphosed strata. In some instances, the decrease in metamorphism away from the core is gradual. More often, the changes are abrupt across major, steep faults. Thrust faulting usually moves strata away from the core of the range. Igneous intrusions are most abundant in or near the crystalline core, but they are not very closely restricted in either space or time. Sometimes they intrude a large part of the mountain system and are emplaced from the early to the final stages of orogeny.

The crystalline core of this region lies in the central and western part of the Columbia Mountains. It contains several structural elements. One is the Shuswap Metamorphic Complex, immediately west of the Columbia River. It consists of thoroughly recrystallized schists and gneisses formed by metamorphism of Late Precambrian (Windermere) and Paleozoic strata. The age of metamorphism has not been determined precisely since evidence is found for several different epochs of deformation and recrystallization. However, the major metamorphism occurred in Middle Jurassic time.

A regional structure known as the Kootenay Arc lies east of the Shuswap terrain in the Selkirk and Purcell Mountains. It curves from Revelstoke south and then southwest along Kootenay Lake into the eastern part of the Okanogan Highland of Washington. The strata range in age from Windermere to Early Jurassic, correlative with those in the Shuswap Complex but only slightly metamorphosed. They have been very intricately folded and faulted during several periods of deformation.

6-5 *Folded gneiss in the Crooked Lake area, west of the Cariboo Mountains, produced by recrystallization of strata of probable Late Paleozoic age. Similar outcrops are found throughout the crystalline core of the Columbia Mountains and Okanogan Highland. (Photograph courtesy of K. V. Campbell.)*

The Kootenay Arc was intruded by numerous granitic intrusions, the largest of which are the Kuskanax and Nelson Batholiths. The latter has been dated by radiometric methods as approximately 160 million years old (Middle Jurassic). Since it cuts across the major folds, the climax of deformation must have occurred before Middle Jurassic time.

The Purcell-Belt strata form most of the outcrops between the Kootenay Arc and the Rocky Mountain Trench. The rocks are not much metamorphosed. They have been deformed into large folds cut by numerous steep faults, but these structures are not as complex as those in the Kootenay Arc. Some of this deformation occurred in the Paleozoic Era. The Purcell-Belt strata were involved also in Rocky Mountains thrusting late in the Mesozoic Era.

The earliest Mesozoic thrusting probably occurred in Late Jurassic time in the Columbia Mountains. However, with time, the zone of thrusting migrated eastward. Successively younger thrust faults formed to the east, ending with displacements in the foothills of the Rockies early in the Tertiary Period. The net effect of all the thrusting was a cumulative displacement of as much as 125 miles and a thickening of the crust by as much as 5 miles. (The origin of thrust faults is discussed in Chapter 7.)

The Kuskanax and Nelson Batholiths were merely the oldest of a host of similar granitic intrusions. Cretaceous intrusions include several batholiths in the Shuswap Metamorphic Complex, the Colville–Loon Lake Batholith of the Okanogan Highland of Washington, the Kaniksu Batholith in northern Idaho, and the large and complex Idaho Batholith of central Idaho (Figure 6-4). Radiometric dating of these rocks indicates that most of them crystallized early in the Late Cretaceous, approximately 85 million to 100 million years ago. This corresponds to

6-6 *Looking north at a peak in the Canadian Rockies that exposes a west-dipping thrust fault. Dark Paleozoic shale has been thrust eastward over younger Paleozoic limestone. Similar thrusts are the dominant structures in the Canadian and Montana Rockies. (Photograph courtesy of James Crawford.)*

the time of major batholithic emplacement in the Coast Mountains of British Columbia, the Blue Mountains of northeastern Oregon, and the Sierra Nevada of California. Somewhat smaller intrusions crystallized in these regions up until Middle Tertiary time, especially in the North Cascades of Washington (Chapter 7).

The batholiths of the Columbia Mountains and the Northern Rocky Mountain Province have not provided much information on the formation of magma. They do reveal, however, many of the relationships between igenous intrusions and metallic mineral deposits. Most of the very considerable mineral wealth of the region occurs at or near the margins of stocks and batholiths.

LEAD AND ZINC DEPOSITS

Eastern British Columbia, northeastern Washington, and northern Idaho have witnessed a long and profitable mining history. Included are some very famous gold, silver, and copper mines. Lead and zinc, however, stand one-two in the all-time dollar-value list for British Columbia and also currently lead Washington's metallic mineral production. The geology of the lead and zinc deposits is everywhere pretty much the same. The two metals occur together in sulfide minerals—the lead as grey, shiny crystals of galena and the zinc as honey-colored sphalerite. The host rock is usually a limestone. Galena and sphalerite are disseminated widely in the rock as a result of solutions that have migrated through cracks, dissolving the lime and precipitating the sulfides.

The limestone host rock is of different ages throughout the region, indicating that the composition rather than the age of the strata is important. In the famous Metaline District of northeastern Washington, the galena and the sphalerite are found in the massive Metaline Limestone of Cambrian age, concentrated especially near the top of the formation. There, rising ore-bearing solutions were trapped under the impermeable Ledbetter Slate of Ordovician age. In addition to the large Pend Orielle Mine, there are many separate workings and prospects. The tonnage of available ore is not known precisely. It may be immense, for there are ore showings over hundreds of square miles. The district extends across the border into southeastern British Columbia. It contains the Sullivan Mine, which has produced 93 percent of British Columbia's lead and almost 85 percent of its zinc. The host rock in the Sullivan Mine is the Precambrian Purcell Group. The ore is concentrated in about 300 feet of thin-bedded shales between more massive beds of quartz-rich sandstone and conglomerate. Other lead-zinc properties in the province are in rocks that range in age from Late Precambrian to Jurassic. Most of the ore deposits lie close to granitic intrusions.

THE TERTIARY RECORD

Strata of Tertiary age are neither thick nor widespread in the Northern Rockies and Columbia Mountains. The Cordilleran Geosyncline had risen to form mountain ranges by the end of the Cretaceous, and the seas never returned to the continental interior. The mountains that we see today are of Late Cenozoic age, however, and not the remnants of Cretaceous ranges. We know very little about the morphology of the older mountains.

The Tertiary deposits that have been preserved consist of nonmarine sedimentary and volcanic rocks, the latter including both lava flows and beds of volcanic ash. Many of the Tertiary formations have been preserved by having been faulted down into the older crystalline rocks. The Republic District of north-central Washington is typical. Most of the rocks in this part of the Okanogan Highland are a part of the Colville-Loon Lake Batholith or are metamorphic schists and gneisses of uncertain age. A fault basin, known as the Republic Graben, trends north-northeast and extends more than 60 miles. The average width of the graben is about 10 miles. The strata within this trough have a total thickness of almost 3 miles and consist of Oligocene and Eocene (Lower Tertiary) sedimentary and volcanic rocks. Evidence for contemporaneous volcanism and fault movement is found, implying that perhaps the sinking of the graben was tied to withdrawal of underlying support due to the volcanic extrusion. (This sort of tie between eruptions and collapse is well documented in modern volcanic areas, such as the East African Rift and Kilauea on the island of Hawaii.) Certainly some of the sediment within the trough was derived from the fault scraps at the edge of the graben. Elsewhere in the region, similar nonmarine sediments and volcanics accumulated in basins of nonfault origin.

Upper Tertiary strata are not extensive. The valleys at the south edge of the Okanogan Highland were filled by floods of Miocene basalt lava that inundated the Columbia Plateau area to the south. Post-Miocene uplift and erosion have removed all but a few patches of this basalt north of the Columbia and Spokane Rivers. Similar sheets of fluid basalt lava covered 300,000 square miles in the Interior Plateau Province of central British Columbia (Chapter 8). They lap eastward onto older rocks of the Shuswap and Quesnel Highlands.

6-7 *The northern part of the Cariboo Mountains, looking south. The glaciers that flank the distant peaks are small remnants of a great ice cap that covered all but the highest summits 15,000 years ago. (Photograph courtesy of K. V. Campbell.)*

GLACIERS COVER THE LAND

Virtually all of British Columbia and the northern parts of Washington, Idaho, and Montana were covered by glacial ice at one or more times in the Quaternary Period. The scene 15,000 years ago was almost unimaginable. Ice covered everything below an elevation of about 8,000 feet in the southern interior of British Columbia. Only the highest peaks of the Rockies and Columbia Mountains stood above this sea of ice, providing a scene not unlike parts of Greenland or Antarctica today. Glaciation had a profound effect on the topography, sharpening it in the high mountains and generally smoothing it in the lower ranges and valleys that were covered by ice.

Today, glaciers exist on many high peaks in the northern part of the Cordillera. In British Colum-

bia, they occur in the Coast, Cariboo, Purcell, Selkirk, and Rocky Mountains. To the south, they are found in the Olympics and the Cascades, especially on the young volcanic cones that rise above the general level of the range (Chapter 13). These alpine glaciers provide a link with the past, a laboratory in which geologists can study the processes of glaciation and can gain fuller understanding of past glaciations.

What caused the Ice Ages? The question has intrigued scientists for more than a century, ever since the first evidence of former continental ice sheets was discovered. Certainly, the principal reason was climatic, but just what brought on a worldwide change in climate is not firmly established. If we understood this, we would have a firmer basis for predicting future climate variations, including the possibility of renewed continental glaciation. (See Chapter 19.)

The great ice sheets of the recent past evolved from the expansion of alpine glaciers such as those preserved at present in the Cordillera. A glacier is born when, over a period of many years, snowfall exceeds melting. This is most likely to occur high on the north or northeast (in the northern hemisphere) flank of a mountain, where the snow is shaded from the warm midday and afternoon sun. The weight of new snow compacts underlying snow layers. Gradually a sort of metamorphic recrystallization occurs at depth. This transforms the delicate snowflakes to compact granules and finally to solid ice, with individual crystals many inches long. Ice is a type of solid known as a plastic substance, that is, capable of slow flowage under conditions of stress. For glacial movement, the important stress is gravity. When ice builds to some critical thickness on a sloping surface, it starts to flow downslope; thus a glacier is created. For the glacier to continue to enlarge, snow accumulation must continue to exceed melting or wastage of the ice. As the glacier extends downvalley to lower elevations where warmer conditions prevail, wastage may begin to exceed snow accumulation. Still the glacier would persist, for the flow is like a conveyor belt that constantly supplies ice from above to be melted below. The position of the lower end, or *terminus,* of the glacier is determined by the opposing forces of melting and the rate of ice flow. If the terminus remains fixed, the system is in equilibrium (Chapter 1). If rate of flow exceeds wastage, the glacier will enlarge; and, if wastage exceeds flow, it will diminish. In the latter instance, the terminus retreats upvalley; and geologists refer to this as *recession* of the glacier. Note, however, that recession can occur even while the ice continues to flow downvalley, for the term is applied only to the relative position in the valley of the terminus.

The great ice cap of the northern part of the Cordillera was created by the continued enlargement of alpine glaciers that began initially high on the flanks of major peaks. Many individual glaciers merged in the lower valleys to form larger glaciers, which advanced finally beyond the range fronts and spread out across nearby lowland areas. The alpine glaciers enlarged until finally they formed continuous ice caps that covered entire ranges, with only the highest peaks still protruding. The lowland ice also expanded until the plateaus and major valleys lay buried beneath thousands of feet of ice.

What determines the direction of movement of a glacier? At first glance the question seems frivolous—obviously ice flows downhill. But this is not necessarily true for large glaciers; at least for them, the direction of slope of the underlying ground is not always critical. The important factor is the direction of slope of the ice surface itself. Glacial ice will flow away from the point of maximum surface elevation of the ice. Thus the explanation for the southward flow of ice into Washington and Idaho from south-central British Columbia was not due to the slope of the ground, for in many instances south was uphill. Rather, the southern flow was away from the center of ice accumulation.

The Cordilleran ice cap attained its maximum elevation along a line extending southeast from the central peaks of the Cariboo Mountains to Chilko Lake in the Coast Mountains. This then was a sort of ice divide. South of this line glacial flow was to the south into Washington and Idaho, while on the opposite side glaciers flowed north toward Prince George. Glacial erosion was particularly intense in the large valleys south of the divide.

6-8 *A glacially scoured cirque basin in the Crooked Lake area, west of the Cariboo Mountains. The layered strata capping the cliff to the left are mica schists belonging to the Shuswap Complex. They were derived from Windermere strata by metamorphic recrystallization. (Photograph courtesy of K. V. Campbell.)*

Troughs such as those occupied by the Okanogan, Columbia, and Kootenay Rivers, and also the Rocky Mountain Trench, were enlarged and smoothed by the passage of major glacial lobes. Erosion was not uniform. In places, the glaciers plucked large quantities of bedrock from the valley floor, producing such deep lakes as Okanagan, the Arrow Lakes on the Columbia, and Kootenay Lake. Glacial erosion also gave these valleys the U-shaped profile characteristic of most glaciated valleys and formed impressive grooves and striations on smoothed rock surfaces of the valley sides.

Erosion is most characteristic of the active stage of glacial advance. During the waning stages of glaciation, as the ice melts away, deposition is the dominant process. Meltwater from the wasting glaciers forms very large rivers. These would be capable of considerable erosion if the melting process did not also liberate tremendous quantities of rock debris carried in or on the ice. This sediment is normally more than the meltwater rivers can handle. The result is, inevitably, that the valleys become floored with outwash sands and gravels, which in some cases are many hundreds of feet thick. This has the effect of evening out the valley floor and accentuating the steepness of the glacially scoured valley walls. The alluvial fill on the valley floor also makes good farm and orchard land, especially if it is mantled by silt deposited in lakes marginal to the stagnating glaciers. Buried glacial gravels, in turn, often contain large quantities of relatively high-quality groundwater and are the principal sources of well water in the region.

The brief period of stagnant valley ice, after glacial motion has ceased but before the complete disappearance of the ice, produces some characteristic landforms. Most prominent often are terraces of sand and gravel, deposited by streams and rivers that flow along the depression that exists between the rock walls of the valley and the stagnant ice. These are known as *kame terraces*. They are particularly prominent in the Okanogan Valley and along the lower reaches of the Methow Valley in north-central Washington.

Most glacial effects identified in the Cordillera date from the last major epoch of glaciation, 15,000 to 20,000 years ago. In certain areas, there is evidence of as many as three older glaciations. Such evidence is quite subtle, normally found only by careful scrutiny of glacial sediments. The last glaciation wrought such profound changes in the landscape that little geomorphic evidence remains from more ancient glacial epochs.

The influence of the Cordilleran ice sheet extended far beyond its actual terminus. The changes brought about on the Columbia Plateau were especially spectacular (Chapter 17).

7 the north cascades

■ The more we learn, the less we know. Scientists live with this truism, for it seems almost axiomatic that in our quest for solutions to particular problems we discover more new ones along the way. The path traveled by a scientist has many branches and side paths. At best,an investigator can recognize the new regions for exploration as he journeys to his particular goal, fully realizing that some other path might prove more profitable. The dilettante will try them all, thus gaining perhaps a measure of understanding of all the problems but not really solving any of them. The scientist must be willing to commit himself—to make judgments and frame conclusions—knowing that his opinions may be challenged someday by someone with more information at hand. If proven wrong, he must content himself with the knowledge that his efforts sparked additional research, that is helped to find the best path. Pioneer work seldom goes unchallenged.

PIONEER WORK ON CORDILLERAN GEOLOGY

The geologic understanding of any region must begin with reconnaissance investigations of large areas. Much pioneer work of this sort was done in the American West late in the nineteenth century, primarily by government geologists charged with inventorying the economic potential of particular regions. Problems of access and travel were enormous, so that some of the areas were virtually unexplored. Often the opportunity for revisiting critical localities was not there—the geologist had to base his conclusions on very limited data.

The results were astonishingly good. The mistakes made are easily dismissed when balanced against the essential correctness of the general conclusions. Naturally, subsequent workers have changed some lines on the maps, have filled in some of the missing detail, and have modified some of the interpretations. But reading the reports of the early reconnaissance studies of regions like the North Cascades fills a contemporary geologist with admiration for the scientific soundness of his predecessors.

What was written about the North Cascades? It was said to consist of a large granitic intrusion—the Chilliwack Batholith—flanked by sedimentary and volcanic strata of varied ages. Evidence of metamorphism was cited, but it was assumed that most of this was caused by the heat and solutions given off by the cooling batholith. The region provided good support for the classic model of a mountain range evolved from a geosyncline. What do we know now? The Chilliwack Batholith is still there, but it is now known to consist of a great number of separate intrusions, plus some large tracts of crystalline schists and gneisses that are very much older than the igneous activity. (This same complex mixture of rock types exists in the Coast Mountains "Batholith" of British Columbia and the Sierra Nevada "Batholith" of California.) Metamorphism occurred not once but many times, as did folding and faulting. Furthermore, the displacement on some of the major faults has been so great that it is almost impossible to reconstruct the original positions of the various pieces. Thus the more we learn, the less we know. We can only look back longingly on the old days when our relative ignorance made the geologic history of North Cascades seem comparatively straightforward.

Geologic studies extending over the past several decades have made the North Cascades one of the better-known mountainous regions in the Cordillera. The pattern emerging is quite comparable to that found in other Northwest ranges—the Klamath Mountains (Chapter 10), the Blue Mountains (Chapter 14), the Columbia Mountains (Chapter 6), and the Coast Mountains of British Columbia (Chapter 8). All these ranges contain an important record for the Paleozoic and Mesozoic Eras, thus providing clues essential to an understanding of the evolution and ultimate destruction of the Cordilleran Geosyncline.

GEOGRAPHY OF THE NORTH CASCADES

The North Cascades are bounded on the north by the Fraser River, the mountains north of there being part of the Coast Mountains. The Puget and Fraser Lowlands lie to the west. The southern edge of the region crosses the range near Snoqualmie Pass east of Seattle. This boundary is defined more on a geologic than on a geomorphic basis. The rocks to the north are primarily crystalline-intrusive granites and a full spectrum of metamorphic rock types. To the south, younger (Cenozoic) volcanic and sedimentary rocks are virtually the only type found all the way to the Sierra Nevada in northern California. The southern end of the North Cascades is bounded on the east along the Columbia River by the young basalt lavas of the Columbia Plateau. North of Brewster, Washington, near the confluence of the Columbia and Okanogan Rivers, the North Cascades merge with lesser ranges in the Okanogan Highland. The Okanogan River is used as the boundary in this area.

The geography within the range is complex. From a distance, the range looks quite uniform and massive, with the average peak elevation generally rising to the north, from an average of 5,000 feet near Snoqualmie Pass to about 7,000 feet at the Canadian border. Standing above the general

7-1 *The North Cascades, looking north from the summit of Eldorado Peak. Pickett Range in the far distance. Most of the higher peaks in this part of the range rise to about 7,000 feet above sea level. (Photograph courtesy of Peter Misch.)*

elevation of the range are the two young volcanoes, Mount Baker and Glacier Peak (Chapter 13). Rivers have cut very deeply into the range, but none have cut through. Thus the Cascade crest is a pronounced divide between east-flowing drainage that reaches the Pacific Ocean via the Columbia River and shorter, west-flowing streams that run directly to salt water. All the valleys within the range have been heavily glaciated in the past 20,000 years. They have relatively flat floors and precipitous sides rising thousands of feet.

The biggest valleys on the west side are those of the South Fork of the Snoqualmie River, containing Interstate Highway 80; the Skykomish River containing U.S. Highway 2; and the Skagit River, followed by the North Cross-State Highway, still under construction in the early 1970s. The major rivers on the east side are the Wenatchee, the Stehekin (including Lake Chelan), and the Methow Rivers. The latter, with its many tributaries, drains much of the northern half of the range in Washington. The Fraser River handles the runoff from the Canadian part of the range. At the higher elevations, rock exposures are excellent, except where covered by permanent snowfields and glaciers. Outcrops below timberline are fair to poor because of the dense vegetation.

THE STRUCTURAL ARCHITECTURE OF THE RANGE

Studying the Cascades is like studying the anatomy, a square inch at a time, of an animal that has very thick, and in some places impenetrable, fur. Outcrops along the road or trail give a very good picture of the types of rock that make up the range. However, the spatial relations or structural architecture of the North Cascades is not immediately apparent.

The distribution of rock formations in the North Cascades is strongly controlled by large, steeply dipping faults of Late Mesozoic and Early Cenozoic age. The range is thus divided by these fractures into a series of distinct structural blocks, most of which trend north-northwest almost parallel to the trend of the mountain chain. Lithologies and structures are commonly quite similar within any one of these blocks but wholly different among adjacent blocks.

In general, the most severely metamorphosed and deformed rocks are found in the center of the range. They form a crystalline core of schists and gneisses of unknown age. Paleozoic, Mesozoic, and Lower Cenozoic sedimentary and volcanic rocks underlie the western foothills of the North Cascades. To the east, an enormous thickness of Upper Mesozoic strata occurs in the Methow Graben, a downdropped fault block that extends north-northwest from near the Columbia River far into south-central British Columbia. Another graben, the Chiwaukum, cuts almost to the Cascade crest in the headwater region of the Wenatchee River. The sediments within this graben are entirely of Early Cenozoic age, as they are in the Republic Graben far to the northeast (Chapter 6). Some recognizable Upper Paleozoic and Mesozoic marine formations crop out in the highlands east of the Methow Graben and west of the Okanogan River. Granitic intrusions are found within all the major blocks of the North Cascades. Some of these clearly postdate the faulting, as they have intruded across the fault planes.

THE HISTORY IN OUTLINE

The age of the oldest rocks in the North Cascades—thoroughly metamorphosed gneisses—is unknown. They may be as old as Precambrian; they are certainly older than Middle Devonian. Late Paleozoic strata, only slightly metamorphosed, crop out in the western foothills of the range and near the Okanogan River to the east. They consist of marine sedimentary and volcanic rocks and constitute one of the best (but still fragmentary) records in the Northwest of the Cordilleran Geosyncline. Rocks of Triassic age are not common. However, episodes of marine sedimentation and volcanism occurred in the succeeding Jurassic and Cretaceous Periods, producing thick sequences of strata. The severe orogeny that finally destroyed the geosyncline reached a climax in the North Cascades in the Cretaceous Period. This

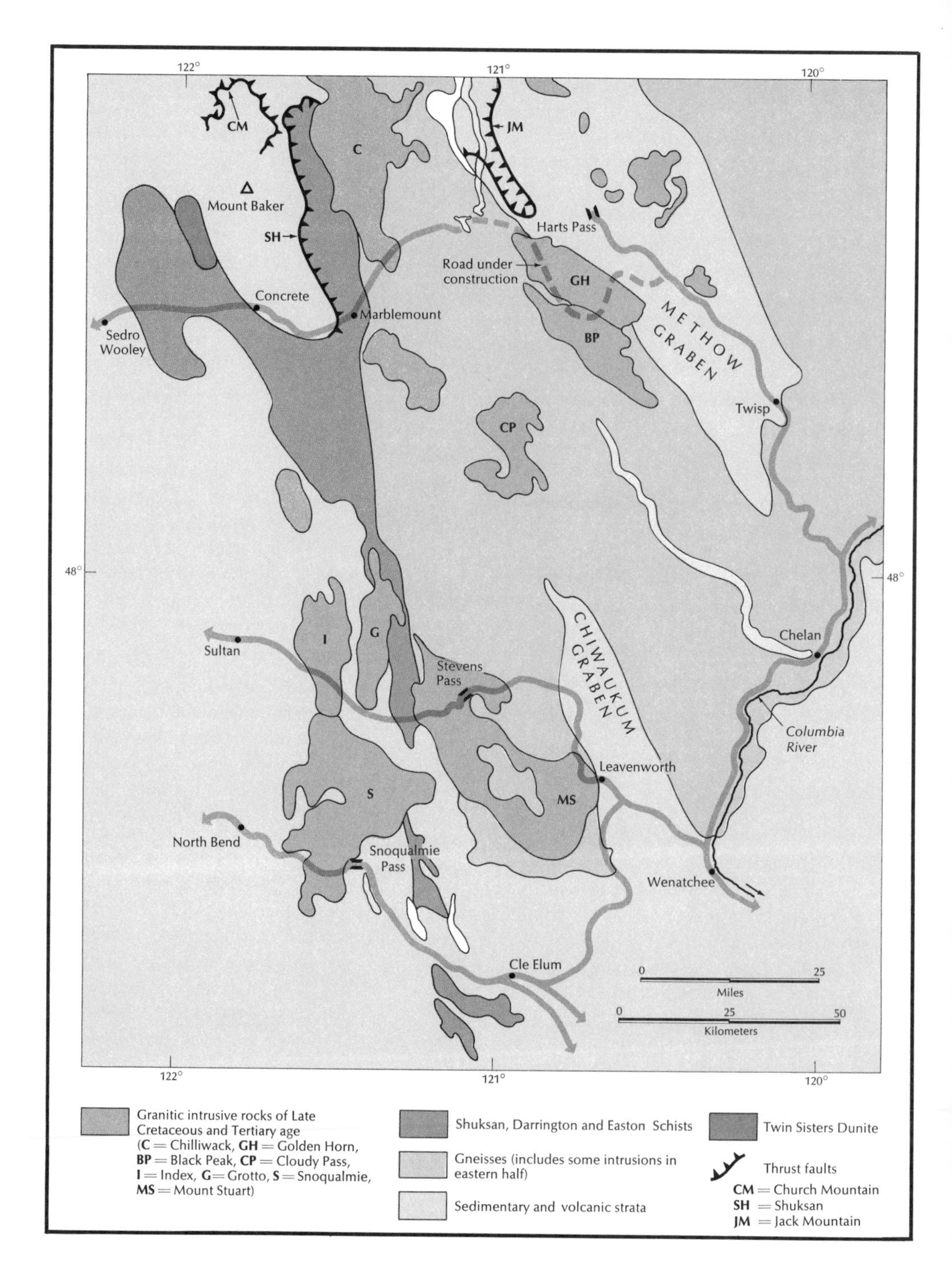

7-2 *A generalized map of the North Cascades of Washington, showing the distribution of crystalline rock. (After Huntting et al., 1961.)*

orogeny involved folding, faulting (including the displacement of great sheets of rock, tens of miles to the west on nearly flat glide planes or *thrust faults*), and severe metamorphism in the core of the range. The first of the granitic batholiths was emplaced in the Cretaceous Period. Intrusion continued, however, far into the Tertiary Period. The Cretaceous Period also saw the final disappearance of salt water from the country that became the North Cascades.

The mountains that rose in the Cretaceous are not those that we see today. By the middle of the Tertiary Period, the older mountains in the Cascades region had been eroded down to relatively low, smooth hills. The present range was built by a new uplift, principally a broad, north-south arch, during the past 10 million years. However, little evidence for post-Cretaceous events is found in the North Cascades proper, for the Cenozoic stratigraphic record is rather sparse. Mount Baker and Glacier Peak are Quaternary volcanoes that seem unconnected to the earlier history of the region, although they must relate somehow to the uplift of the range.

The high peaks of the North Cascades spawned large Quaternary alpine glaciers. These moved down the valleys on either side and merged with immense glaciers that occupied the adjoining lowland regions. Eventually ice covered all but the highest peaks. Glaciation was a mixed blessing, geologically. On the plus side, the glacial history of the Northwest is as fascinating as any in the world. Moreover, the principal mineral resource currently produced in Washington is sand and gravel, much of it a direct glacial inheritance. Finally, glaciation has stripped away the cover and provided the geologist with some fine rock exposures, especially in the high mountains, where otherwise there might be none. Balancing these blessings is a thick cover of glacial sediments dropped in the lowlands that conceals most of the bedrock and structure there.

7-3 *A schematic cross section through Mount Baker and Ross Lake in the North Cascades. Numbers below the section refer to the following major blocks: (1) Western Foothills region, consisting of folded and faulted Paleozoic marine strata of the Chilliwack Group, plus Mount Baker, a Quaternary volcano; (2) Crystalline Core, composed of granitic rock of the Chilliwack Batholith plus highly deformed schist and gneiss; (3) The Methow Graben, containing folded Upper Mesozoic strata cut by Cenozoic granitic intrusions; (4) The Eastern Highland, underlain near the Methow Graben by schist, gneiss, and granite.*

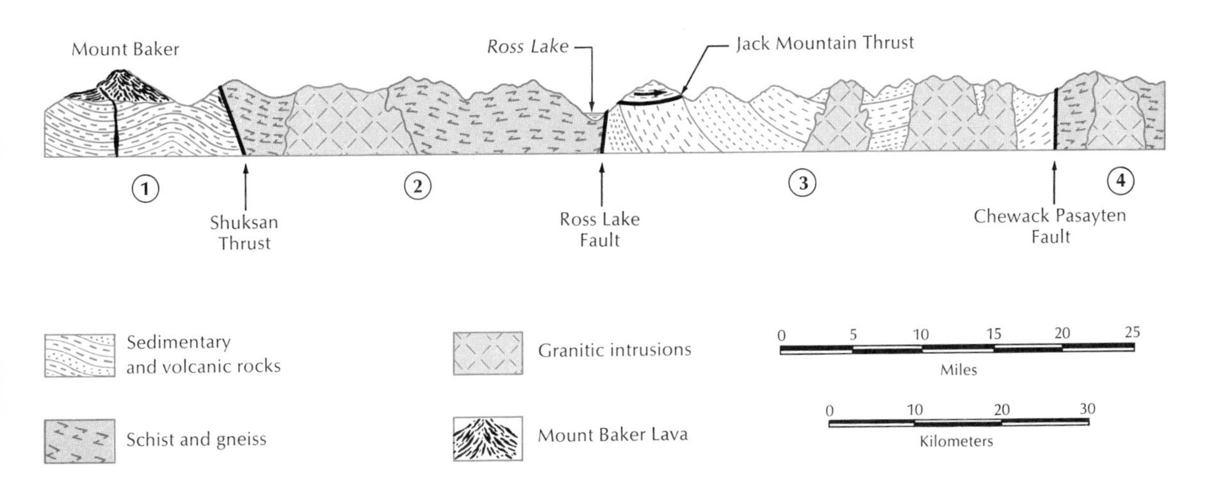

THE WESTERN FOOTHILLS

Three different formations dominate the geology of the western foothills of the North Cascades. These are the Chilliwack Group, the Nooksack Group, and the Chuckanut Formation (Table 7-1). The oldest is the Chilliwack Group. It consists of a typical eugeosynclinal assemblage—dark sandstones, black shales, thin beds of chert, and submarine lava flows. The Chilliwack also contains some rather thick layers of limestone, lenticular in form and seldom extending for more than a few miles. They are locally of importance as host rocks for ore deposits and as a source for a once-thriving cement industry. The limestone contains most of the fossils found in the Chilliwack Group—fossils as old as Middle Devonian and as young as Middle Permian. Certainly, the stratigraphic record is incomplete. It contains many gaps, or unconformities, including an apparent absence of any strata of Mississipian age. The Chilliwack Group has not been studied in detail; its thickness exceeds 10,000 feet.

The Nooksack Group consists of dark sandstones and shales made up for the most part of fragments of volcanic rock. It has been intensely folded and sheared locally by some of the large Cretaceous thrust faults that cut through the foothills. This deformation and the lithologic monotony of the Nooksack make thickness determinations very difficult. Certainly, the section is more than 5,000 feet thick. It contains marine invertebrate fossils that prove a Late Jurassic and Early Cretaceous age.

TABLE 7-1 **Stratigraphy of the western Cascades, northern Washington. (From Misch, 1966, and personal communication.)**

AGE	FORMATION	DESCRIPTION	THICKNESS, FEET
Oligocene (unconformity)	Hannegan Volcanics	Dacite and andesite flows, breccias, and tuff. Minor clastic beds	>3,000
Middle to late Eocene (unconformity)	Huntingdon	Arkosic sandstones and shales. Minor conglomerate. Nonmarine	4,000
Paleocene to Eocene	Chuckanut (Swauk)	Arkosic sandstones, siltstones, shale, conglomerate. Local coal beds. Nonmarine	>15,000
Late Jurassic to Early Cretaceous	Nooksack Group	Graywacke sandstones and siltstone. Minor conglomerate. Marine	5,000
Middle to Late Jurassic	Wells Creek Volcanics	Andesite and dacite flows and breccias. Graywacke sandstone and shale interbeds. Marine	3,500
Late Triassic to Early Jurassic	Cultus	Shale and siltstone; minor sandstone. Marine	>3,000
Middle Devonian to Middle Permian	Chilliwack Group	Graywacke sandstones and shales, limestone, chert; basaltic to andesitic flows, breccias, tuffs. Marine	>10,000
Pre-Middle Devonian (Precambrian ?)	Yellow Aster Basement Complex	Gneisses derived mostly from plutonic rocks	?

——— Major unconformity

The Chuckanut Formation consists of nonmarine sandstones and shales of Early Tertiary age. They postdate the Cretaceous orogeny and are discussed later in this chapter.

CRYSTALLINE ROCKS IN THE RANGE CORE

Schists, gneisses, and granites compose the backbone of the North Cascades. They can be seen along the Skagit River east of Marblemount, along the Fraser River near Hope, and along the west side of the Columbia River from Rocky Reach Dam north to the Okanogan River. (There are also fine exposures along the Stevens Pass Highway, but stopping on this busy route is extremely hazardous.) The ages of the metamorphic rocks are not known, except that they are older than the Cretaceous Period. The oldest may be granitic gneisses found around Marblemount and included as slices in the thrust sheets farther west. These constitute the Yellow Aster Complex or more generally, the basement complex; they are metamorphosed intrusions that appear to be older than the Chilliwack Group and hence are pre-Middle Devonian. Similar granitic gneisses underlie large parts of the Entiat Mountains southwest of Lake Chelan, the south end of the Methow Graben and the western part of the Okanogan Highland. Gneisses almost certainly correlative with the Yellow Aster Complex are found on Orcas Island in the San Juan Islands, where they are referred to as the Turtleback Complex. Once again, a stratigraphic position below Middle Devonian strata is demonstrable.

The Yellow Aster Gneiss is quite uniform. In contrast, the Skagit Gneiss to the east is strikingly heterogeneous. It contains distinct layers of quite dissimilar composition and a great variety of irregular veins cutting through it. Continuous exposures in road cuts along the Skagit River from Newhalem east past Diablo Dam offer a unique opportunity to study this formation. It is a classic example of the type of "mixed" igneous and metamorphic rock known as a *migmatite*. The origin of migmatites is a subject that generates considerable debate among geologists. The Skagit Gneiss is closely associated with the Cascade River Schist. This latter unit formed by metamorphism of sedimentary and volcanic strata of unknown age. Some evidence suggests that the Skagit Gneiss may have been derived by metamorphism from the Cascade River Schist. Similar schist east of Stevens Pass is referred to as the Chiwaukum Schist. Migmatites very similar to the Skagit Gneiss are well exposed at Chelan Falls on the Columbia River and in the Cascades of British Columbia. There they are called the Custer Gneiss.

A thin fault block in the western part of the core contains metamorphic rocks that are quite unlike anything else in the Northwest. What makes these strata so unusual is not their original composition but the type of metamorphism that affected them. Two formations are recognized. The lower one consists of as much as 10,000 feet of shale. It has been highly sheared and transformed by metamorphism to phyllite, here the Darrington Phyllite (see also Chapter 3). Overlying the Darrington Phyllite are about 5,000 feet of submarine basalt flows. These were metamorphosed with the sediments and changed to *greenschist*. Mount Shuksan east of Mount Baker is composed mostly of this rock, which is known as the Shuksan Greenschist. The Darrington Phyllite and Shuksan Greenschist together form the Shuksan Suite.

The presence of the blue amphibole minerals, crossite and glaucophane, distinguishes the Shuksan Suite from other low-grade phyllites and greenschists of the Northwest. These minerals are found only in certain parts of the world, most notably in places around the Pacific rim. They are known to form under rather unusual metamorphic conditions—moderate to high pressure but low temperature. The Shuksan Suite has been correlated with a similar formation at the south end of the North Cascades. Known as the Easton Schist, it too contains minor amounts of glaucophane.

7-4 *Eldorado Peak, south of the Skagit Valley, looking south. Glacier Peak, a Quaternary volcano, is in the distance to the right. Eldorado Peak is composed of crystalline rock of the Yellow Aster Complex. (Photograph courtesy of the National Park Service.)*

The Shuksan-Easton unit may be very critical to our understanding of the geologic history of the North Cascades. The concepts of sea-floor spreading and subduction have made geologists try to recognize zones where ancient underthrusting occurred. Certain rock types and structures have been predicted as likely to occur in "fossil" underthrust zones. These include marine strata that have undergone glaucophane metamorphism and that are cut by many steeply dipping shears or faults. The Shuksan-Easton unit thus qualifies. Perhaps it does represent an ancient subduction zone, but this is highly speculative. No fossils have been found in the unit, and its age is unknown. How it correlates with the Chilliwack Group to the west or the Skagit-Custer Gneiss to the east, is also uncertain. If the Shuksan Suite does represent an ancient subduction zone, then the Chilliwack and Skagit units must belong to separate plates that somehow drifted together.

7-5 *Highly deformed Skagit Gneiss, east of Gorge Dam on the Skagit River. The rock, a complex mixture of metamorphic and igneous-appearing material, is a typical migmatite. (Photograph courtesy of Peter Misch.)*

CRETACEOUS OROGENY

The North Cascades region was deformed severely between the accumulation of Early Cretaceous sediments and the resumption of widespread, nonmarine sedimentation early in the Tertiary Period. The principal deformational events were minor metamorphism, batholith intrusion, movement on very large thrust faults, and uplift of the entire area above sea level. The metamorphism of the Upper Jurassic–Lower Cretaceous Nooksack strata was at low temperatures and pressures. This was certainly a Cretaceous event. The age of the higher-grade metamorphism of the range core is less certain. It was probably pre-Cretaceous.

The most noteworthy Cretaceous deformation was the formation of very large, nearly horizontal thrust faults, which are very characteristic of major mountain chains. The thrusting involves the displacement of large plates of strata from the axial region of the range outward tens of miles into less-deformed marginal zones. In the Cordilleran region, the largest of these thrust faults, or *over-*

thrusts, are found along the front of the Rocky Mountains (Chapter 6), in the eastern part of the Basin and Range Province, in the Klamath Mountains, and in the North Cascades. The largest displacements in the Cascades were to the west; slices of the crystalline core of the range were transported more than 30 miles. The biggest thrust faults preserved are the Church Mountain Thrust on both sides of the Canadian border, and the Shuksan Thrust. The latter involved mostly rocks of the Shuksan Suite, but it also contains numerous slices of gneiss from the core of the range. The Shuksan Thrust extends from the vicinity of Mount Baker south at least as far as the Skykomish River and perhaps all the way to Easton.

A smaller structure, the Jack Mountain Thrust, carried Late Paleozoic strata eastward over Cretaceous rocks in the Methow Graben (Figure 7-3). The total demonstrable displacement of this thrust is about 6 miles, considerably less than that suggested for the Shuksan and Church Mountain Thrusts.

The origin of overthrusts is highly debatable. Some geologists regard overthrusts as essentially gigantic landslides in which plates of rock have traveled many tens of miles off the flanks of rising mountains. For many years the topography implied in this theory seemed wholly unrealistic. However, recent theoretical calculations have demonstrated that gravity sliding of large rock plates down

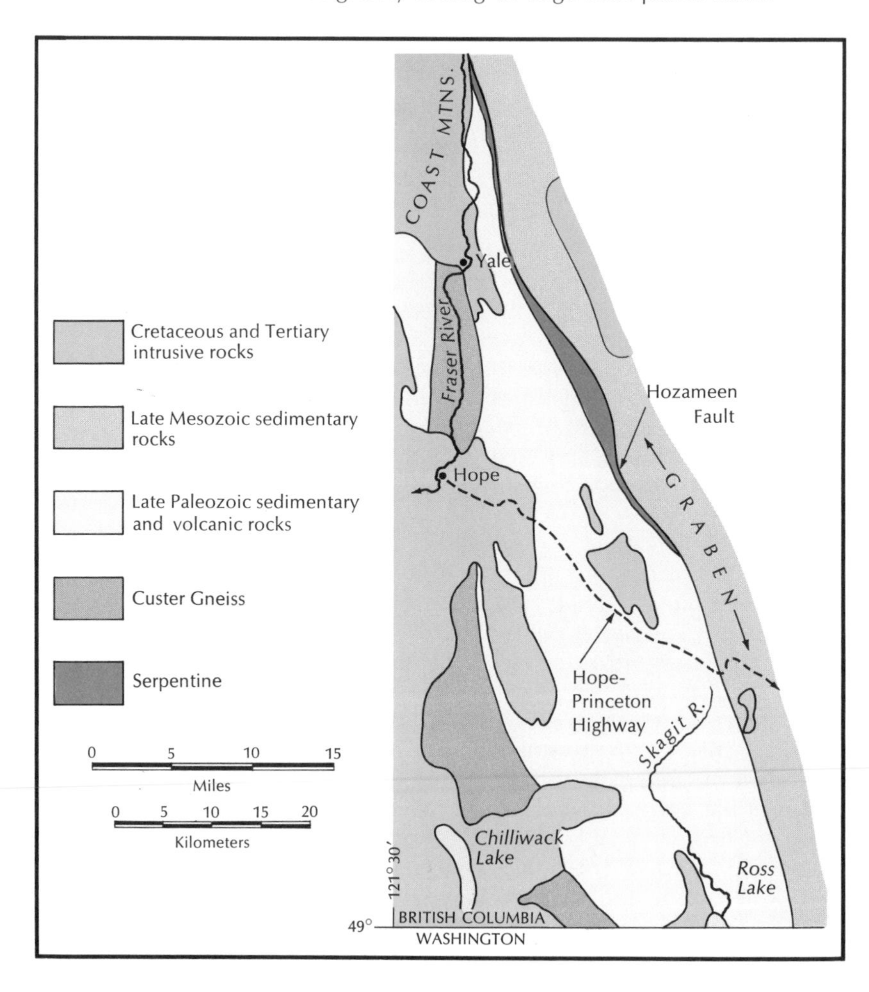

7-6 *Generalized geologic map of a part of the Cascade Range in British Columbia. The Custer Gneiss is continuous with the Skagit Gneiss south of the border. (After McTaggart and Thompson, 1967. Reproduced by permission of the National Research Council of Canada.)*

slopes of only a few degrees is possible. This type of deformation is sometimes referred to as thin-skinned tectonics, for it does not involve structures deep within the crust. Convincing arguments have been made for gravity sliding as a mechanism to explain overthrusts in the Appalachians and in the Alps, for example.

An older view holds that thrusting away from the axis of a range represents marked compression and shortening of the crust. This implies that thrust faults should steepen and pass into the core of the range. Gravity gliding suggests that they should disappear out into the air, like the sliding surface at the head of a landslide (Figure 7-8). The critical evidence is often hard to find. Near the center of major mountain ranges other types of faulting, plus folding, metamorphism, and intrusion all combine to vastly confuse the record. Moreover, erosion following range uplift may remove much of the evidence.

The overthrusts of the North Cascades appear to dip steeply back toward the range core. This suggests that the thrust plates were squeezed up out of the ground. The principal zones of steep dip are the Shuksan zone on the west and the Ross Lake zone on the east. Both have been modified by subsequent faulting and batholithic intrusion. The role of gravity in keeping the thrust plates moving once they were extruded may have been significant. However, severe compression rather

7-7 *Severely sheared sedimentary and volcanic strata of the Chilliwack Group, 2,000 feet below the principal movement plane of the Shuksan Thrust. (Photograph courtesy of Peter Misch.)*

than simple uplift and spreading seems to have been involved in the deformation of the range core.

A REMARKABLE SECTION IN THE METHOW GRABEN

The thick stratigraphic section in the Methow Graben further complicates our interpretation of Cretaceous orogeny. In the range to the west, Cretaceous sedimentation seemed to end with the deposition of the Nooksack Group. This would imply that deformation and uplift occurred for the rest of the Cretaceous Period. However, if the Cordilleran Geosyncline disappeared early in the Cretaceous, how can we explain the extraordinary thickness of later Cretaceous sediments, many of marine origin, that fill the Methow Graben?

The graben trends north-northwest and is bounded by the Ross Lake Fault on the west and the Chewack-Pasayten Fault on the east. The faults have been partly obscured by Tertiary intrusions that have risen along the fracture zones. There are some older crystalline rocks at the south end of the graben, between Twisp and the mouth of the Methow Valley. These probably represent the floor of the fault basin. They are cut by a number of high-angle faults of presumed Tertiary age.

The various sedimentary and volcanic units found within the Methow Graben are listed in Table 7-2. Several features of the section should be emphasized, as follows:

1. The Newby Group at the bottom of the section is equivalent to the marine Nooksack Group west of the crystalline core of the North Cascades. It rests unconformably beneath the overlying formations and has been more severely deformed.

2. All units below the Winthrop Sandstone are marine and most contain fossils. The total thickness of this marine section is approximately 8 miles—a rather astonishing thickness.

7-8 *Two possible mechanisms to explain thrust faults. Section A shows large-scale gliding down a low slope due to gravity. Note the stratigraphic gap near the head of the slide. In Section B, thrusts are produced by compression, the end blocks moving together and the right side riding up and over the left side. Here the sliding surfaces dip underground to the right. In Section A, they reach the ground surface to the right. Note that the left halves of both sections are the same, which is one problem in the interpretation of such features.*

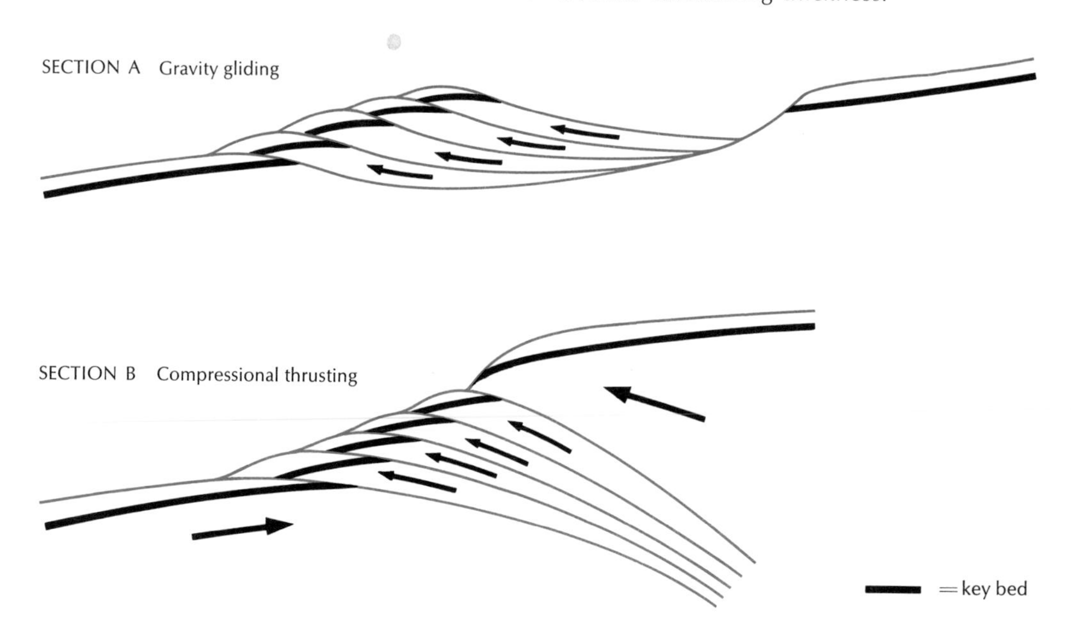

3. As much as 4 miles of nonmarine sediments and volcanics were deposited on top of the marine section. The entire sequence was then folded, faulted, and intruded locally by granitic rocks.
4. The change from marine to nonmarine conditions after the deposition of the Virginian Ridge Formation reflects the final emergence of this area above sea level.

The Methow Graben extends almost 100 miles north of the border (Figure 7-6). The names of formations within the graben change but the rock types remain the same. The Dewdney Creek Group is a thin (approximately 1,000 feet) equivalent to the Newby, without the volcanics. This is overlain by 14,000 feet of marine sedimentary strata, known as the Jackass Mountain Group. They correspond to the Goat Creek, Panther Creek, Harts Pass, and Virginian Ridge Formations of Washington. The Canadian equivalent of the nonmarine Winthrop and Ventura Formations is called the Pasayten Group. It has a maximum thickness of about 10,000 feet. The fault along the west side of the graben in British Columbia is called the Hozameen Fault, and it contains a significant band of serpentine. This fault apparently merges to the south with the Jack Mountain Thrust and its root, the Ross Lake Fault.

TABLE 7-2 Late Mesozoic stratigraphy of the Methow Valley, North Cascades. (From J. D. Barksdale, personal communication.)

AGE	FORMATION	THICKNESS, FEET*	DESCRIPTION
Early Tertiary	Pipestone Canyon	2,300	Granitic conglomerate, arkosic sandstone, and shale. Fossil leaves. Nonmarine
Late Cretaceous	Midnight Peak	10,000	Andesite flows, breccias, tuffs. Minor red sedimentary interbeds. Nonmarine
	Ventura	2,100	Reddish-purple sandstone, siltstone, and shale. Nonmarine
	Winthrop	6,500 (13,500)	Massive arkosic sandstone. Fossil leaves. Nonmarine
Early Cretaceous	Virginian Ridge	7,200 (11,600)	Chert, conglomerate and sandstone, black shale. Marine fossils
	Harts Pass	7,900	Massive arkosic sandstone, minor black shale. Marine fossils
	Panther Creek	4,500	Black shale with minor sandstone and conglomerate. Marine fossils
	Goat Creek	5,100	Arkosic sandstone and black shale. Minor conglomerate. No fossils. Presumed marine
	Newby (upper)	14,500	Andesite flows, breccias, tuffs, in lower one-third. Volcanic sandstone and shale. Marine fossils
Late Jurassic	Newby (lower)	5,000	Black shales and volcanic sandstones. No fossils. Presumed marine

▬▬ Major unconformity
*Thickness in parentheses denotes maximum for notably variable units.

7-9 *The Methow Graben northwest of Harts Pass, looking southwest toward Azurite Peak. All the strata from the foreground through Azurite Peak are Lower Cretaceous sedimentary rocks within the graben. More distant peaks belong to the crystalline core of the range. The Barron Mine in the right foreground is a gold mine (not currently in operation) adjacent to a granitic intrusion. (Photograph courtesy of the Washington State Department of Commerce and Economic Development.)*

What interpretation seems reasonable for this extraordinarily thick Cretaceous section in the Methow Graben? One possibility is that the graben is on the site of a former narrow "arm" of the sea that somehow covered this area while pronounced uplift and deformation took place in the immediate surroundings. This theory has problems. For one thing, there is little in the stratigraphic record to suggest a Cretaceous shoreline near the edge of the graben. For another, similar Cretaceous sediments are preserved in other grabens or large downfolds in the Cordillera. The number of "arms" required to explain all these makes for pretty complex geography. Furthermore, it is easy to overlook the vast amount of time (60 million years) represented by the Cretaceous Period. That we can demonstrate a Cretaceous age for thrusting, metamorphism, intrusion, and marine sedimentation does not demand that it all happened at just the same time. The Jack Mountain Thrust, for example, clearly postdates the marine sedimentation in the Methow Valley. Paleozoic strata in the upper plate of the thrust moved eastward over the uptilted and eroded edge of the Harts Pass and older formations. In the San Juan Islands to the west, for example, Upper Cretaceous marine sediments were deposited after movement of Cretaceous thrust faults.

The Cretaceous Period was the last time that the seas covered much of the North American Continent. Perhaps Cretaceous marine sediments were once much more widespread in the Cascade region than is suggested by present-day exposures. Only where faulting or folding placed these beds down into the hard crystalline core of the range were they able to survive the great erosional stripping of the past 60 million years.

7-10 *The intrusive contact between the Tertiary Chilliwack Batholith and overlying altered Paleozoic lava of the Chilliwack Group. Note inclusions of the country rock in the granodiorite to the left of the hammer. East Lake Ann Butte, North Cascades, Washington. (Photograph courtesy of Peter Misch.)*

GRANITIC INTRUSIONS

Granitic rocks underlie a large area of the North Cascades. Some of these, such as the Yellow Aster Complex, are demonstrably old, but most are of Late Cretaceous or Tertiary age. None of these batholiths is uniform; rather they consist of multiple intrusions of magma of slightly different ages and compositions. The principal intrusive complexes in the North Cascades are shown in Figure 7-2. The Golden Horn is the only batholith that is mineralogically a true granite; the rest are principally granodiorite and quartz diorite (see Appendix A).

The age of intrusion of some of these batholiths is fairly certain, but for others it is quite conjectural. Close dating depends upon identifying the youngest units intruded by the magma and the oldest formations that postdate it. Radiometric age dating techniques have also been very helpful. The Mount Stuart Batholith is probably the oldest. It has a radiometric age of 80 million to 90 million years (early Late Cretaceous) and was exposed and supplying sediment to nearby basins in the Eocene Epoch (Early Tertiary). The Black Peak intrusion is of latest Cretaceous or Early Tertiary age. It postdated Late Cretaceous thrusting since it intruded along the Ross Lake Fault. The Chilliwack Batholith had at least four intrusive pulses, ranging in age from perhaps Eocene to early Miocene. The main (second) phase was Eocene. The Golden Horn intrusion cuts (and hence postdates) the Black Peak Batholith; it is of probable Eocene or early Oligocene age. The Snoqualmie Batholith is the youngest of all. It intrudes sedimentary and volcanic rocks as young as early Miocene and has yielded radiometric dates of 16 million to 18 million years (middle Miocene). No accurate date for the Summit Creek Batholith is available. It is probably of Early Tertiary age, based on the ages of comparable intrusions to the west.

Dates as young as Miocene for large batholiths are rare and of considerable geologic interest. This does not mean that granitic batholiths have not been emplaced in the past few tens of millions of years or that granite is not crystallizing today underground in certain parts of the world, including perhaps the Northwest. However, batholiths crystallize at depths of several miles or more. Thus when we see granitic rocks exposed at the surface, this implies that very considerable erosion must have occurred to uncover them. Young batholiths exposed at the surface mean recent vigorous mountain building—entirely compatible with our interpretation of the Cascades as one of the youngest mountain ranges in the world.

TWIN SISTERS—A PIECE OF THE MANTLE?

The mountain mass known as Twin Sisters, 10 miles southwest of Mount Baker, is one of the most interesting parts of the North Cascades. The mountain itself is rather stark. It consists of massive rock that weathers reddish-brown and does not support as much vegetation as surrounding slopes. The fresh rock is green, consisting primarily of the mineral *olivine* (a magnesium-bearing silicate) with scattered grains of black chromite. Locally, the chromite is concentrated in nearly pure layers several inches to several feet thick. A rock rich in olivine is known as a *dunite*. Although small pods of dunite are found in fault zones in many mountain ranges, this is the largest single mass in the western hemisphere. Detailed study of the Twin Sisters has shown that the dunite was emplaced by faulting as a solid mass sometime in the Early Tertiary.

Dunite is of great interest to geologists because the physical properties of the rock are very similar to those determined for the mantle—that part of the earth that underlies the crust. We will not know if the mantle contains dunite until we succeed in drilling through the crust (a feat that should be accomplished within the next few years). It seems probable. The occurrence of dunite (and its common alteration product, serpentine) along major steeply dipping faults in deformed mountain belts supports the concept that this rock may have been derived from beneath the crust. If this turns out to be correct, the North Cascades contains the hemisphere's largest unaltered piece of exposed mantle.

TERTIARY STRATIGRAPHY

The North Cascades does not provide one of the better Tertiary stratigraphic records in the Northwest. Tertiary strata are locally quite thick. How-

7-11 *Mount Index, a Middle Tertiary granodiorite batholith that towers above the Skykomish River and U.S. Highway 2. (Photograph courtesy of Washington State Department of Commerce and Economic Development.)*

ever, like the unusual Methow section or the Tertiary of the Okanogan Highland, downfaulting or folding was generally necessary for the preservation of the Tertiary section.

The best record is for the Eocene Epoch. Strata of this age are widespread in the southern part of the North Cascades and in the western foothills around Bellingham. The dominant lithology is a light-colored *arkosic* (feldspar-rich) sandstone, with interbeds of medium grey siltstones and shales and locally thick conglomerate. Fossil wood, plant stems, and leaves are abundant. For some unknown reason, vertebrate fossils have not been found. However, the fossil flora is extremely varied. It consists of ferns, conifers, and many deciduous species, but unfortunately the longevity of most botanical species prevents precise paleobotanical dating. This flora does indicate, however, that there was a warm, semitropical climate, with one of the most common fossils being a relative of the modern palmetto.

The beds around Bellingham are known as the Chuckanut Formation. They are very well exposed along Chuckanut Drive and in cuts along Interstate Highway 5 near Lake Samish. Correlative strata in the North Cascades are assigned to the Swauk Formation. Its type area is the Swauk Pass–Blewett Pass region southeast of Mount Stuart. Neither formation has been studied in great detail, in part because each is rather monotonous in character and contains few distinctive beds that can be traced for any great distance. The sediments were deposited by rivers and in lakes on broad, alluvial floodplains. Some of the basins must have been subsiding to allow the accumulation of such thick sections. One clear example is the Chiwaukum Graben, northwest of Wenatchee. This graben is bounded on the west by the nearly vertical Leavenworth Fault and on the east by the equally steep Entiat Fault. The faults bordering the Chiwaukum Trough were active during sedimentation, so that the basin subsided as sediment was deposited. The Swauk sediments have been folded parallel to the long axis of the graben; the folds are exposed along the Wenatchee River and along U.S. Highway 2 between Leavenworth and Wenatchee. Conglomerates are common along the edges of the graben. Cobbles and boulders are composed largely of the same types of crystalline rocks that compose the highlands adjacent to the graben. The stratigraphic thickness of the Swauk Formation northeast of Leavenworth is at least 23,000 feet—a rather remarkable figure. Unfortunately, this is the only place where a serious attempt has been made to measure the thickness of the Swauk. The correlative Chuckanut to the northwest is in excess of 15,000 feet thick.

What did the landscape of the North Cascades look like in the Eocene? It was once presumed that there were no mountains here at this time and that the entire region was a relatively featureless coastal plain with many rivers meandering westward across it toward the sea. The abundant quartz and feldspar grains were presumably derived from the primarily crystalline terrain of the Okanogan Highlands and the eastern edge of the North Cascades. The recognition of active faults and conglomerates of local derivation within the Cascades region complicates this picture, but perhaps not seriously. The older Mount Stuart Batholith probably was exposed. It contributed arkosic sand and also pebbles from its cap of volcanic rock. (The only evidence of this cap is an abundance of volcanic pebbles in Swauk conglomerate along the west edge of the Chiwaukum Graben. The Stuart Range itself has been eroded too deeply to preserve any lava in place.) A flat coastal plain environment also seems unlikely in view of the suggested Eocene age of emplacement of some of the other batholiths in the North Cascades. These must have been accompanied by local uplift and volcanic eruptions.

The Swauk and the Chuckanut Formations are the most extensive Early Tertiary units in the North Cascades, but there are others. Arkosic sediments, 2,500 feet thick, crop out northeast of Bellingham. These have been correlated with the middle to upper Eocene Huntingdon Formation of southwestern British Columbia (Table 7-1). Stratigraphically above these sediments are approximately

3,000 feet of lava flows of andesitic to dacitic composition, with lesser amounts of interbedded ash beds and arkosic sediments. This formation is known as the Hannegan Volcanics and is of Oligocene age.

Sedimentation also occurred along the east side of the Methow Graben. Swauklike sediments, including very coarse boulder conglomerates, were deposited in this trough adjacent to the Chewack-Pasayten Fault. Exposures are restricted to three small downfaulted blocks (thus, grabens within a graben) and the total section is only a few thousand feet thick. The unit is known as the Pipestone Canyon Formation; its fossil flora has been interpreted as Paleocene in age.

Post-Swauk volcanism and sedimentation occurred in the Cle Elum area south of Mount Stuart. The record is preserved in the east-west trending Cle Elum Syncline. (A *syncline* is a downfold.) The stratigraphic relations are apparent in a cross-sectional view (Figure 7-12). The Stuart Range to the north consists of Cretaceous granodiorite, surrounded by serpentine and older metamorphic rocks. The Swauk Formation rests on these crystalline rocks, which composed the floor of the Swauk basin in this vicinity. (Unlike the east side of the range, the contact between the Swauk and the pre-Tertiary rocks is here an unconformity rather than a fault.) Volcanism occurred after thousands of feet of Swauk sediments had accumulated. This resulted in as much as 5,000 feet of andesitic and basaltic lava flows, breccia beds, and thin interbeds of dark, volcanic sandstone and ash being piled on top of the Swauk. The flows were fed by long, straight fissures that cut through the underlying units; these were subsequently filled by lava that congealed as dikes. The formation is known as the Teanaway Basalt. Its many dark dikes cutting through light-colored Swauk sediments are very striking. There are thousands of these dikes, mostly trending northeast, with each one typically a few tens of feet thick.

7-12 *Generalized cross section across the Cle Elum Syncline from Mount Stuart to Cle Elum Ridge. The town of Cle Elum and Interstate Highway 90 are located in the valley of the Yakima River.*

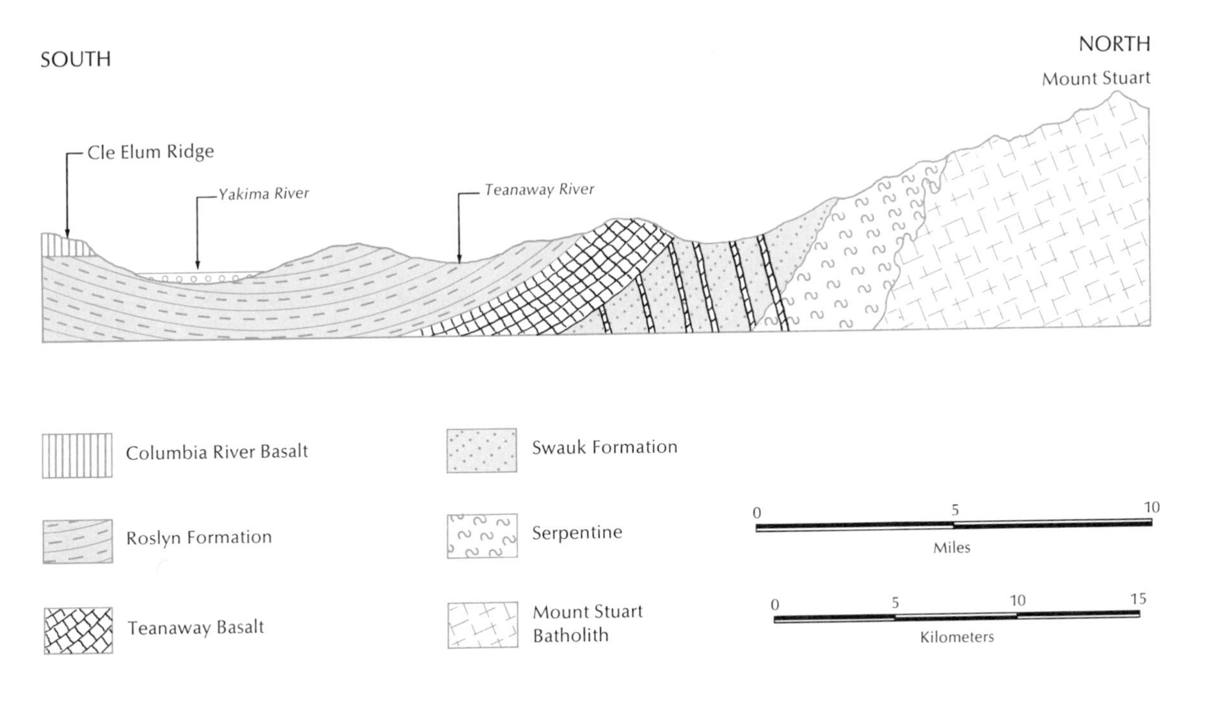

Arkosic sedimentation resumed in the late Eocene after this volcanic interlude. It produced the Roslyn Formation, which consists of Swauklike sandstones and shales. The Roslyn also contains some coal beds, which formed the basis for the now defunct coal-mining industry of Kittitas County. (The coal resources are not exhausted. The beds, however, dip rather steeply in places as a result of regional folding, and excavation costs are not competitive. A similar problem exists in the coal beds of King and Pierce Counties in the western Cascade foothills.)

The post-Eocene record of the North Cascades is relatively poor. The folding of the Swauk, Teanaway, and Roslyn Formations occurred prior to the Miocene eruption of the Columbia River Basalt, which rests unconformably on the upturned and eroded edges of the Eocene section. The Snoqualmie Batholith, also of Miocene age, intruded the Swauk beds along the Cascade crest. The heat of the invading magma caused local recrystallization of the sediments so that they look very much like the granodiorite of the batholith. These metamorphosed sediments form such rugged peaks as Big Four, west of Monte Cristo, and those at the head of Gold Creek immediately east of Snoqualmie Pass. The Snoqualmie and other Tertiary batholiths produce many deposits of gold, silver, copper, and other metals, but no significant mines are operating currently. Upper Tertiary strata have not been found in the North Cascades.

THE QUATERNARY PERIOD—GLACIERS AND VOLCANOES

The Tertiary history of the North Cascades was quite different from that of the rest of the range. The north end of the range was mostly uplifting and being eroded. The region south of Snoqualmie Pass witnessed an extraordinary sequence of volcanic eruptions and the accumulation of a very thick section of lava, ash beds, mudflows, and volcanic sandstones and breccias. In the past few million years the histories have come together. The principal events of the Late Cenozoic—uplift of the modern range, development of the Quaternary volcanic cones, and sculpturing by glaciation—have occurred throughout the Cascades. The uplift of the range is treated in Chapter 12, for the evidence that documents the evolution of the Cascade Arch lies mostly south of Snoqualmie Pass. The young volcanoes are discussed in Chapter 13, for the two cones of the North Cascades (Mount Baker and Glacier Peak) are not sufficiently unlike the other Cascade volcanoes to merit separate treatment. This leaves the glaciation for discussion here.

The first geologists who did reconnaissance mapping in the higher ranges of the Northwest recognized that today's glaciers are only tiny remnants of alpine ice that once covered a much greater area and extended to much lower elevations. The evidence was obvious—the characteristic U-shaped cross profile of the valleys; the semicircular basins (called *cirques*) with nearly vertical headwalls at the upper end of the valleys from which the glacier flowed; the grooves or striations of exposed bedrock surfaces produced by abrasion by stones transported in the moving ice; the erratic stones on the valley floor; and the embankments of glacial material along the sides and looping across the floor of the valley. These embankments are called *moraines,* and the one farthest down the valley (the terminal moraine) marks the farthest extent of the former glacier. The embankment consists of *till*—a mixture of clay, silt, sand, pebbles, and even boulders, deposited directly by the glacier and not size-sorted by running water.

The pioneer geologists merely noted the past existence of large alpine glaciers. Only in the past several decades have geologists attempted to work out the details of glacial events in the Cascades. Our knowledge is distinctly spotty. The glacial sequence has been worked out in detail for a few valleys, such as that of the Yakima River on the east flank of the Cascades and the several valleys that head on Mount Rainier. Reconnaissance mapping of glacial features has been done in many other areas, as well. Several periods of advance and retreat (or disappearance in some instances) of alpine glaciers have been recognized. Radiocarbon age data, though not abundant, have en-

7-13 *View northwest up the Stehekin Valley at the head of Lake Chelan. Glacial ice covered all but the highest peaks in the right distance. The smooth, rounded appearance of the slopes is the result of glacial abrasion. Parts of Lake Chelan are almost 2,000 feet deep, attesting to the efficiency of glacial erosion. (National Park Service photograph, courtesy of Washington State Department of Commerce and Economic Development.)*

abled workers to make preliminary correlations between alpine and lowland glacial events. We know more about the most recent glaciation than about previous ones, primarily because (as in the lowland) each glaciation destroyed much of the evidence for preceding ones.

A valley-by-valley treatment of Cascade glaciation would be inappropriate here. Instead, a few of the more interesting or better-studied areas will be noted, beginning with the Methow Valley. This is Washington's answer to Yosemite Valley, with steep valley walls and a broad flat floor. The glacier that moved down the Methow was derived in part from ice that flowed south from British Columbia up over 7,000-foot-high Harts Pass and in part from alpine glaciers that formed in the surrounding Cascades. Methow glaciers extended all the way to the Columbia Plateau at Brewster and Pateros. Impressive kame terraces (Chapter 6), deposited between the stagnating glacier and the valley wall, line the sides of the lower Methow Valley.

Lake Chelan is a classic example of a glacial basin carved out by a thick valley glacier (Figure 7-13). The ice formed near the crest of the Cascades and extended all the way to the Columbia River. The Chelan glacier had receded well up the lake by the time of farthest advance of lowland ice across the Columbia Plateau. The latter dammed the valley mouth at Chelan and raised the level of the lake hundreds of feet. Finally water spilled across the south flank of the basin and into the Columbia River via Knapp and Navarre Coulees. These coulees are now dry, but their large size suggests that this escape overflow was very considerable. The very prominent terraces found on the hillsides at the southeast end of the lake were formed by sediment being deposited upvalley from the lowland ice that occupied the valley of the Columbia River. A comparable relationship existed along the west front of the Cascades where the Puget Lobe dammed Cascade valleys (Chapter 18).

Various valley galciers coalesced to flow far down the valley of the Yakima River, reaching finally almost to the Kittitas Valley. Tributary glaciers carved out basins now occupied by three "finger" lakes of the central Cascades—Lakes Keechelus, Kachess, and Cle Elum. Downvalley from Cle Elum, the Yakima Valley glacier split around Lookout Mountain. One lobe extended a few miles farther down the Yakima River, and the other flowed east across Swauk Prairie, building a large terminal moraine complex across its eastern end. Prominent recessional moraines were left in the Yakima Valley during the retreat of the ice. They include the large one under Lake Keechelus Dam and smaller moraines well exposed in road cuts between Snoqualmie Pass and the west end of Lake Keechelus.

Establishing the glacial chronology in one valley may not be too difficult. Correlating from one valley sequence to another is much more of a problem. Certainly a region's climate has a major influence on the size and state of health of its glaciers, and one might expect that all of the glaciers in an area would wax or wane in unison. This is not always the case; and, even at the present time, the fronts of some glaciers are advancing while nearby ones recede. The problem is compounded for past glaciations because there were several periods of major alpine glaciation, and each was characterized by numerous short intervals of glacial advance or retreat. Glacial geologists are therefore reluctant to generalize about the glacial history of a region as large as the Cascade Range. It will probably be many years before enough detailed work has been done to establish a glacial chronology for the entire range.

8 central and western british columbia

■ At first glance, the geology of western British Columbia looks relatively simple, since individual formations seem to cover large areas. This impression is incorrect. It merely reflects how relatively little we know about the geology of this vast region. In much the same way, the older geologic maps of parts of the United States, now known to be very complex, show large areas of uniform lithology. Consider the Coast Mountains "Batholith." It extends 1,000 miles from the Fraser River north into Alaska, and has been cited as the world's largest batholithic intrusion. In fact, it consists of many diverse intrusions surrounded by or including gneissose remnants of the country rock.

Western British Columbia is a magnificent place for "do it yourself" geology. It is no great trick to find areas that have not been mapped—where your geologic observations may be the first ever made. The greatest problem for geologic investigation throughout the province has been the very limited access provided by roads. With a good boat, the immense and intricate coastline can be reached, and it affords generally excellent rock exposure. However, the mountains rise so steeply out of the fiords and their sides are so heavily timbered that movement inland from the coast is nearly impossible.

The incentive for geologic investigation is great because of the potential mineral wealth of the province. Reconnaissance mapping by airplane and helicopter is being done at an accelerating pace, by mining concerns, the British Columbia Department of Mines, and the Geological Survey of Canada. The task is so large that it will be many years before any sort of "honest" geologic map of the province will be available. At present, we can only make generalizations that we hope will stand up to the detailed studies that someday must follow.

8-1 *Geologic field party working on the Coast Mountains project of the Geological Survey of Canada. Small boats, such as this, are essential for work along shore, but helicopters are needed for exploration inland. (Photograph courtesy of J. A. Roddick, Geological Survey of Canada.)*

THE PHYSIOGRAPHY OF THE REGION

Mile for mile, western British Columbia has probably more breathtaking scenery than almost any comparable region in the world. In detail, the physiography seems a complex mix of innumerable jagged peaks, overpowering cliffs, immense fiords, and lovely islands. The overall physiographic plan is not so complex (Figures 5-2, 5-3). From west to east there are, first, a partly submerged mountain range (Vancouver Island and the Queen Charlotte Islands), then a submerged lowland (the Strait of Georgia, Queen Charlotte Strait, and Hecate Strait, plus coastal lowlands), and finally the Coast Mountains. The latter merge with the Interior Plateau Province of south-central British Columbia and various highlands and interior ranges in the north-central part of the province.

The highest peaks on Vancouver Island, rising more than 7,000 feet, are located near the middle of the island. The mountains diminish in height rather gradually from there to the northwest and southeast ends of the island. Surrounding lowlands include the Nanaimo Lowland on the east side, the Nahwitti Lowland at the northwest end of the island, and the narrow Estevan Coastal Plain along the central part of the west coast.

The mountains on the Queen Charlotte Islands are not as high (maximum elevation around 3,700 feet) as those on Vancouver Island, but they too have been glaciated and are every bit as steep. They lie along the western side of the island group, including all of Moresby Island. On Graham Island, the range grades eastward into the Skidegate Plateau, which in turn slopes down to the Queen Charlotte Lowland.

The Coast Mountains are 1,000 miles long. They extend from the valley of the Fraser River northwest along the entire length of the province and 50 miles into the Yukon. They are breached by a number of rivers that originate along the western edge of the Interior Plateau Province and the various interior highlands and plateaus that lie east of the range in northern British Columbia. The largest of these rivers are (from south to north) the Klinaklini, the Bella Coola, the Dean, the Skeene, the Nass, the Stikine, and the Taku Rivers. All these valleys were occupied by alpine glaciers in the past, and all empty into the magnificent fiords along the Pacific coastline. The positioning of a river across the axis of a major mountain chain raises a question as to the origin of the river. In these instances, it seems as if the rivers were there before the mountains formed. The Coast Mountains, like the Cascade Range to the south, were raised in Late Cenozoic time. The rivers were able to maintain their courses across the rising range by downward erosion at a rate at least equal to that of the uplift. This is a testimonial to the efficiency of stream erosion and also an indication of the gradual rate of mountain building. Rivers that are able to maintain their courses across rising mountains are known as *antecedent rivers.*

The tallest peak in the Coast Mountains is Mount Waddington, which rises to an elevation of 13,177 feet. It lies near the head of Knight Inlet, about 100 miles east-northeast of the north end of Vancouver Island. This southern part of the Coast Mountains is the loftiest part of the chain, with many peaks above 9,000 feet and numerous alpine ice fields and glaciers.

The Interior Plateau Province covers more than 100,000 square miles of the central and southern interior of British Columbia. To the west, it is bounded by the Coast Mountains and Cascades, to the north by the Skeena and Omineca Mountains, and to the east by the Columbia Mountains. Most of this immense region is drained by the Fraser River and its tributaries (Figure 5-2). Much of the Interior Plateau Province lies at elevations of 4,000 to 5,000 feet, with some highlands rising above this. A plain surface at only 2,000 to 3,000 feet is widespread around Prince George.

THE GEOLOGIC OVERVIEW

Geologically, the Coast Mountains are composed essentially of metamorphic rocks and many granitic intrusions of Late Mesozoic or Early Cenozoic age. In this respect, they much resemble the North Cascades of Washington and the Shuswap Metamorphic Complex in the Columbia Mountain Province (Chapter 6). A major part of the Interior Plateau Province is underlain by Tertiary volcanic rocks. These generally are inclined slightly to the east away from the Coast Mountains uplift. The province also contains exposures of pre-Tertiary rock, especially Late Paleozoic and Early Mesozoic sedimentary and volcanic rock. They constitute a most valuable record because strata of comparable age in the Coast Mountains to the west have been thoroughly metamorphosed. Vancouver Island and the Queen Charlotte Islands, with an especially good Mesozoic record, also add much to our knowledge of the Cordilleran Geosyncline.

The Cenozoic record of central and western British Columbia is not very informative. Granitic intrusions occurred at various times, but these are volumetrically rather minor compared to the great Mesozoic emplacements. The Cenozoic strata of the mainland are nonmarine volcanic and sedimentary rocks. Except for large sheets of Miocene basalt lava on the Interior Plateau, these strata are neither extensive nor very thick. The record on the islands is a little better, but not much. Tertiary strata crop out locally along the southwest coast of Vancouver Island and throughout much of Gra-

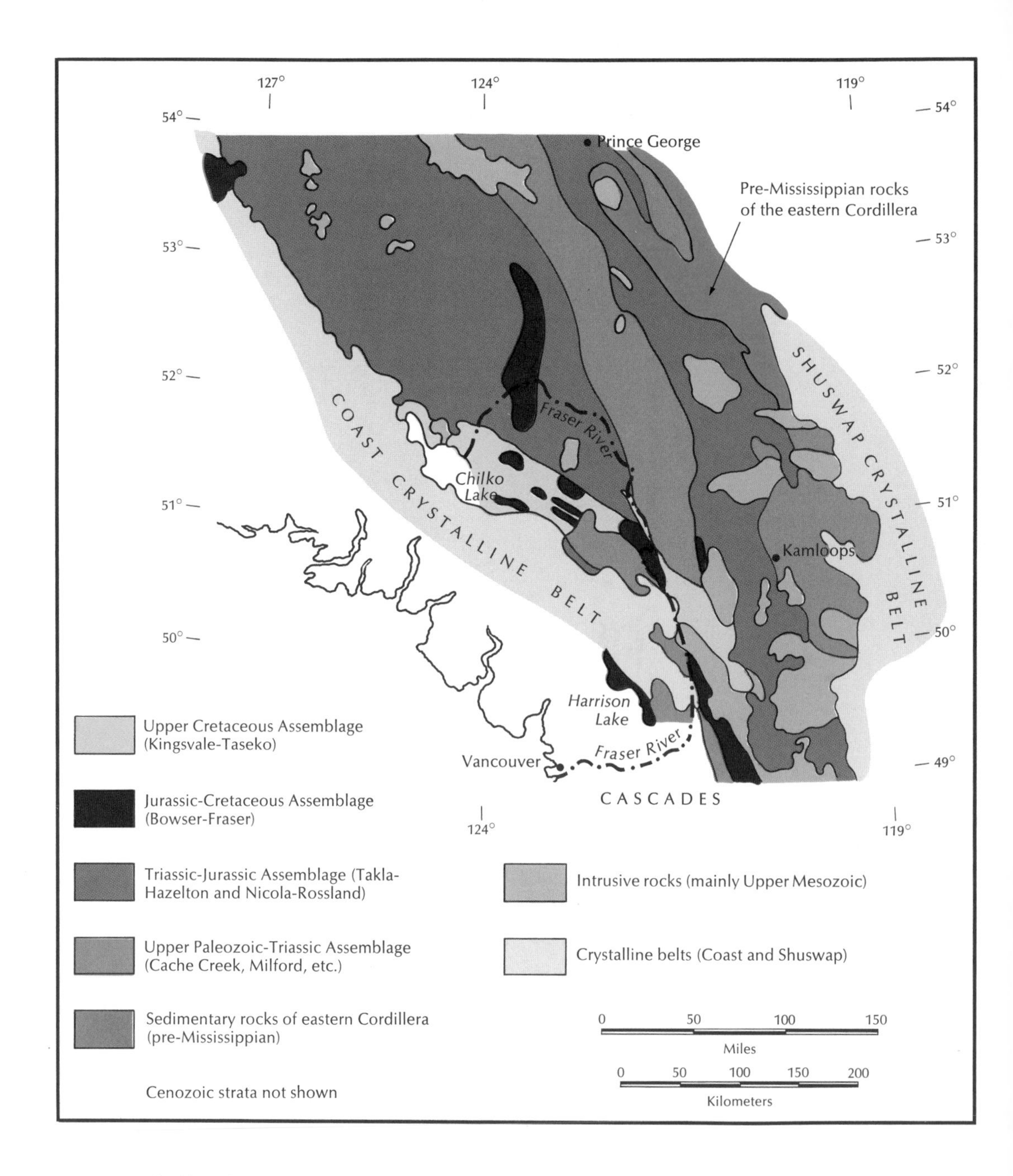

8-2 *Generalized geologic map of the southern and central part of the Interior Plateau Province of British Columbia. (After CIM Sp. Vol. 8, 1966, Fig. 4-1.)*

ham Island at the north end of the Queen Charlotte Islands. The present distribution of these strata is controlled by large and relatively young faults. Tertiary marine strata undoubtedly were once more extensive along the coast.

The mountains of western British Columbia rose late in the Cenozoic and are probably still rising today. Young volcanoes, which are so much a part of the recent history of Washington and Oregon, are found also in western British Columbia. Like the Cascade volcanoes, these developed after the main uplift of the underlying mountain chain.

Much of the spectacular scenery of British Columbia is due to Pleistocene glaciation. Glaciers that began as small alpine glaciers grew until finally much of the province lay buried beneath thousands of feet of ice. This happened perhaps as many as four times. The total change to the landscape produced by glaciation was enormous.

THE EARLY RECORD

The age of the oldest rocks in western British Columbia is unknown. The oldest fossiliferous strata are Devonian in age—just as they are in western Washington. Rather thoroughly recrystallized schists and gneisses that underlie these formations could be as old as Precambrian. These metamorphic rocks do not crop out extensively. They are most easily seen in the Victoria area at the southeastern end of Vancouver Island.

The record for the Late Paleozoic is much better. It includes the Sicker Group of southeastern and central Vancouver Island (Table 8-1), and the Cache Creek Assemblage east of the Coast Mountains (Figure 8-2). The Sicker Group consists of approximately 10,000 feet of submarine lava flows and volcanic ejecta, mostly of basaltic composition, with interbedded dark marine sandstones, shales, and beds of chert. These pass upward into beds of fossiliferous limestone and shale of earliest Permian age. The Cache Creek rocks are much the same type as the Sicker Group—marine volcanic rocks, shale, chert, and some limestone. They range in age from Mississippian to Early Triassic. The volcanics have been altered to greenstone and the shale to phyllite, but metamorphism was not severe. In many areas, the Cache Creek strata have been folded and faulted rather markedly. This has discouraged detailed study of the stratigraphy. No estimate of the total thickness of this assemblage is available, but it must exceed 10,000 feet. Thus the Paleozoic Era ends with seas apparently covering much of western British Columbia. There was active volcanism that created islands, but there were also relatively stable regions in which limestone was deposited.

MARINE LAVA AND SEDIMENTS

The general environment of the Paleozoic Era persisted through the first half of the Mesozoic Era. It gave way then to the great cataclysmic events that destroyed the Cordilleran Geosyncline. The Triassic record is not too extensive, although it is better here than in Washington. In the Interior Plateau Province, the principal Triassic units are the Nicola and Rossland Groups in the south and the Takla and Hazelton Groups in central British Columbia. All these are predominantly marine basalt or andesite flows and breccias, with interbedded conglomerates, sandstones, and shales rich in volcanic fragments. They also contain Lower and Middle Jurassic strata in their upper part. Fossils are uncommon.

To the west, the story is the same. Starting in Middle Triassic time, great outpourings of submarine basalt occurred on what is now Vancouver Island. This lava pile, the Karmutsen Formation, is more than 2 miles thick and may once have had a total volume of perhaps 100,000 cubic miles. As soon as the eruptions ceased, the organic and inorganic precipitation of lime began. The rest of the Triassic saw the formation of probably the most extensive and the thickest (up to 3,000 feet) limestone formation in the entire northern part of the eugeosyncline. This is the Quatsino Formation, and it correlates with the Kunga Formation in the Queen Charlotte Islands (Table 8-2). These limestones served as the principal host rock for ore deposits that were emplaced later in the Mesozoic Era. Volcanism resumed once again in the Jurassic Period. It was more violent than the Karmutsen eruptions had been, producing andesitic breccias and tuffs, interbedded with marine volcanic sandstones and shales. These units comprise the Bonanza Formation of Vancouver Island and the Yakoun Formation in the Queen Charlotte Islands.

A break or unconformity exists in this position in the stratigraphic record of western British Columbia and elsewhere in the eugeosyncline. This mid-Jurassic period of erosion was followed by renewed sedimentation and minor volcanism, beginning in the Late Jurassic and lasting into the early part of the Cretaceous Period. In the northern part of the province, the strata deposited at this time are called the Bowser Assemblage. They

TABLE 8-1 Stratigraphy of Vancouver Island. (From Sutherland-Brown, 1966.)

AGE	FORMATION		MAXIMUM THICKNESS, FEET	DESCRIPTION
Early to Middle Tertiary	Carmanah and Sooke		8,000	Sandstone, shale, and conglomerate. Marine. Locally along west coast
Early Tertiary (Eocene)	Metchosin		>7,500	Basaltic flows with minor sedimentary interbeds. Pillows. Marine. South edge of island
Late Cretaceous	Nanaimo Group		10,000	Arkosic sandstones, siltstones, shales, conglomerates, coal beds. Shallow marine and nonmarine. Nanaimo and Suquash Basins
Latest Jurassic to Early Cretaceous	Unnamed		3,600	Graywacke sandstone, siltstone, conglomerate. Marine. Patches around and south of Quatsino Sound
Jurassic	Granitic intrusions			Widespread
Late Triassic to Early Jurassic	Vancouver Group	Bonanza	12,000	Andesitic breccias and tuffs, explosive volcanism. Minor graywacke sandstone and shale. Marine. Nootka Sound to north. Local elsewhere
Late Triassic	Vancouver Group	Quatsino	3,000	Limestone, minor shale. Marine. Discontinuous bands from Nootka Sound north
Permian(?) to Late Triassic	Vancouver Group	Karmutsen	>10,000	Massive basaltic flows, local pillows. Marine. Widespread throughout entire island
Devonian(?) to Early Permian	Vancouver Group	Sicker Group	10,000	Basaltic flows, breccias, tuffs; graywacke sandstone, chert, limestone, shale. Marine. Discontinuous band south-center, Saltspring Island to Hesquiat Inlet
?	"Basement Complex"		?	Dioritic gneiss, migmatites. Discontinuous band along south side from Victoria to Hesquiat Peninsula

——— Major unconformity

8-3 *Steeply dipping quartzites on Addenbroke Point, 18 miles south of Namu, British Columbia. These are probably derived from Middle to Late Paleozoic chert beds correlative with the Sicker and Chilliwack Groups to the south. (Photograph courtesy of J. A. Roddick, Geological Survey of Canada.)*

8-4 *Submarine pillow basalt of Mississippian age near Little Fort on the North Thompson River, British Columbia. This formation, the Fennel, belongs to the Cache Creek Assemblage. (Photograph courtesy of K. V. Campbell.)*

consist of conglomerate, sandstone, and shale; some of the beds are nonmarine. In the southern Coast Mountains and to the east, rocks of comparable age are known as the Fraser Assemblage. It resembles the Bowser Assemblage except for the presence of volcanic rocks near Harrison Lake.

The action was to the west, where the first of the great Mesozoic batholiths was emplaced. Radiometric ages from some of the larger granitic intrusions on Vancouver Island are as old as 167 million years, or approximately mid-Jurassic. This is appreciably older than most of the intrusions in the Coast Mountains but approximately synchronous with the Nelson Batholith and others to the east (Chapter 6). Granitic pebbles produced by the erosion of these intrusions are found in some of the conglomerates of the Bowser and Fraser Assemblages.

THE ORIGIN OF BATHOLITHS

The great deformation that finally destroyed the Cordilleran Geosyncline is as well documented in British Columbia as it is in the North Cascades. Once again, it was not a spectacular, short-lived event. In a sense, the orogeny has continued to the present day and has produced the very young mountains that we now see. In northern Washington, the most notable aspect of the orogeny was perhaps the great westward moving thrust plates (Chapter 7). In western British Columbia, the formation of the Coast Mountains crystalline complex was the most significant event. We have little detailed information about the complex. Clearly, it contains many separate granitic intrusions of various sizes, shapes, and ages, including some as young (18 million years) as the youngest intrusions of the Cascades.

The total volume of intrusion was immense. Geologists have long debated the fate of country rock that is displaced or replaced by batholiths. (This problem is commonly referred to as the room problem.) One explanation, the *forceful injection* hypothesis (Figure 8-6), states that the rising magma forces the rocks aside or upward. Lateral compression should produce severe folding and faulting in the adjacent country rock. Upward displacement is convenient, for erosion would strip away all of the evidence rather quickly. Another hypothesis, known as *stoping,* maintains that the country rock sinks down into the rising magma. In this scheme, the intrusion in a sense "eats" its way upward, incorporating the rocks above. We might expect more pieces of the country rock frozen into the solidified granite than we usually see. The theory's proponents often explain this by the hot magma's melting and mostly obliterating these inclusions.

The related theory postulates formation of the batholiths by *anatexis.* In this process magma is formed by melting of the country rock itself, the idea being that heat (presumably of radioactive origin) concentrates in the rock faster than it can dissipate. The temperature ultimately rises to the point at which the rock starts to melt. This solves the room problem rather neatly; for, in this scheme, no rocks are missing—they have simply melted and recrystallized. Chemically, it works rather well in certain instances. The compositions of some batholiths are often quite similar to that of the older sedimentary and volcanic rock of the

TABLE 8-2 Stratigraphy of the Queen Charlotte Islands. (From Sutherland-Brown, 1966.)

AGE	FORMATION		MAXIMUM THICKNESS, FEET	DESCRIPTION
Middle to Late Tertiary	Skonum		>6,000	Sandstone, shale, and conglomerate. Marine and nonmarine. Eastern Graham Is.
Early Tertiary	Masset		18,000	Basalt flows and breccias, rhyolitic ash flows. Nonmarine. Central Moresby Is., most of Graham Is.
Late Cretaceous	Queen Charlotte Group	Honna and Skidegate	4,000	Conglomerate and sandstone. Marine? Skidegate Inlet area and northwestern Graham Is.
		Haida	3,000	Sandstone and shale. Marine. Skidegate Inlet area and northwestern Graham Is.
Early Cretaceous	Longarm		>4,000	Graywacke sandstone and siltstone, local conglomerate. Marine. Eastern Moresby Is. and southern Graham Is.
Jurassic	Granitic intrusions			Western Moresby Is. and southwestern Graham Is.
Middle Jurassic	Yakoun		6,000	Andesitic breccia and tuff. Volcanic sandstone and shale. Local coal. Shallow marine and nonmarine. Moresby and southern Graham Is.
Late Triassic to Early Jurassic	Kunga		2,000	Limestone and shale. Marine. Widely scattered
Permian(?) to Late Triassic	Karmutsen		14,000	Basaltic flows, local pillows. Marine. Moresby Is.
Late Paleozoic	(Sicker Group equivalent)		?	Volcanic and sedimentary rocks. Marine. Eastern Moresby Is.

▬▬▬ Major unconformity

8-5 *Migmatitic gneiss on the southeast face of Remote Peak, 12 miles northwest of Mount Waddington. The rock is typical of the Coast Crystalline Complex. (Photograph courtesy of J. A. Roddick, Geological Survey of Canada.)*

host geosyncline. However, there are also rather sophisticated chemical arguments against this hypothesis.

Yet another theory is that of *granitization*—a process whereby the granitic rocks are formed by metamorphism rather than by crystallization from a magma. The possibility of granitization is no longer seriously questioned, but its significance relative to other batholith-producing processes is in doubt. Like the theory of anatexis, granitization solves the room problem by transformation of the country rock.

The Coast Mountains of British Columbia are potentially one of the world's greatest laboratories for the testing and refinement of these important questions. They contain a great variety of batholiths and unlimited rock exposure above timberline. However, the laboratory has barely been unlocked. Substantial masses of the older country rock still remaining in the range have undergone a moderate to high degree of metamorphic recrystallization—sufficient in most cases to prevent any routine identification of the parent formation. Many of these old remnants have been metamorphosed several times, and this further complicates their interpretation. Future work should confirm that the Coast Mountains "Batholith" consists of many different intrusives—some igneous, some metamorphic, some complex. They formed at different times, but principally during the Cretaceous and early Tertiary Periods. Remnants of almost all the older formations of western British Columbia will be found in and around the many granitic bodies. Seeing the original stratigraphy through the effects of metamorphism, however, will necessitate the application of rather sophisticated techniques.

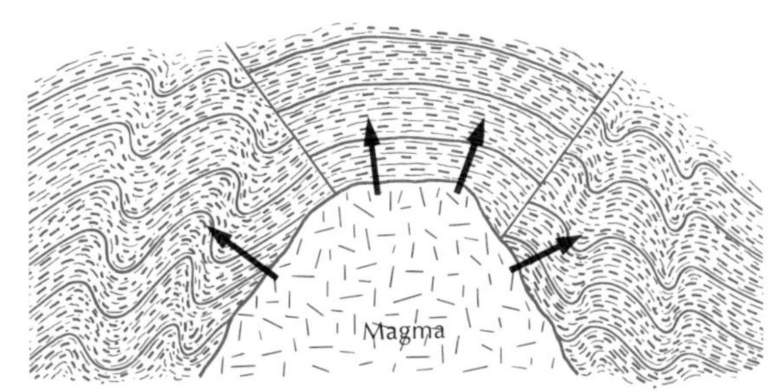

FORCEFUL INJECTION
Magma pushes strata of country rock up and aside

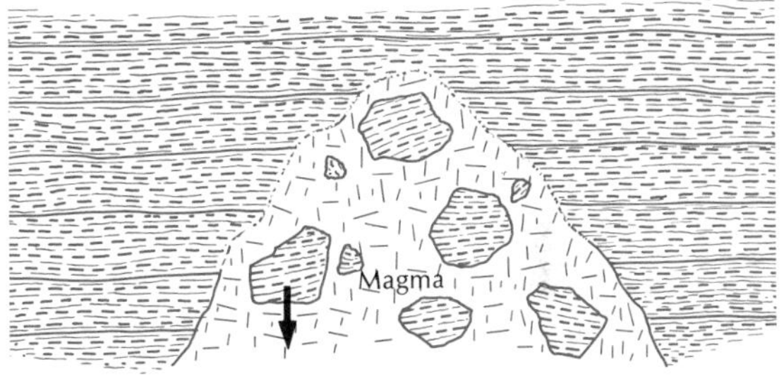

STOPING
Pieces of country rock fall off and sink into magma

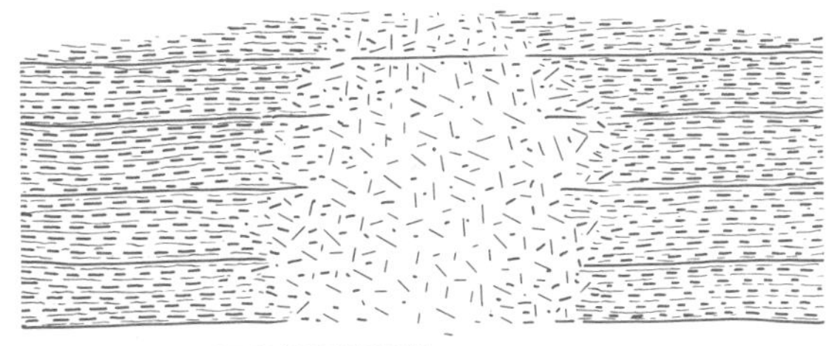

ANATEXIS OR GRANITIZATION
Country rock melts to form magma (anatexis) or recrystallizes without melting to form magmatic-appearing rock (granitization)

8-6 *Various mechanisms of granitic emplacement.*

METALLIC RESOURCES

The potential for the development of mineral resources in the Coast Mountains seems almost unlimited. The relative inaccessability of the area means that any prospect must be large to be profitable. As of 1965, there were 160 metallic mineral deposits in British Columbia that had produced 10,000 tons or more of ore and countless thousands of smaller prospects. The principal metals produced in the province have been, in decreasing dollar value, lead, zinc, copper, gold, silver, iron, mercury, and tungsten. The two leaders, lead and zinc, are mined primarily in eastern British Columbia (Chapter 6). The association of ore deposits with large intrusions is clear throughout the province. Typically, the ore is concentrated in the country rock rather than in the intrusion itself.

The actual processes of ore genesis are much debated. One argument centers on whether the metallic elements are derived from the intrusion, or whether they were present in the older rock and concentrated by the heat and solution action associated with the intrusive process. The theory of the host rock as a source for the metal is well supported by some of the large iron deposits in western British Columbia, such as those on Texada Island and at Zeballos in west-central Vancouver Island. The ore is magnetite. It occurs as massive replacements of the Triassic marine limestone, the Quatsino Formation, near Jurassic batholiths. The intrusions themselves are not unusual. Neither is the

8-7 *Contact between light-colored quartz diorite intrusion (on left) and older amphibole gneiss, in the Coast Mountains near Butedale. Note the pervasive penetration of the quartz diorite into the country rock. (Photograph courtesy of J. A. Roddick, Geological Survey of Canada.)*

limestone much different from limestone formations elsewhere that have been intruded but not mineralized. The source of the iron may have been the volcanic strata of the Karmutsen Formation, which underlies the Quatsino. It contains appreciable, but not profitable, amounts of iron. The intrusions may have mobilized the iron, which moved in some form through the country rock and replaced the limestone.

A similar origin has been proposed for some small copper deposits on Texada and Vancouver Islands and in the Coast Mountains. Copper is another element concentrated in some basaltic lava, both in British Columbia and elsewhere. (Most of the world's finest examples of native copper, such as those from the Lake Superior region, are found in cavities in basalt.) In many of British Columbia's copper deposits, however, the intrusion rather than the country rock appears to have been the source for the metal.

The prospector is not much interested in these arguments. His aim is to find the ore, not to debate its origin. He will continue to search along the margins of the province's many intrusions for evidence of ore mineralization. Clearly, if we knew more about the origin and the controls on mineralization, we could better focus our search and identify particularly favorable areas. The prospector can hardly be expected to stay home and wait for these arguments to be resolved. Furthermore, the scientist can often oversimplify and set up too restrictive a set of guidelines. A classic example was the search for uranium. If the prospectors had read the textbooks and confined their investigations to the favorable environments (in veins around intrusions), the great finds would not have been made. They showed instead that many of the world's largest deposits are disseminated through sandstones and shales many miles from the nearest intrusion. So the textbooks change as the "amateurs" make new finds.

THE LAST INVASION BY THE SEA

The Cretaceous Period presents a most enigmatic record. The period witnessed the end of the Cordilleran Geosyncline. The orogeny was a climax to crustal instability that had characterized the entire history of this great trough. By the beginning of the Cenozoic Era the Cordilleran region had been firmly added to the continent. Thereafter, the Pacific Ocean covered only the westernmost edge of the Cordillera.

We might predict that Cretaceous deposits would be fairly restricted in distribution and that the Jura-Cretaceous marine sediments and volcanics would represent the last strata deposited in the geosyncline. This is not the case. Some of the thickest accumulations of marine Cretaceous strata found anywhere in the world crop out in the North Cascades (Chapter 7). Similar assemblages are widely scattered throughout British Columbia. The rocks are principally sandstones, shales, and conglomerates. They contain marine fossils in the lower part of the section, but they grade upward into nonmarine beds with abundant plant fossils. The character of the sediment itself also changes. Younger formations have a higher proportion of the minerals feldspar and quartz. They are also "cleaner" (better sorted) than are units lower in the section. These features suggest that the Jurassic and Cretaceous batholiths of the Cordillera, which are high in quartz and feldspar, were being progressively uncovered and were providing sediment to Cretaceous basins. Pebbles, cobbles, and boulders of granitic rock are much more common in the conglomerates of Late Cretaceous age than they are in older conglomeratic beds.

West of the Coast Mountains, Upper Cretaceous sediments, the Nanaimo Group, accumulated in a trough situated between the Coast Crystalline Complex and Vancouver Island. These were mostly fossiliferous marine sandstones, shales, and conglomerates, generally light in color and containing much quartz and feldspar. Coal beds have formed from plant material that accumulated in coastal swamps. The Nanaimo Group is at least 10,000 feet thick in places. It is well exposed in the Gulf Islands and along the east side of Vancouver Island from Saltspring Island to Campbell River. The elongated form of many of the Gulf Islands reflects the structure—a series of northwest-trending folds and faults. The islands and reefs are underlain by relatively resistant beds of sandstone and conglomerate. Intervening channels are floored by shale. The Suquash Basin in the Fort Rupert–Albert Bay area of Vancouver Island is another remnant of the Nanaimo Trough.

To summarize, the Cordilleran Geosyncline died slowly. Major intrusions began in the Jurassic, especially on Vancouver Island and east of the Coast Mountains. Intrusion reached something of a climax in the Cretaceous, but it continued into the Cenozoic Era. Major thicknesses of marine sediment accumulated in Late Jurassic–Early Cretaceous time, and also locally later in the Cretaceous Period. Meanwhile, much folding, faulting, metamorphism, and uplift occurred. Reconstructing

8-8 *View southeast along Broughton Strait, a few miles south of Alert Bay. Vancouver Island is on the right. The land in the foreground is underlain by sedimentary rocks of the Nanaimo Group in the Suquash Basin. Steep slopes on far right consist of resistant volcanic strata of the Karmutsen Formation. (Photograph courtesy of James Crawford.)*

the geography of the Cretaceous is difficult. There must have been major mountains present; but, at the same time, quite thick sections of marine strata were deposited. The problem is not unique. The same situation prevailed throughout the Cordillera. The Cretaceous Period was a long one (approximately 60 million years), and the land did not have to go up and down at any spectacular rate to accommodate all of the geological history suggested by the record.

THE REGION DURING THE TERTIARY PERIOD

British Columbia's Tertiary history seems comparatively simple, perhaps because the great uplift and erosion of the past few million years have removed complicating evidence. Except for the west coast of Vancouver Island, the seas had retired once and for all. Lower Tertiary strata consist of local, but thick, accumulations of nonmarine sandstone and shale, with some interbedded coal deposits. The section also contains a varied assemblage of volcanic rocks, including sheets of silicic ash that were once very extensive. The most continuous covering of Tertiary rocks is found on the east side of the Coast Mountains in central British Columbia. It consists of great sheets of basaltic lava of Miocene age, very similar to, and probably genetically related to, the more famous Columbia River Basalt of Washington, Oregon, and Idaho (Chapter 17).

Marine sediments and volcanic rocks are found along the outside of Vancouver Island and in the Queen Charlotte Islands. The Metchosin Formation, which consists of at least 7,500 feet of submarine basalt flows of Eocene (Early Tertiary) age, covers the south end of the island. It correlates with the Crescent Formation of the Olympic Mountains (Chapter 11). Farther north on Vancouver Island are the Carmanah and Sooke Formations. Together they are more than 8,000 feet thick on the Hesquiat Peninsula. The restricted distribution of these Tertiary beds is due in part to large faults that parallel the coast and that have been active for a long time.

The Tertiary history of British Columbia, although imperfectly known, bears much resemblance to that of Washington and Oregon. The record to the south is more complete and has been studied more thoroughly. Investigation of the faults and folds along the west coast of Vancouver Island and in the Queen Charlotte Islands will produce most interesting data, for here we probably have a direct interaction between continental and oceanic crust.

LATE CENOZOIC MOUNTAIN BUILDING

The principal events in western British Columbia during the past several million years have been (1) the rise of the mountains, (2) volcanic activity in the Coast Mountains, and (3) the formation of immense ice sheets that at times buried all but the highest peaks.

The rise of the Coast Mountains is documented by the eastward dip of the Miocene basalt of the Interior Plateau Province. These great "flood" lavas spread almost like water, forming initially perfectly level surfaces. Thus the tilting of the layers to the east reflects deformation of the past 10 million to 15 million years. The total amount of uplift was more than a mile, since the basalt rises from an elevation of 2,500 feet in the interior to about 8,500 feet in the Coast Mountains. The Coast Range appears to have formed as a large but rather simple arch, now much modified by erosion. The range is probably still rising. The mountains of Vancouver Island and the Queen Charlotte Islands are presumably as young and as active as the Coast Mountains. Unfortunately, there is no convenient datum surface, such as the Miocene flood basalt, from which to measure this.

QUATERNARY VOLCANISM

Quaternary volcanoes are surprisingly numerous in British Columbia. Most of them lie in three connecting belts. One starts at Squamish in Howe Sound north of Vancouver and extends north for more than 100 miles. Another runs east-west through the center of the province. The third belt extends north from southeastern Alaska across the northwest corner of British Columbia into the Yukon (Figure 8-10). Most of the cones are small and were formed primarily by the eruption of basaltic cinders and small blocky flows. At least twenty larger composite volcanoes were also constructed. The best known of these, including Mount Garibaldi, are in the southern belt. The cones are younger than the Coast Mountains, since the first lava flows that erupted filled canyons cut into a mature mountain range. Compared to the more famous volcanoes of the Cascade Range, those of the Garibaldi group are not very large, and they do not tower above the surrounding peaks of crystalline rock. Otherwise, they are rather typical "Cascade" volcanoes. They consist of several types of lava (basalt, andesite, and dacite) interbedded with rubble and ash beds. (See Chapter 13 for a discussion of volcanic products and processes typical of Cascade volcanoes.) The total volume of lava extruded in the eruptions of Mount Garibaldi and its neighbors has been estimated to have been about 6 cubic miles—less than half the volume of any one of the larger volcanoes of the Cascades. Perhaps the most interesting feature of these cones was the interaction between erupting lava and glaciers. Much of the volcanic activity occurred during and immediately following the period of maximum glaciation of the range. This produced some interesting structures in the lava and undoubtedly some rather spectacular floods.

8-9 *Miocene basaltic lava flows of the Interior Plateau Province, British Columbia. (Photograph courtesy of James Crawford.)*

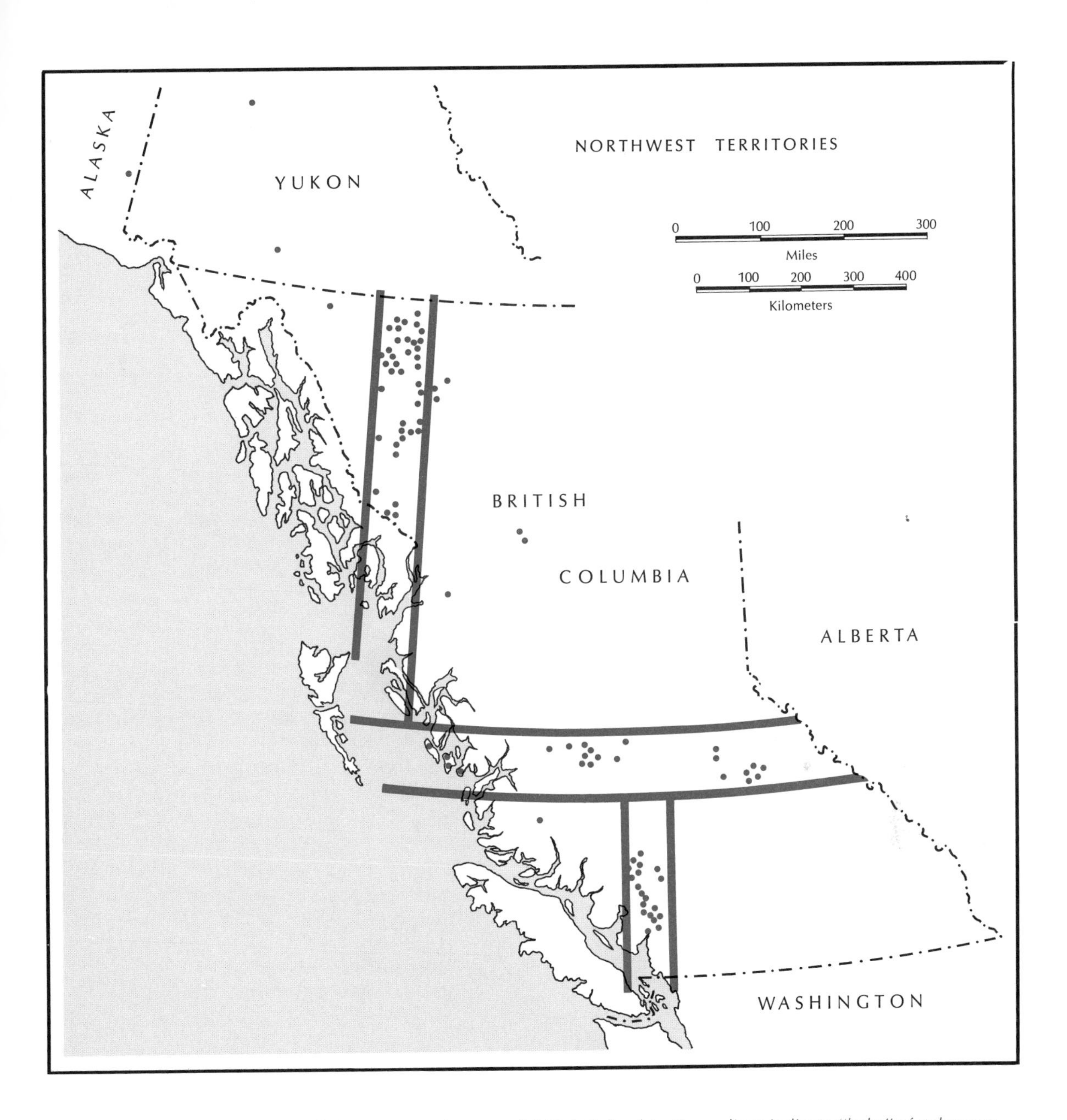

8-10 *The distribution of principal Quaternary volcanoes of British Columbia. Green lines indicate "belts" of volcanoes. (After Souther, 1970, p. 565. Reproduced by permission of the National Research Council of Canada.)*

Mount Edziza in the northern belt is also impressive. It is located just east of the Coast Mountains, approximately 200 miles north of Prince Rupert and 130 miles east-southeast of Juneau. The cone rises to more than 9,000 feet in elevation, far above the surrounding plateau. It began to form more than 4 million years ago (Pliocene Epoch) by first building a broad platform of initially rather fluid basaltic lava. In the Pleistocene Epoch, the erupting lava changed to more viscous ("sticky") andesite, dacite, and rhyolite. These flows built a series of domes. Some thirty or more basaltic cinder cones have formed recently in the immediate vicinity of Mount Edziza. Interplay between glaciers and volcanic eruptions, such as occurred on Garibaldi, has been documented also on Mount Edziza (Figure 8-12).

The reasons for volcanism in British Columbia are not at present clear. The province's volcanoes

8-11 *Dacite and rhyolite flows near the summit of the central dome of Mount Edziza, British Columbia. (Photograph courtesy of J. G. Souther, Geological Survey of Canada.)*

are, of course, a part of the "circle of fire" around the Pacific Ocean. Chapter 4 suggested that this volcanism is related in some way to processes that consume the oceanic crust beneath the margins of continents. If this is true for British Columbia, the processes must be complex, for there is no oceanic trench adjacent to its shore, there is no zone of earthquakes descending beneath this part of the continent, and the volcanoes do not form chains parallel to the coastline. Instead, the cones seem to lie along major north-south and east-west fracture zones in the continental plate. Very likely these relate to spreading axes and transform faults in the Pacific plate to the west, but the relationships are as yet unclear. Geological and geophysical studies now in progress, especially those being made offshore in the northeast Pacific, should provide answers to many important tectonic problems.

8-12 *Pillow structure in a basalt flow that erupted beneath a glacier at the 8,000-foot level of Mount Edziza, British Columbia. (Photograph courtesy of J. G. Souther, Geological Survey of Canada.)*

FIORDS AND GLACIERS

Fiord; fjord: A long, deep arm of the sea, occupying a portion of a channel having high, steep walls, a bottom made uneven by bosses and sills, and with side streams entering from high-level valleys by cascades or steep rapids. (American Geological Institute, *Dictionary of Geological Terms,* 1962)

Writing about fiords is as frustrating as trying to describe in words a great masterpiece of art—("There's this woman called Mona Lisa standing in front of a weird sort of landscape and she has this enigmatic expression on her face"). Clearly a photograph can help, but fiords, like paintings, really must be seen to be fully appreciated. For one thing, a photograph simply cannot give the proper feeling of scale—the immensity of the cliffs rising up straight out of the water into (too frequently) the clouds. On clear days, ice-mantled peaks many tens of miles distant seem near at hand. Sometimes only the sight of an occasional boat, a mere speck in this immensity of lonely space, brings the splendor of the scene to proper focus.

The fiords along the coast of British Columbia are some of the finest in the world. They were carved by large alpine glaciers that formed during the Ice Ages on the high peaks of the Coast Mountains and the ranges of Vancouver Island and the Queen Charlotte Islands. The glaciers flowed down river valleys to the sea, widening and deepening them as they passed. Much of the bedrock in the mountains is hard, crystalline rock, but it proved to be no match for the power of the moving ice. The glaciers grew larger and larger until finally they filled the valleys completely and merged with glaciers in adjoining valleys. The ice extended out to sea as shelf ice, much like that surrounding Antarctica. Large icebergs broke from its margin and drifted out into the North Pacific. Stones picked up by and incorporated in glaciers in the mountains were dropped to the sea floor as the icebergs melted. Some have been recognized in dredge hauls and drill cores taken from the sea bed hundreds of miles from land.

The fact that most of the mass of an iceberg lies submerged beneath the surface of the sea is familiar to all. This reflects the relatively small difference in density between glacial ice and seawater

8-13 *A cinder cone, approximately 1,400 years old, on the Tahltan Highland north of Mount Edziza. (Photograph courtesy of J. G. Souther, Geological Survey of Canada.)*

8-14 *View northwest up Sydney Inlet, a typical fiord along the central part of Vancouver Island's west coast. The northwest trend of this fiord, nearly parallel to the coast, is along a fracture zone. (Photograph by Bates McKee.)*

(about 10 to 15 percent), which means that ice barely floats in water. This has considerable significance in the origin of fiords. A glacier which is, for example, 2,000 feet thick, requires a depth of water of almost 1,800 feet in order to float. This explains how glaciers, unlike rivers, can erode below sea level and why some of British Columbia's fiords are several thousand feet deep. The sea, of course, occupied the overdeepened valleys after the glaciers melted away—in fact, the melting of the great Pleistocene ice caps of the northern hemisphere probably raised the sea level as much as 400 feet above its level during maximum glaciation.

Many of the fiords on the mainland of British Columbia have very long and straight reaches, then turn sharply. A somewhat rectilinear pattern is apparent, with one trend essentially at right angles to the coast (that is, northeast) and a second trend parallel to the coast (that is, northwest). The explanation is a structural one. The glaciers followed river valleys. The valleys, in turn, were positioned partly by the existence in the Coast Mountains of northwest- and northeast-trending fracture zones, along which the bedrock is somewhat more crushed and more easily eroded. Thus the presence of two fracture sets in the Coast Crystalline Complex is accentuated in the topography by two dominant trends to fiord segments.

Not all the glacial ice from the high ranges could escape directly to the ocean. Glaciers at the south end of the Coast Mountains were blocked off from the sea by Vancouver Island. Consequently a very large lowland glacier filled the Georgia Depression. It was nourished by tributary glaciers flowing in from the Coast Mountains and off the northeast flank of the Vancouver Island Ranges. This lowland lobe of ice flowed both ways around Vancouver Island—northwest out the Queen Charlotte Strait and south via the Strait of Georgia. The ice divide lay probably near the many islands that almost fill the lowland opposite the mouth of Bute Inlet. The southern ice lobe divided around the northeast flank of the Olympic Mountains of Washington. One branch continued west out the Strait of Juan de Fuca to the sea. The other (the Puget Lobe) flowed south along the Puget Lowland for another hundred miles. The glaciation of the Georgia Depression and Puget Lowland is treated in Chapter 18.

9
the san juan islands

■ A cynic, when seeing an entire chapter devoted to the San Juan Islands, might wonder if I own property there or am particularly familiar with its geology. I plead innocent to both counts. The San Juans are a kind of separate world, related geologically to surrounding regions but sufficiently apart from mainland Washington and Vancouver Island so that it has its own stratigraphic section and formational names. The islands are also unique in providing outcrop of pre-Tertiary rock along the axis of this part of the Puget-Fraser Lowland. Considering that the entire area of the San Juan Islands is only a few hundred square miles, the variety of rock types and structures seen is quite astonishing. Finally, so many people visit the islands that they deserve special attention.

The bare hills and rocky shores of the San Juan Islands provide excellent exposures of the bedrock. Considering the obvious advantages of doing field work in these pleasant surroundings, surprisingly little attention has been given to the geology of the islands. The most complete treatment is that done by R. D. McLellan (see reference list at end of book). His work was done almost fifty years ago for a Ph.D. dissertation. At that time, little was known about the geology of the entire group, although limestone quarrying had been going on for many years. McLellan investigated all the islands and larger rocks, making extensive use of a rowboat, and synthesized the general stratigraphy and structure of the San Juans. In recent years, his synthesis has been modified somewhat by more detailed studies (largely unpublished) of particular areas within the island group, especially parts of the two largest islands, Orcas Island and San Juan Island proper.

THE GEOGRAPHY AND GENERAL GEOLOGY

The San Juan Island group consists of many separate islands and rocks, but more than 80 percent of the total area is derived from the three largest islands—Orcas, San Juan, and Lopez Islands. Orcas Island is almost cut in half by East Sound, a deep, glacially scoured channel that lies west of the highest point of the San Juans, Mount Constitution. Orcas Island has the greatest variety geologically of the group. Lopez Island, to the south, is covered largely by Pleistocene glacial deposits, and the only extensive bedrock outcrops are at the south end. San Juan Island to the west has particularly good exposures of bedrock along its northern and western shores. All the many smaller islands have some rock exposed. The many submerged rocks and reefs contrast markedly with the relatively deep, unobstructed passages in Puget Sound to the south, where rock outcrops are uncommon and glacial deposits mantle everything. (In a sense we can think of the great lowland glaciers as giant earth movers that eroded large quantities of rock and soil in Canada, carried them south, and dumped them in Washington. The farther south you look, the thicker the mantle of glacial sediment and the less bedrock exposed. The San Juan Islands, near the midpoint of the journey, received an aesthetically pleasing balance of both glacial erosion and deposition.)

The geologic history in outline looks familiar. Most of the major elements of the Cascades are here—an old, crystalline basement rock; a Late Paleozoic and Early Mesozoic eugeosynclinal sequence of marine volcanic and sedimentary rocks; thrust faults of Cretaceous age; and Lower Tertiary sandstones and shales. Missing are any significant intrusions of Cretaceous or Tertiary age. Otherwise, the history of the San Juan Islands confirms impressions gained in studies to the east and to the north.

THE TURTLEBACK COMPLEX

The oldest rocks in the San Juans are a series of gneisses and granitic rocks. These crop out on several of the islands but most prominently in the Turtleback Range in the western half of Orcas Island. They are referred to as the Turtleback Complex. McLellan thought that these crystalline rocks formed as igneous intrusions in the Mesozoic Era, but now it is clear that Paleozoic strata as old as Middle Devonian were not intruded by the Turtleback Complex. Instead, these beds were deposited on top of the crystalline rocks. Furthermore, conglomerate beds at the base of the Paleozoic section locally contain pebbles apparently derived from the Turtleback gneisses; so, clearly, the crystalline rocks cannot intrude the sediments. Finally, radiometric dating of zircon crystals from Turtleback granitic rocks have a *minimum* age of 360 million years (that is, Devonian). The Turtleback thus compares directly with the Yellow Aster Complex of the North Cascades in that it is pre-Middle Devonian and possibly Precambrian in age. Furthermore, both units consist principally of granitic intrusive rocks (technically, quartz diorite and diorite), contain only minor amounts of older schists and gneisses, and were metamorphosed prior to the deposition of Middle Devonian strata.

9-1 *San Juan Islands, looking northeast toward Orcas Island and the Turtleback Range. Shaw Island is in the lower foreground, and Crane Island lies between the ferry and Orcas Island. The flank of Mount Constitution is in the distance to the right. (Photograph courtesy of Washington State Department of Commerce and Economic Development.)*

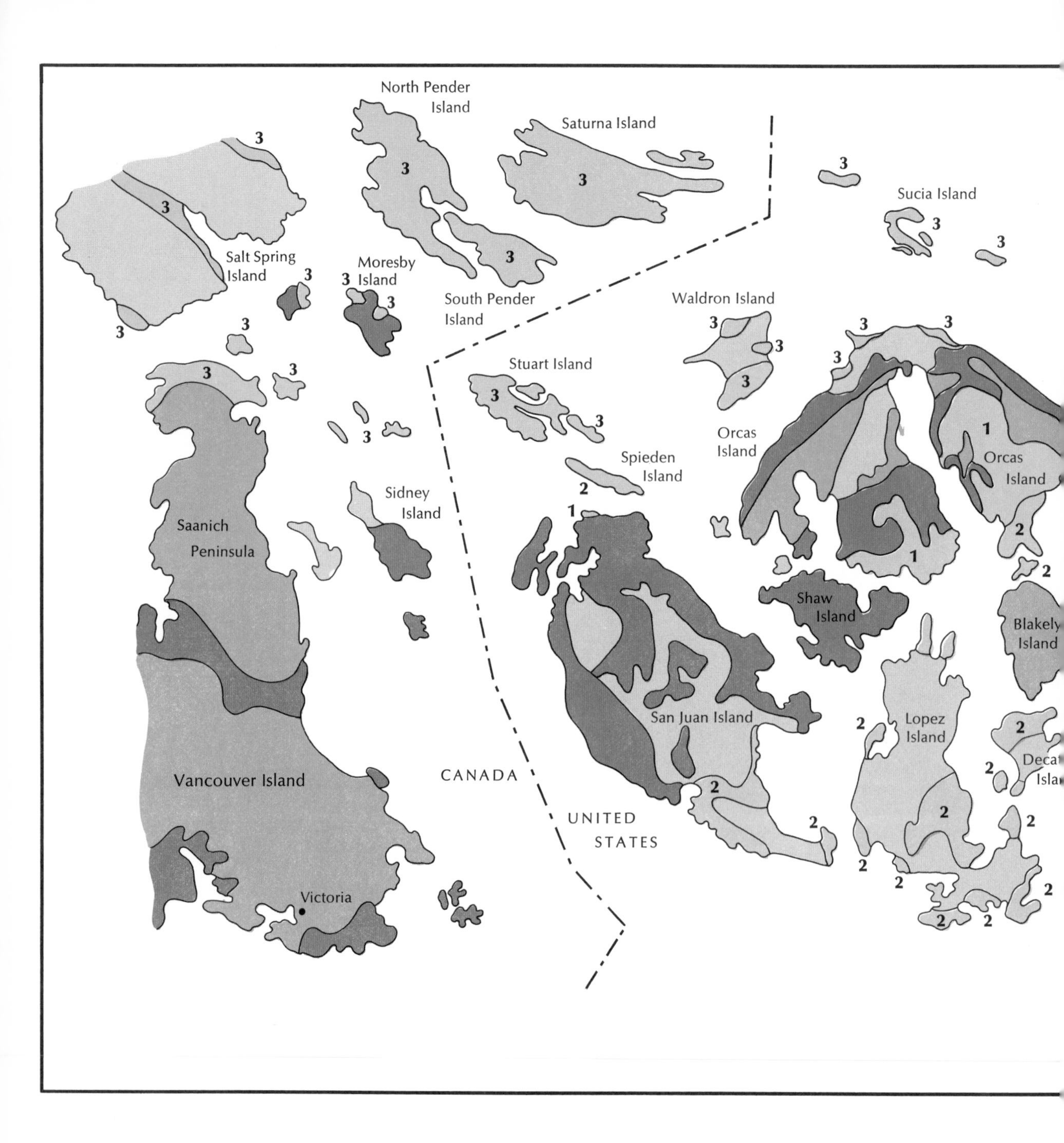

9-2 *Generalized geologic map of the San Juan Islands and environs. (Modified from Huntting* et al., *1961, and unpublished maps by J. A. Vance and W. R. Danner.)*

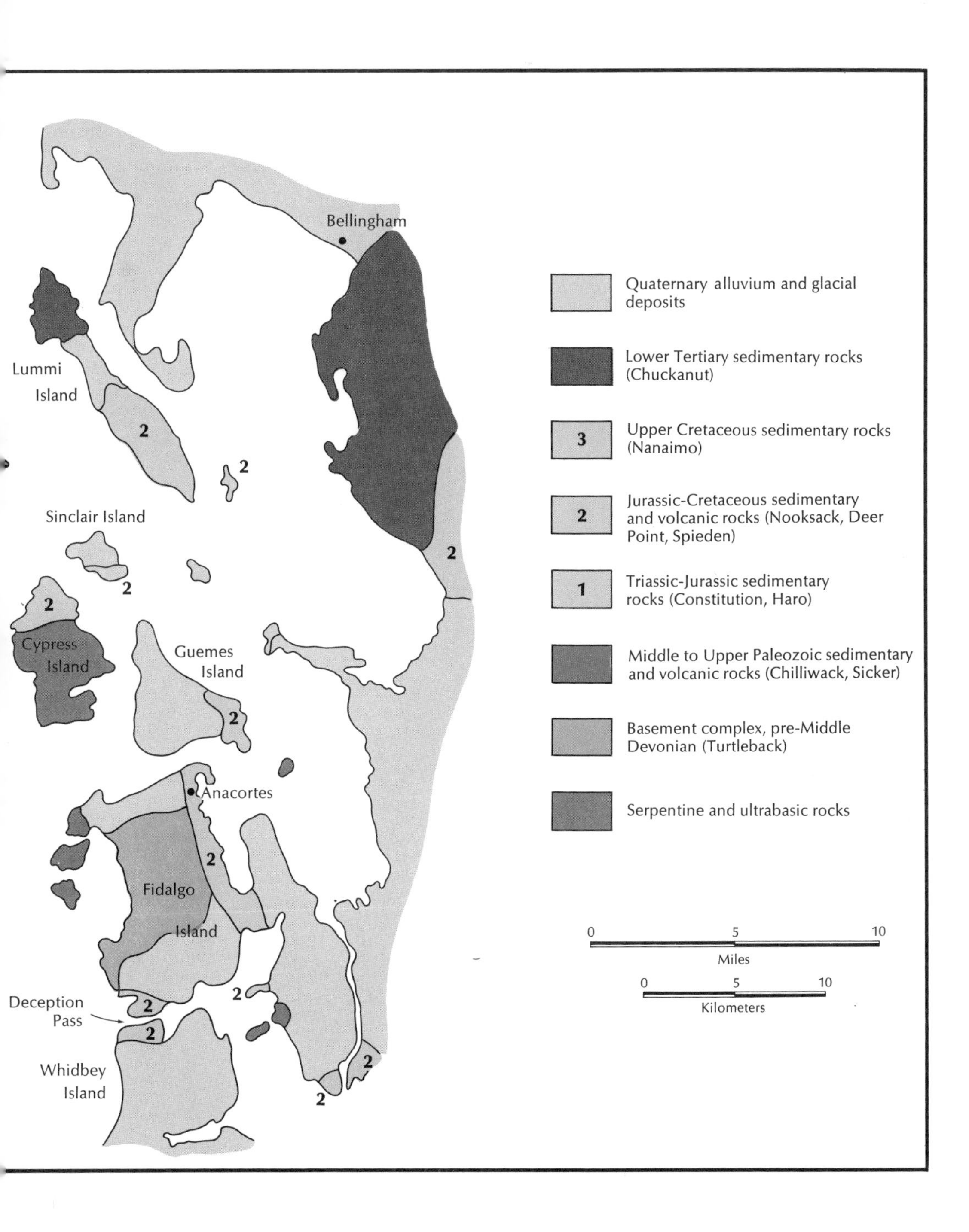

Bellingham
Lummi
Island
Sinclair Island
Cypress
Island
Guemes
Island
Anacortes
Fidalgo
Island
Deception
Pass
Whidbey
Island
Quaternary alluvium and glacial deposits
Lower Tertiary sedimentary rocks (Chuckanut)
3
Upper Cretaceous sedimentary rocks (Nanaimo)
2
Jurassic-Cretaceous sedimentary and volcanic rocks (Nooksack, Deer Point, Spieden)
1
Triassic-Jurassic sedimentary rocks (Constitution, Haro)
Middle to Upper Paleozoic sedimentary and volcanic rocks (Chilliwack, Sicker)
Basement complex, pre-Middle Devonian (Turtleback)
Serpentine and ultrabasic rocks
0
5
10
Miles
0
5
10
Kilometers

9-3 *Rocks of the Turtleback Complex along the west shore of Fidalgo Island. Light-colored layers are composed of granitic veins that intruded the older and darker gneiss. Both units are displaced by faults. (Photograph courtesy of J. A. Vance.)*

A recent maritime mishap has extended the known distribution of the Turtleback Complex. In 1961, an American freighter, the *Island Mail,* ran aground on Smith Island. This is located at the east end of the Strait of Juan de Fuca, 8 miles south of the south end of Lopez Island. All surface exposures on Smith Island are of glacial sediments. The freighter grounded on solid rock, however, some of which was torn loose and wedged in the ruptured hull of the ship. The rock is identical to the Turtleback Complex and can be considered as a southern extension of that unit. (This is clearly a prohibitively expensive sampling method and not a routine technique for field investigation.) Rocks essentially the same as the Turtleback Complex also underlie much of the Victoria-Saanich Peninsula area at the southeast end of Vancouver Island and the Deception Pass–Fidalgo Island area east and southeast of Lopez Island.

PALEOZOIC ROCKS AND LIME QUARRIES

The Paleozoic section is only a few thousand feet thick. It has been deformed quite severely in places by later folding and faulting, and this makes accurate thickness estimates impossible. The strata apparently correlate well with the Chilliwack Group in the Cascade foothills and in southwestern British Columbia, but local formational names are sometimes used in the islands. Fossils are not common. Most identifiable species have been collected from the many limestone lenses and pods that crop out locally. The fossils indicate that sedimentation occurred during the Middle and Late Devonian, Early Pennsylvanian, and most of the Permian Periods. The pre-Permian formations consist mostly of volcanic breccia and ash of andesitic composition, with interbeds of marine sandstones, shales, limestone, and minor amounts of chert. Chert is more common in the Permian sequence. It occurs both as thin interbeds in limestone and as more extensive units of relatively pure chert. Marine sandstone, shale, and volcanic rock also are common, especially toward the top of the Permian section.

Paleozoic limestone in the San Juan Islands and the North Cascades has been the source rock for virtually all of western Washington's lime industry. Limestone is used by a variety of manufacturers, including those of cement, fertilizers, paper and pulp, and building stone. At the present time, limestone quarrying in western Washington has all but ceased, in part due to the impurity of the limestone and in part because of the small size and relative inaccessibility of most deposits. Local users of high quality limestone are getting it from Texada Island (150 miles northwest of Vancouver, British Columbia) or elsewhere.

The largest lime-producing operation in the Northwest was at Roche Harbor at the northern end of San Juan Island, where quarrying began perhaps as early as 1857. The Roche Harbor Lime and Cement Company operated continuously from 1882 until 1956 and at one point was shipping rock as far as the Hawaiian Islands and South America. The area was sold as a resort in 1956, by which time the principal quarries (fifteen in number) were largely depleted. In recent years, several of them have been used for water storage. In a sense, this depicts the economic future of the islands, for they are now much more valuable for recreational and homesite development than for any mineral wealth they contain. Furthermore, the availability of fresh water is critical to the homesite and recreational development of the islands, especially the smaller ones.

THE SEA RETURNS

No strata of latest Permian to Late Triassic age have been found in western Washington. This gap (unconformity) in the stratigraphic record indicates that the region was uplifted above sea level during at least the latter part of this interval. Triassic deformation was apparently not severe since the older Chilliwack strata are neither appreciably more folded and fractured nor more metamorphosed than are the Mesozoic strata lying above the unconformity.

Marine sedimentation resumed late in the Triassic Period and continued intermittently into the Cretaceous Period. Fossils are exceedingly rare in these strata, and the limestones that were so useful for dating the Paleozoic formations are no longer found in the section. (This is not true to the north, where the very thick Late Triassic and Jurassic section of Vancouver Island contains fossiliferous limestone.)

The principal formations of Middle Mesozoic age in the San Juan Islands are the Haro, Constitution, "Deer Point," and Speiden Formations. The Haro Formation underlies Davison Head at the northern tip of San Juan Island and contains the only certain Triassic fossils in all of western Washington. It consists of conglomerates, sandstones, and siltstones with abundant volcanic rock fragments, and several very thin beds of limestone and chert. The limestone contains shells of the distinctive Late Triassic clam, *Halobia,* as well as fragments of a few other organisms.

The Constitution Formation covers much of the eastern half of Orcas Island (including the type area, Mount Constitution) and much of the central part of San Juan Island. It consists of generally fine-grained sandstones and siltstones—a rather monotonous nonfossiliferous section. It is overlain on the southeastern tip of Orcas Island by the Deer Point Formation, a series of sandstones and shales containing many volcanic fragments. Deer Point strata are well exposed also at the south end of Lopez Island. They resemble the Nooksack Formation of the North Cascades, a marine unit with both Late Jurassic and Early Cretaceous fossils; but the Deer Point beds are nonfossiliferous. If the two formations are the same age, the Constitution Formation must be older than Late Jurassic. (The Constitution Formation unconformably overlies Permian beds, but its stratigraphic relationship to the Haro Formation is uncertain.)

Tiny Speiden Island, immediately north of San Juan Island, is the type area of the Speiden Formation. This unit consists mostly of conglomerate, with minor amounts of fossiliferous sandstone, siltstone, and silty limestone. The fossil fauna is quite varied. It includes clams, ammonites (coiled cephalopods, related to our modern pearly nautilus), and straight-shelled cephalopods known as belemnites. The age of this fauna is Early Cretaceous, so that the Speiden Formation is equivalent to the upper parts of the Nooksack and Newby Groups of the North Cascades. Sedimentary rocks similar to the Speiden Formation are interbedded with volcanic strata on Lummi, Eliza, Samish, and Fidalgo Islands in the eastern part of the San Juan group.

CRETACEOUS THRUSTING

Thrust faulting, an important part of the mid-Cretaceous orogeny in the North Cascades, oc-

TABLE 9-1 Stratigraphy of the San Juan Islands. (From Danner, 1966, and J. A. Vance, personal communication.)

AGE	FORMATION	THICKNESS, FEET	DESCRIPTION
Quaternary	Glacial deposits and alluvium	500	Gravel, sand, silt, clay, peat.
Early Tertiary(?)	Cypress Island serpentine	?	Serpentine and ultrabasic rocks, correlative perhaps with Twin Sisters dunite (Chap. 7). Cypress and Fidalgo Is.
Late Cretaceous	Nanaimo	10,000 (in B.C.)	Arkosic sandstone, conglomerate, and siltstone with marine fossils. Orcas, Waldron, Sucia, and Stuart Is., and Gulf Is. of B.C.
Early Tertiary	Chuckanut	15,000 (N. Casc.)	Arkosic sandstone, siltstone, and conglomerate. Leaf fossils. Nonmarine. On mainland, Lummi Is., and possibly Sucia Is.
Early Cretaceous	Speiden	1,000(?)	Conglomerate with minor sandstone, siltstone, and limestone. Marine fossils. Speiden Is. Equivalent to upper part of Nooksack Group
Late Jurassic or Early Cretaceous(?)	"Deer Point"	3,000	Graywacke sandstone and siltstone with abundant volcanic rock fragments. Nonfossiliferous but presumed marine. Throughout southeastern part of islands
Late Triassic or Jurassic(?)	Constitution	3,000	Monotonous fine-grained sandstone and siltstone. Presumed marine, but no fossils. Orcas Is.
Late Triassic	Haro	1,400	Conglomerate, sandstone, siltstone, and minor chert and limestone. Marine. Davison Head on San Juan Is.
Devonian, Pennsylvanian, and Permian	Chilliwack	7,500	Andesitic breccia and tuff, graywacke sandstone, shale, chert, limestone. Several unconformities. Marine. San Juan, Orcas, and Shaw Is.
Precambrian or Early Paleozoic	Turtleback Complex	?	Gneiss intruded by diorite and quartz diorite veins. Low-grade metamorphism. Orcas, Blakely, and Fidalgo Is.

——— Major unconformity

curred also in the San Juan Islands. The Turtleback, Chilliwack, and Constitution Formations on Orcas Island have been sliced by low-angle thrusts. This deformation may have coincided with thrusting in the Cascades; but, if so, a real problem in interpretation is posed by the disparate directions of movement. The Cascade thrusting involved westward displacements of strata some tens of miles; the thrusts are traceable almost to salt water. The thrusting in the San Juans, in contrast, was to the north—essentially at right angles to the Cascade direction. This suggests that the two systems were not related directly to any simple stress field. Perhaps two unrelated stress systems were in close proximity at the same time, or possibly the thrusts are of different ages.

This kind of problem may be solved only with a better understanding of plate tectonic history. Perhaps it is a mistake to assume that the present proximity of the San Juan Islands to the North Cascades region existed in the Cretaceous Period. Stratigraphic similarities might suggest that it did, but this is not very powerful evidence—especially since lithologic patterns are so similar over such large parts of the Cordillera. Suppose that the San Juan Islands belong to a separate crustal plate that arrived in its present position sometime during the Cretaceous orogeny. Then the disparity in direction of thrusting between Orcas Island and the nearby Cascade system might be explained.

At present, this is highly speculative, of course. It is, however, the kind of speculation that must be tested in the years to come—not just by imagining possible plate configurations for the past but by detailed study in the field and laboratory. Do the paleomagnetic data from the rocks in the San Juans differ from results obtained in the North Cascades? What is the significance of the large serpentine mass on Cypress Island (Figure 9-5)? Does it, like similar bodies of serpentine in the Klamath Mountains (Chapter 10) or the Coast Ranges of California, mark the location of a possi-

9-4 *Dark sandstone beds of the Deer Point Formation, southeastern Orcas Island. Light-colored spots are living barnacles. (Photograph courtesy of J. A. Vance.)*

9-5 *View east from the summit of Mount Constitution, Orcas Island. Cypress Island, smoothed by glaciation and composed largely of serpentine, is in the middle distance. Mount Baker volcano is on the left horizon with Twin Sisters (Chapter 7) in front and to the right. (Photograph courtesy of Washington State Department of Commerce and Economic Development.)*

ble plate boundary? Until these types of questions can be answered, reconstruction of ancient geography is an uncertain business.

SUCIA ISLAND AND THE NANAIMO GROUP

Sucia Island, which is perhaps the most picturesque and unusual island in the entire San Juan group, consists primarily of Upper Cretaceous sandstone, siltstone, and conglomerate. The beds have been bent into open folds that trend northwest-southeast and are inclined to the southeast. Erosion by waves and glaciers has carved out the less resistant siltstones, so that the harder layers of sandstone and conglomerate form prominent ridges, which project offshore as submerged reefs. Weathering and wave erosion have accentuated the differential concentration of cement in the sandstones, so that locally near the tide line the rock is markedly honeycombed. Fossil Bay on Sucia Island is one of several good fossil localities; clams, snails, and ammonites are found. The beds on Sucia belong to the Nanaimo Formation (Chapter 8). It underlies most of the Gulf Islands to the north, plus Stuart, Johns, Waldron, Patos, and Matia Islands in the San Juans. Nanaimo strata also crop out along the north shore of Orcas Island opposite the head of East Sound.

The sediments of the Nanaimo Formation are much lighter in color than any of the older beds and are of arkosic composition (that is, high in feldspar and low in rock fragments). They accumulated in a shallow marine trough that developed late in the Cretaceous on the deformed and eroded remains of the Cordilleran Geosyncline. Probable sources for the arkosic sediment were the exposed Jurassic granitic intrusions of Vancouver Island and the older gneisses of the North Cascades and the Coast Mountains of British Columbia.

Arkosic sandstones and shale also make up the Chuckanut Formation in the western foothills of the North Cascades (Chapter 7). Generally, the Chuckanut arkoses are lighter in color and not as firmly cemented as are the Nanaimo beds. Moreover, the Chuckanut Formation contains only plant and leaf fossils, and these suggest an Early Tertiary rather than a Late Cretaceous age. Undoubted Chuckanut strata occur on the north end of Lummi Island only 8 miles east of Sucia. They may also overlie Nanaimo beds on Sucia Island. Tertiary units younger than the Chuckanut Formation have not been found in the San Juan Islands, and in all probability the region has been eroding during much of the Cenozoic Era.

THE GLACIAL RASP

The San Juan Islands lay directly in the paths of the great lobes of glacial ice that formed in the Pleistocene Epoch in the mountains of western British Columbia and flowed south out of the Georgia Depression. There was no stopping the ice. Even the summit of Mount Constitution, at 2,409 feet the highest point in the islands, lay beneath ice more than $\frac{1}{2}$ mile thick just 15,000 years ago. Glacial erosion was so pronounced that geologists have little idea what the San Juans looked like before the first glaciation. Channels oriented parallel to the direction of ice movement, such as East Sound and Haro Strait, were very markedly deepened by the passage of the ice. Most of the landscape was smoothed and rounded, and this effect is seen clearly in the Turtleback Range (Figure 9-1), Cypress Island (Figure 9-5), the western side of San Juan Island, and on Blakeley Island. Smaller islets and rocks likewise were shaped by glaciation. Large glacial grooves were scraped in the bedrock surface exposed along the south shores of San Juan and Lopez Islands.

Erosion was not the only effect of glaciation, for the ice deposited sediment as it wasted away.

Rock particles ranging in size from clay to gigantic boulders were lying on or incorporated in the glacial ice. Some of this came from the Coast Mountains or the ranges of Vancouver Island, some undoubtedly from the San Juan Islands themselves. (Material eroded from beneath the ice likely would remain close to the base of the glacier.) When the terminus of the glacier melted back north of the islands the rock fragments in the ice either dropped to the ground beneath or were carried off and deposited by meltwater streams. Very thick beds of outwash sand form the cliffs immediately west from Cattle Point at the south end of San Juan Island.

Further discussion of lowland glaciation will be deferred to Chapter 18, for the glacial history of the San Juan Islands constitutes only a small part of a much grander picture. If geologists had been forced to confine their glacial studies to just the San Juans, they would not have found evidence for more than one period of glaciation, nor would they have been able to estimate the true thickness of the ice.

10 the klamath mountains of oregon

■ Here is a genuine geologic nightmare. Few regions in the Cordillera have yielded their geologic secrets as unwillingly as has the Klamath Mountains Province of southwestern Oregon and northwestern California. A geologist who spends just a few days in these mountains without a reasonably good picture of the regional structure and stratigraphy must come away in a somewhat confused state. Yet this is not virgin country. The Klamaths have been the site of considerable mining activity for more than a hundred years, and there are numerous roads (none of them very good) and trails through the region. The principal problems blocking geologic synthesis have been the dense vegetation throughout the region, the steep terrain (which does not necessarily mean good rock exposure), the lithologic monotony of many of the formations, a paucity of fossils, and extreme structural complexity. The Klamath Mountains geologists must think longingly sometimes of the glacial erosion that has stripped the soil cover from ranges in the northern part of the Cordillera. He might even wish for a good southern California-type brush fire!

REGIONAL GEOGRAPHY

The Klamath Mountains are defined partly on a geomorphic and partly on a geologic basis. They are surrounded on three sides by other mountain ranges, from which they are distinct only in that they contain older rocks. To the north lies the Coast Range of Oregon and to the south the Coast Range of California. Unlike these ranges, the Klamaths are not bounded on the east by an interior valley but merge with the south end of the Cascade Range.

The limits of the Klamaths are drawn arbitrarily at the edge of the younger strata that crop out extensively in the surrounding mountains—Lower Tertiary marine strata to the north, nonmarine Lower Tertiary volcanic rocks to the east, and Cretaceous marine beds to the south. Geographically, the contact with the Oregon Coast Range extends southwest from the Bandon area to the Rogue River, then northeast to Roseburg. The limit of younger volcanic rocks of the Cascades runs south from Roseburg to Medford, then southeast over the mountains near Siskiyou Pass, and finally south to Redding, California. The contact thus approximates the position of Interstate Highway 5. Cretaceous strata across the north end of the Sacramento Valley flank the Klamaths for 50 miles southwest from Redding. From there, the contact with the Coast Range follows a steeply dipping fault that trends northwest almost to the coast near the California-Oregon border.

The topography of the Klamath Mountains is rugged throughout. Slopes are steep; rivers run in narrow canyons with no consistent trends; and peaks generally rise to around 5,000 feet (although the highest ones reach 7,500 feet in Oregon and 9,000 feet in California). Separate named ranges within the Klamath Mountains are the Trinity, Salmon, Scott, Scott Bar, Trinity Alps, South Fork, and Marble Mountains in California, and the Siskiyou Mountains along the state line. North of the Siskiyous, the Klamaths are generally lower and the topography is more subdued. Drainage of the area is westward to the Pacific Ocean except in the southeastern part of the range, which drains into the Sacramento River.

THE HISTORY IN BRIEF

The stratigraphic record of the Cordilleran Geosyncline is more complete and less metamorphosed in the Klamath Mountains than it is in the nearby Sierra Nevada Province of California. The most complete sequence is in the eastern part of the Klamaths of California where approximately 25,000 feet of Paleozoic and 15,000 feet of Lower Mesozoic strata occur. The section is markedly eugeosynclinal in aspect, consisting primarily of volcanic rocks (flows, beds of ash, and breccia), marine sandstones and shales, and minor amounts of conglomerate, chert, and limestone. Apparently, all periods from Ordovician to the Jurassic are represented by some strata; but unconformities (breaks in the record) are numerous, and contacts between some of the units are faults. The overall picture is one of general marine sedimentation, frequent volcanic eruptions, and periods of local uplift and erosion. Certainly, much of the sediment was produced by weathering and erosion of nearby volcanoes. To the west, in the core of the Klamaths, the rocks have been more highly deformed and metamorphosed. They tell us much less about the environment of the geosyncline.

The Paleozoic and Early Mesozoic history of the Klamaths north of the California-Oregon border is very poorly known. Schists south of Medford were derived from sedimentary and volcanic strata, but all of the original features of the beds, including fossils, have been obliterated by metamorphism. The Applegate Group, somewhat metamorphosed, covers a much larger area. It is in part Triassic in age, but it may also include Upper Paleozoic strata. The Applegate Group is composed of marine sedimentary and volcanic rocks.

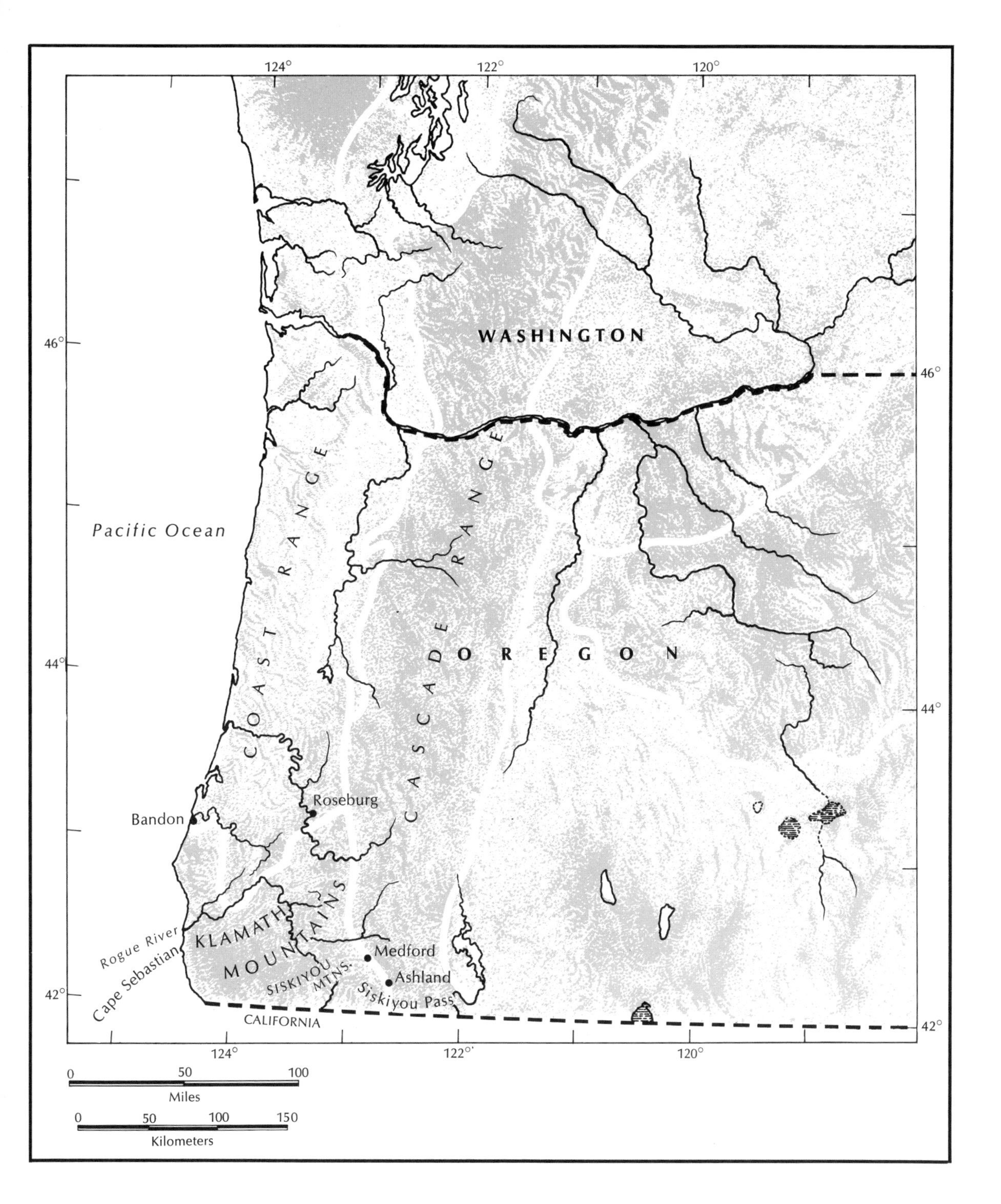

10-1 *Physiography of the Klamath Mountains of Oregon. (From Raisz, Landform Map of Oregon, 1955.)*

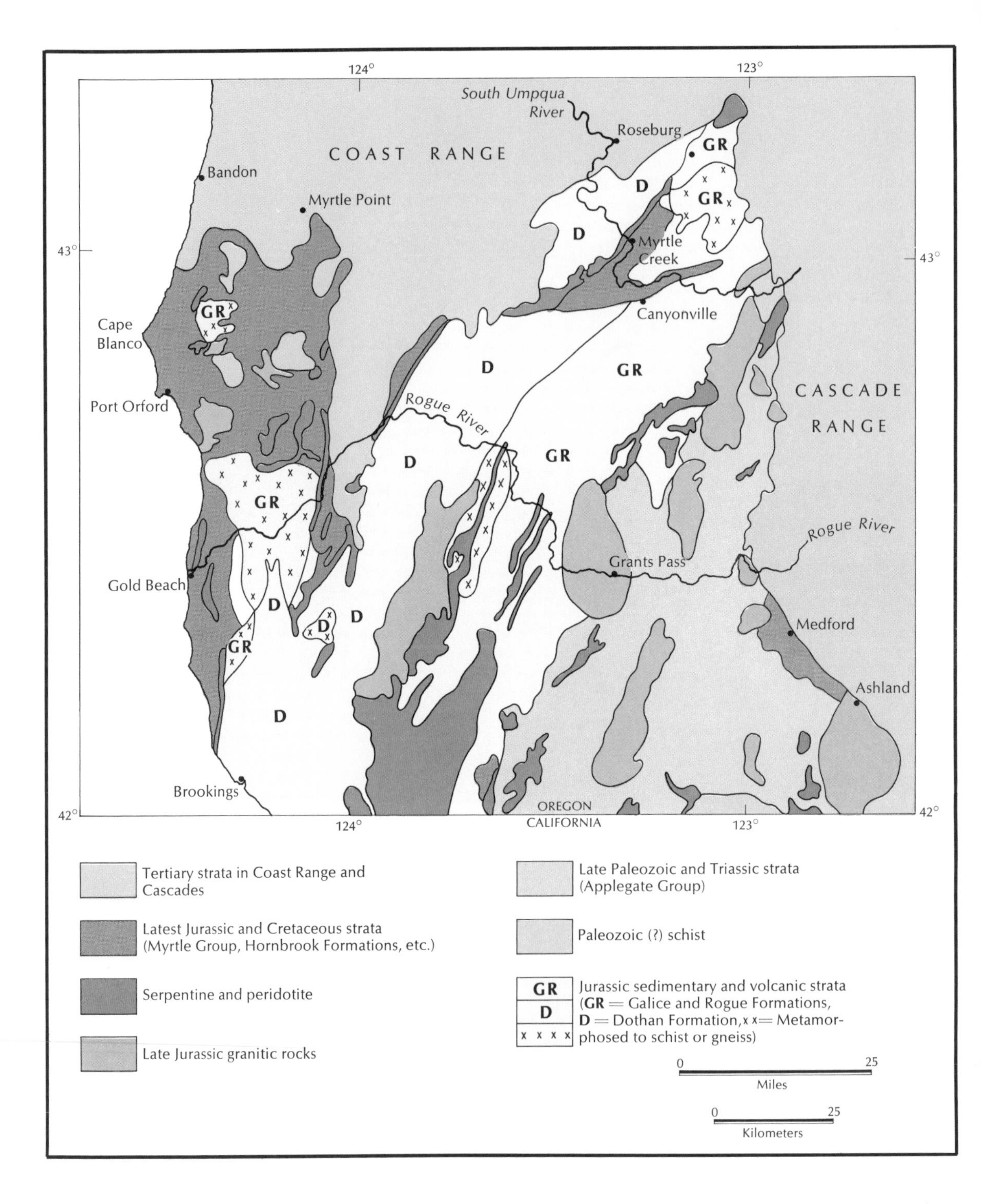

10-2 *Distribution of major rock sequences in the Klamath Mountains of Oregon. (Generalized from Wells and Peck, 1961.)*

The Late Mesozoic record is quite complete in the Oregon Klamaths. It consists of (1) very thick sections of marine volcanic sandstone, shale, and lava of Late Jurassic age; (2) igneous intrusions, also of Late Jurassic age; and (3) various shallow-water marine sandstones and shales of latest Jurassic and Cretaceous age. The Cretaceous sediments represent a sort of "last gasp" of the Cordilleran Geosyncline, quite comparable to the Methow Valley section of the North Cascades and the Nanaimo Group of Vancouver Island and the San Juan Islands.

No extensive outcrops of Tertiary rocks occur within the range, although Late Cenozoic uplift and erosion may have removed some strata. Alpine glaciers formed on the highest peaks of the Klamaths during the Pleistocene Epoch. They never grew very large, presumably because the range was not high enough to spawn glaciers this far south.

PRE-JURASSIC ROCKS

The frustrations inherent in the interpretation of schistose regions have been apparent in other chapters. Metamorphic recrystallization masks so much of the depositional history of the rocks that we can only guess at their original character. The fossil record is destroyed, and radiometric dates are difficult to interpret. Normally, they suggest the *minimum* age of the strata but not their true age.

Schists at least as old as Paleozoic crop out in Oregon near the California border, 20 miles south of Medford. They are in apparent fault contact with the Applegate Group and continue into California where they underlie much of the core of the Klamath Mountains. Two distinct units are recognized in California—the Salmon Hornblende Schist and the Abrams Mica Schist (recently renamed the Grouse Ridge Formation). Radiometric dates indicate a Late Pennsylvanian or Early Permian age of the metamorphism. The original rock was probably a rather thick section of sedimentary and volcanic strata, which might have been deposited in the Cordilleran Geosyncline or could even be Precambrian.

Applegate strata underlie the eastern third of the Klamath Mountains of Oregon. They have been metamorphosed, but the degree of recrystallization was not great except near some large igneous intrusions. As much as two-thirds of the rocks were volcanic, consisting of flows and fragmental ejecta of mostly basaltic and andesitic composition. The sediments were the typical eugeosynclinal assortment of sands and silts, with lesser amounts of gravel, limestone (now marble), and chert. Fossils are rare, but a few from the upper part of the group indicate a Late Triassic age. Applegate strata extend south into California, where some of the beds are of Late Paleozoic age.

The Applegate Group has been folded and faulted quite extensively. No detailed stratigraphic studies have been made, but the thickness of this unit must exceed 10,000 feet. The Karmutsen Group of Vancouver Island (Chapter 8) and the Clover Creek and Seven Devils Volcanics of the Blue Mountains (Chapters 14 and 16) are similar to the Applegate in age and lithology.

A THICK JURASSIC SECTION

Three formations that underlie much of the central and western part of the Klamath Mountains in Oregon are all of Jurassic age. They are well exposed along many of the rivers (especially the Rogue). They have not been studied in great detail, although they have received more attention than the Applegate Group. The units are known as the Rogue, Galice, and Dothan Formations. Their relative ages are uncertain because all three have been folded and faulted rather extensively, and this has obscured their original stratigraphic relationships. Low-grade metamorphic recrystallization is a further complication. Fortunately, the three formations tell much the same story, and the details can be left for future study by geologists.

The Rogue Formation is entirely volcanic—a sequence of massive andesitic lava flows and pyroclastic beds, all of which have been altered to greenstone. Presumably these volcanics were extruded in a marine environment, but the evidence is not conclusive. Some geologists regard the Rogue Formation as a part of the Galice, others consider it as underlying both the Galice and Dothan Formations.

The Galice Formation contains the most fossils. These are marine clams and ammonites of early Late Jurassic age. The strata consist of dark, fine-grained sandstones and siltstones, plus ash beds (tuff) and some andesitic lava flows. The Galice Formation is thought to be at least 15,000 feet thick.

The Dothan Formation covers the largest area, a wide band emerging from under the Cenozoic lava cover 10 miles east of Roseburg and extending southwest into California. The principal rock is a dark sandstone, with lesser amounts of shale, conglomerate, chert, and pillow basalt. (*Pillow*

TABLE 10-1 Stratigraphy of the Klamath Mountains of Oregon, (From Baldwin, 1964, and Ramp, 1969.)

AGE	FORMATION	THICKNESS, FEET	DESCRIPTION
Late Cretaceous	Hornbrook	2,000	Arkosic sandstone, siltstone, and conglomerate. Shallow-water marine. Medford-Ashland area, and northern California
Latest Jurassic and Early Cretaceous	Days Creek and Riddle	10,000	Dark siltstone and sandstone; minor basalt and andesite, chert, limestone, conglomerate. Marine. Myrtle Creek area and equivalent strata in coastal belt
Late Jurassic	Nevadan intrusions		Diorite, granodiorite, and quartz diorite stocks and batholiths. Widespread
Late Jurassic	Dothan	18,000	Dark graywacke sandstone, lesser shale, conglomerate, chert, pillow basalt. Marine. Probably equivalent to Galice. Forms western Jurassic belt from near Roseburg past Brookings into California
Late Jurassic	Galice	15,000	Dark gray mudstone, siltstone, and fine-grained sandstone. Marine fossils. Metamorphosed locally, including Colebrooke schist east of Gold Beach. Widespread in eastern Jurassic belt
Jurassic	Rogue	15,000	Submarine flows, breccias, and tuffs of basaltic and andesitic composition, locally weakly altered to greenstone. Widespread in eastern Jurassic belt
Permian and Triassic	Applegate Group	>10,000	Mostly andesitic and basaltic flows and pyroclastic rocks. Sandstone, siltstone, conglomerate, limestone, chert. Marine. Weakly metamorphosed. Widespread in eastern Klamaths
Paleozoic(?)	Salmon and Abrams (Grouse Ridge) Schists	?	Marine sedimentary and volcanic strata, metamorphosed to hornblende and mica schist. Southwest of Ashland

▬▬ Major unconformity

10-3 *Massive volcanic rocks of the Rogue Formation at Hellgate Canyon on the Rogue River, 15 miles northwest of Grants Pass. The strata dip steeply to the east (left) and have been altered to greenstone. Boats in the river indicate scale. (Photograph courtesy of Oregon State Highway Division.)*

10-4 *Thin-bedded sandstone and siltstone of the Galice Formation, southeast of Myrtle Creek. (Photograph by Bates McKee.)*

structure is characteristic of lava that crystallizes under water.) A minimum estimate of the thickness of the Dothan Formation is 18,000 feet.

The Rogue Volcanics form a central belt in the Klamaths, with Galice strata to the east and the Dothan Formation to the west. The aggregate thickness of the three units is at least 40,000 feet—a very considerable total for strata of this restricted an age range. The nonfossiliferous Dothan Formation presents the biggest problem in establishing the relative positions of the three units. It has been assigned to the bottom of the sequence by some geologists, to the top by some, and as equivalent to the Galice by others. At any rate, the Late Jurassic was clearly a time of rapid marine subsidence and sediment accumulation, accompanied by impressive outpourings of submarine volcanics.

THE NEVADAN OROGENY

The term "orogeny" implies an episode of folding, faulting, metamorphism, and batholithic intrusion. "Nevadan Orogeny" was first applied to the deformational and intrusive climax of the Sierra Nevada region, which was thought to have occurred near the end of the Jurassic Period. Subsequently, the term was applied throughout the Cordillera to denote any Late Jurassic deformation. Unfortunately, radiometric dating has now proved that much of the "Nevadan" deformation, including the emplacement of the batholiths of the Sierra Nevada, occurred toward the middle of the Cretaceous Period. Thus the type area for Late Jurassic orogeny was excluded, since by common usage the term "Nevadan" had become more closely tied to the time than to the place. Cretaceous dates were obtained from other large western batholiths, including the Idaho Batholith and many intrusions in the Coast Mountains of British Columbia, the North Cascades, and the Blue Mountains. The concept of a Late Jurassic orogeny waned. However, further dating established intrusions of this age on Vancouver Island, in the Interior Plateau and Columbia Mountains of British Columbia, in the western foothills of the Sierra Nevada, and in the Klamath Mountains of Oregon and California.

The Jurassic intrusions in the Klamaths consist of many separate batholiths and stocks. Compositionally, they are quite varied; the most abundant rock types are diorite, quartz diorite, and granodiorite (Appendix A). Most batholiths in the Klamath Mountains have a radiometric age ranging from about 150 million to 130 million years, which agrees well with the field evidence for Late Jurassic intrusion.

Not all the intrusions in the Klamath Mountains have a Jurassic age. Dark masses of gabbro and peridotite locally cut the Applegate Group and the older schists. Peridotite and its soft, green alteration product, serpentine, also occur along many fault zones. Since some of these faults are Nevadan or younger, the emplacement of the peridotite and serpentine must have been post-Nevadan. (Because serpentine is so soft and easily deformed, it often squeezes up cold along faults. Under pressure it has the consistency of toothpaste.) There are also small granitic stocks of Cretaceous and Early Tertiary age, but their total volume is insignificant compared with that of the large Jurassic batholiths. They do indicate, however, that, as in the North Cascades and the ranges of British Columbia, intrusive activity continued long after the orogenic climax.

The emplacement of Jurassic batholiths is only one indication of the Nevadan Orogeny. Another is the more intense deformation suffered by the Rogue-Galice-Dothan strata in comparison to younger formations in the Klamath Mountains.

ORE DEPOSITS

Intrusions were responsible, directly or indirectly, for most of the many mineral deposits of the region. Old mines and prospects are found throughout the Klamath Mountains. The principal metals produced have been gold, nickel, and chromium, with lesser amounts of antimony, copper, iron, mercury, platinum, silver, tungsten, and zinc. Many of the deposits are related genetically to the granitic rock, the ore occurring most commonly in veins near intrusive margins. Nickel and chromium are found in peridotite and serpentine—the chromium as distinct layers of the mineral chromite and nickel disseminated through the rock. The "black sands" common along certain parts of the southern Oregon coastline contain appreciable amounts of chromite derived from the weathering and erosion of peridotite and serpentine within the range. Mining activity is currently low in the Oregon Klamath Mountains, but mineral exploration is continuing. Especially promising prospects are certain nonmetallic resources such as asbestos, barite, garnet sand, and limestone.

The history of gold mining in the Klamath Mountains is of interest. Native gold occurs with quartz in veins and cavity fillings near the margins of some intrusions. This type of occurrence is known as a lode deposit, and more than 150 lode mines have produced gold in the Klamath Mountains. The largest mines in Oregon were the Greenback Mine, 15 miles north of Grants Pass, and the Ashland Mine near Ashland. The former produced 3.5 million dollars worth of gold from veins in *greenstone* (recrystallized andesite) of the Rogue Formation; the latter, gold worth 1.3 million dollars from veins in metamorphosed sedimentary rock of the Applegate Group. The total value of gold mined from lode deposits in the Klamath Mountains of Oregon has exceeded 11 million dollars.

Gold prospectors in the past have devoted most of their efforts to panning loose sediment from stream beds and beaches. Nature has many ways of concentrating ore minerals. Man must take advantage of these, for frequently the concentration of the desired metal in the primary host rock (for example, a lode deposit) is insufficient for profitable mining. Fortunately, minerals like native gold, platinum, chromite, and diamonds have properties that lead to their concentration at the earth's surface. For one thing, they resist weathering, and consequently remain in soil while less resistant minerals decompose. Furthermore, these minerals are moderately resistant to abrasion, so if they are transported as sediment they do not disintegrate as readily as do other mineral grains. Finally, they are all heavy minerals that tend to be concentrated in small pockets or riffles on a beach or in a stream—left behind as lighter grains are transported on. This type of concentration of heavy minerals is known as a *placer deposit.* It was placer gold that sparked the great California and Alaska Gold Rushes.

More than three-quarters of the total gold mined in the Klamath Mountains of Oregon has come from placer deposits. The first workings were begun around 1850 in stream sediments in Jackson and Josephine Counties and on the beaches of southwestern Oregon. In succeeding years, virtually all of the streams in the Klamath were worked at one time or another in the search for gold. Platinum and chromite also occur in the gold placers and have been produced in the past as by-products of gold mining. In recent years, chromite concentrations in beach sands of southwestern Oregon have been mined on occasion.

In many instances, geologists and prospectors have been led to primary lode deposits by following placer deposits upstream to their source. This happened in the Klamaths, the Sierras, and many other mining districts. The technique is still used extensively today. It is now called "geochemical prospecting," since this sounds more scientific than "panning upstream."

UPPER MESOZOIC STRATA

As in so many parts of the Northwest, marine sedimentation occurred in the Klamath Mountains region during the orogeny that essentially closed the long history of the Cordilleran Geosyncline. This emphasizes once again that the final disappearance of a geosyncline is not a short-lived event. Rather, it involves contemporaneous deformation and deposition over tens of millions of years, until finally the entire region is raised above sea level.

Marine rocks of latest Jurassic and Early Cretaceous age exist along the Pacific side of the Klamath Mountains and in fault blocks inland. Collectively, these are sometimes referred to as the Myrtle Group, although local formational names include the Otter Point, Humbug Mountain, Rocky Point, Riddle, and Days Creek Formations. The Myrtle Group has a maximum thickness of at least 10,000 feet. Most typical rock types are dark siltstones and sandstones, although andesitic to basaltic lava flows, pyroclastic beds, chert, limestone, and conglomerate are all found. Clams are the most common fossils. Myrtle Group strata have been folded and fractured but certainly not to the same degree as have the pre-Nevadan section. Moreover, they have not suffered the low-grade regional metamorphism that characterizes the older section.

The Hornbrook Formation, of Late Cretaceous age, crops out in the Ashland area at the eastern edge of the Klamath Mountains. It consists of fossiliferous shallow-water marine sandstone, siltstone, and conglomerate. These beds are generally light in color and arkosic in composition. Marine sedimentary rocks that resemble the Hornbrook Formation crop out at Cape Sebastian on the Oregon coast. Their fossils indicate that these beds are also of Late Cretaceous age but somewhat younger than the Hornbrook strata.

MESOZOIC GEOGRAPHY

A prediction was made in Chapter 4 that the geosyncline concept, a pillar of past geologic interpretations, would gradually be replaced by more realistic models. The Mesozoic record in the Klamath Mountains Province illustrates the problem. The strata are typically eugeosynclinal, which is to say that they constitute a very thick section of marine volcanic and sedimentary formations. But what sort of mental picture does this give of the actual environment during the Mesozoic Era? Not a very clear one. For the geosyncline concept to have much meaning, it must be related to present-day environments in which comparable strata are accumulating and similar types of structures are forming. When geologists attempt this, they use terms like "continental shelf," "continental slope and rise," "island arc," "oceanic trench," and "abyssal plain"; they substitute all the complex features of a continental margin for the concept of a simple subsiding trough. Furthermore, we have stressed that the relative positions of rock units in a deformed mountain range may have been produced by very large fault displacements, so that present distribution patterns differ markedly from the original ones. How then can we proceed?

Clearly the first step must be to identify all the pieces—to recognize structural blocks that have some continuity and stratigraphic meaning. For the Klamath Mountains, these would include the eastern belt of Paleozoic and Mesozoic strata in California, the Applegate Group, the central core of schists (Salmon and Abrams Schists), the eastern Jurassic belt (Rogue and Galice Formations), the western Jurassic belt (Dothan Formation), and the various patches of latest Jurassic and Cretaceous strata (Myrtle Group, Hornbrook Formation, and so on). The contacts between most of these blocks are faults of probable large displacement; many appear to be west-dipping thrust faults. Having recognized the separate elements, each one must be studied thoroughly for information useful in interpreting its environment of deposition. Formations deposited in shallow water should be more fossiliferous and should perhaps show more evidence of the sediment having been reworked by waves and currents than strata deposited in deep water. Limestone, especially if it contains coral or other reef-forming organisms, would indicate shallow water. Chert, on the other hand, is more indicative of a deep-water environment, possibly even the sea floor; but chert and limestone are sometimes interbedded. The composition of the volcanic rocks in the section is also very important. Assemblages rich in andesite are characteristic of island arcs or volcanic chains along the edges of continents. Volcanic eruptions along oceanic ridges or those that build islands or submarine seamounts far from a continental margin are almost invariably basaltic in composition. Recent studies have suggested a great variety of chemical and mineralogical criteria that are useful in recognizing the environment of accumulation of ancient volcanic rock.

Once the geologist can envision the original geography that accounts for all the structural pieces in a range, he is ready for the next step—unravelling the tectonism that brought about the present configuration. His interpretation must account for the particular structures (folds and faults) and their relative ages. It must also explain such fea-

10-5 *Hypothetical cross section through the Klamath Mountains region late in the Jurassic Period.*

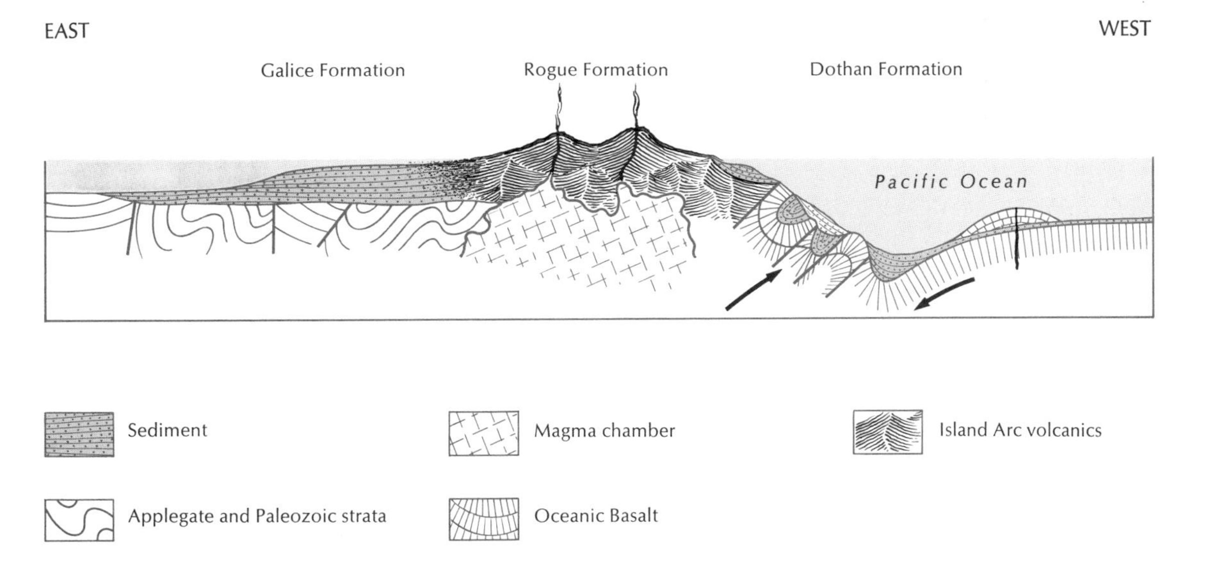

tures as metamorphism of the strata and the intrusion of granitic batholiths. Finally, to complete the picture, the orogenic history of the region should be explained and not just outlined. Was the tectonism a result of oceanic crust underthrusting the margin of the continent or were other processes involved? Are there pieces that do not belong—separate crustal plates that have been rafted in from elsewhere and incorporated in the continental margin? When a geologist can answer these questions with confidence, he really has come a long way from the classical concept of a geosyncline.

Now for the big letdown. When it comes to understanding the Klamath Mountains, geologists have barely reached first base. Hopefully, this brief treatment of the major geologic features of the region suggests some of the possibilities. For instance, by late in the Jurassic Period, we might assume that the eastern belts (Paleozoic and Early Mesozoic of California, the crystalline schists, and the Applegate Group) had become a part of the continental crust. What about the Jurassic belts? The eastern one, consisting of the Galice and Rogue Formations, has the aspect of a volcanic nearshore island chain or perhaps deposits formed on the continental shelf adjacent to a coastal volcanic range. (The island arc fits the present geometry better, for the andesitic Rogue Formation lies on the ocean side of the shallow marine Galice Formation.) Either environment could provide the lava, the volcanic debris in the sediments, and the ammonite and clam fossils.

The western Jurassic belt, consisting primarily of the Dothan Formation, has quite a different aspect. The volcanic rocks are basalts rather than andesites. No fossils are found. Chert is interbedded in the section, and the sandstone layers are massive with little evidence of having been reworked by waves or currents. All this suggests deep water—possibly toward the foot of the continental slope, maybe even in an oceanic trench. The sandstones could have been deposited by turbidity currents. The lava of the Dothan Formation might have been a piece of the oceanic crust, perhaps generated at a mid-ocean ridge or erupted in a volcanic seamount.

This model supposes that compression and thrusting of the continental margin has occurred to account for the close juxtaposition of the various pieces. The compression could have been due to sea-floor spreading and subduction of oceanic crust. The presence of glaucophane minerals in the Colebrook Schist near the coast supports this hypothesis, for a similar metamorphic assemblage of minerals has developed in the California Coast Range where Late Mesozoic subduction has been strongly suggested.

The intrusions are more of a problem. In composition, they are mostly the equivalent of andesite and might be related to the island arc volcanism. They could represent the crystallized chambers from which the andesitic flows and tuffs of the Rogue Formation were derived. The fact that they intrude the Rogue strata is not necessarily damning, for not infrequently a thick volcanic cap has formed above a batholith and has then been intruded by rising magma.

Well, as the saying goes—"I have already told you more than I know." At the risk of offending geologists who have worked in the Klamaths, I have outlined a simplistic model to explain some of the relationships found there. The sole purpose of this exposition is to illustrate the kind of interpretations that global tectonic theory encourages. If something like this actually occurred, it will have to be substantiated by more solid data than are presented here.

UPLIFT IN THE CENOZOIC ERA

The Cenozoic record is, in a word, sparse. As noted, Cenozoic rocks around the Klamaths are excluded from the province by definition. Within the range are found stream gravels and sands, some of which are as old as Miocene. Apparently, they were deposited at a time when the range was lower, and erosion was less vigorous than it is today. The uplift of the Klamath Mountains in the

10-6 *Stacks or erosional remnants of bedrock on a wave-cut platform, near Brookings. The stacks are composed of strata of the Dothan Formation. Note the flat surface of an uplifted wave-cut platform at far right. (Photograph courtesy of Oregon State Highway Division.)*

past few millions of years has raised many of these older gravels to quite high elevations, so that some remnants are still preserved. Some of these were worked for gold by hydraulic mining. Recent uplift is also demonstrated by raised wave-cut benches and terraces along the Pacific shore (Figure 10-6 and Chapter 11); gravels on some of these, as well as some of the alluvial deposits inland, have supported small placer gold-mining operations in the past.

A few of the higher peaks in the Klamath Mountains and perhaps the Coast Range of Oregon to the north contained Pleistocene glaciers, although none is found there today. These were small and located mostly on the relatively protected north- and east-facing slopes, but little is known about them.

11 the coast range province

■ A restatement of the reason for the seemingly curious organization of this book is appropriate here. Chapters 6 through 9 treated northern provinces—everything north of central Washington and north-central Idaho. The preceding chapter jumped down to the southwestern corner of the region covered by this volume and discussed the Klamath Mountains. With this chapter, the intervening area starts to get filled in. Why the jump?

The answer is geologic. Regions studied so far contain a substantial stratigraphic record for the Mesozoic and Paleozoic Eras, but relatively little strata of Cenozoic age. Provinces treated in this and succeeding chapters have good Cenozoic sections but relatively little in the way of an older record. (Exceptions are the Blue Mountains and Idaho, which have substantial amounts of both Cenozoic and older strata. However, the superiority of their more recent record warrants their inclusion in later chapters.) The book is organized, then, in a way that evolves the entire region through geologic time. This results in a sort of geographic hopscotch game, but this is deemed preferable to excessive jumping around in the geologic time scale.

The reasons for the present distribution of strata in the Northwest are complex. The Cenozoic section has a maximum thickness in the Coast Range, the Puget-Willamette Trough, the Cascades south of Snoqualmie Pass, the Columbia Plateau, and the northwestern flank of the Blue Mountains. This region is surrounded by mountains that contain older rocks, but virtually no pre-Tertiary strata crop out within it. Some geologists explain this as the natural result of Tertiary subsidence and deposition in this entire region. Surrounding provinces stood high through much of the period, while in the center extraordinary thicknesses of sedimentary and volcanic strata accumulated. Older crustal rocks are presumed to lie buried beneath the Tertiary section of the Coast Range, Cascades, and others. The great central downwarp in this interpretation is referred to sometimes as the Columbia Embayment.

A different explanation is suggested by other geologists. They believe that the edge of the Columbia "Embayment" is really a great indentation in the edge of the North American continent, brought about perhaps by Cenozoic bending or fracturing of the crust. In this interpretation, there would be no pre-Tertiary continental crust beneath northern and western Oregon and southern Washington. Instead, the Cenozoic strata would have been deposited directly on oceanic crust. Proponents of this hypothesis often use the term "Columbia Arc" when referring to the pattern of distribution of pre-Tertiary rock in the Northwest.

The data required to accept or reject either hypothesis simply are not available at present. Certainly, the area of the "Columbia Embayment" subsided relative to surrounding regions during much of the Cenozoic Era, for how else could it have accumulated such a great thickness of Cenozoic strata? Conversely, an analysis of Cenozoic deformation, both of the Cordillera and of the floor of the Pacific Ocean, supports the possibility of major displacements of the western margin of the continent.

The Coast Range Province lies across the mouth of the "Columbia Embayment." The range itself is a young feature, reflecting uplift at the continental margin during the past few million years. Within the rocks of the Coast Range are exceedingly important clues to the history of the edge of North America during the rest of the Cenozoic Era.

COAST RANGE GEOGRAPHY

The province is bordered on the west by the Pacific Ocean, on the east by the Puget-Willamette Lowland, on the north by the Strait of Juan de Fuca, and on the south by the Klamath Mountains. The border with the Klamaths is placed arbitrarily at the contact between Tertiary and pre-Tertiary strata (Chapter 10). The Coast Range is lowest in northern Oregon and southern Washington—a relationship seen also in the Cascade Range to the east. Separately named subdivisions of the Coast Range include the Willapa Hills in southwestern Washington and the Olympic Mountains to the north. The Olympics are the highest and the most rugged part of the province; peaks tower to elevations of a mile or more above the surrounding lowlands. Glaciers on the highest of all, mount Olympus (elevation, 7,954 feet), extend to lower elevations than they do anywhere else in the "Lower 48." (This is due, of course, to the extraordinary annual precipitation for the higher parts of the range, much of which falls as snow. Accurate figures are not available, but estimates based on stream runoff measurements are as high as 300 inches of water per year.)

The Coast Range is breached by only a few rivers. Most of the region is drained by short streams that flow west to the Pacific or east to the interior lowland. Throughgoing streams include the Chehalis River, which separates the Olympic Mountains from the Willapa Hills; the Columbia River; and the Umpqua and Rogue Rivers of southwest Oregon. None of these channels pro-

11-1 *The south end of the Coast Range Province, looking southeast. Cape Arago in the foreground is composed of steeply dipping sandstone of the Eocene Coaledo Formation. (Copyrighted photograph courtesy of Delano Photographics.)*

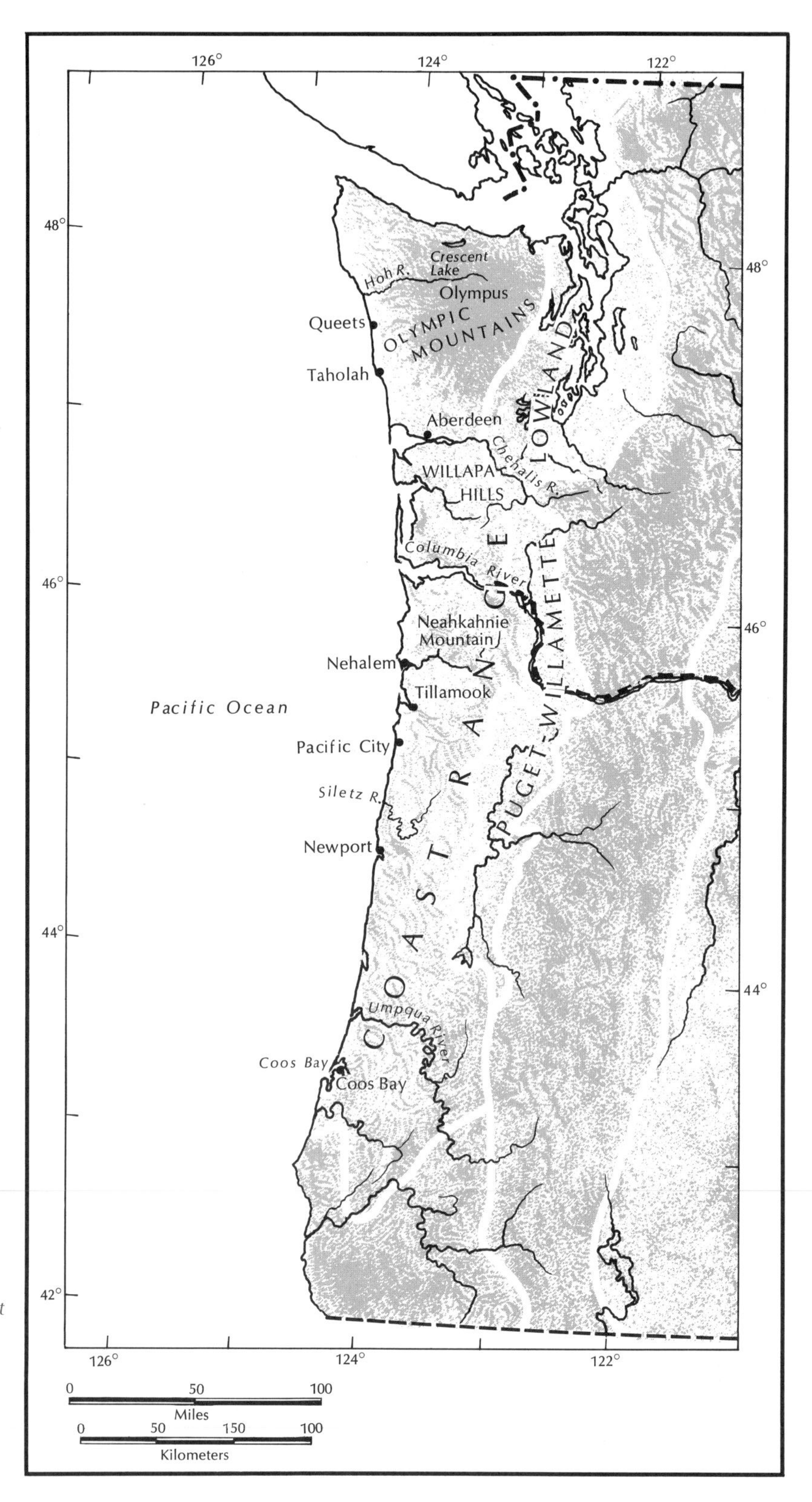

11-2 *The physiography of the Coast Range Province. (After Raisz, 1941.)*

vides complete exposure of the bedrock of the Coast Range. The structure of the region has been pieced together primarily from many small, isolated outcrops found in roadcuts, stream beds, and sea cliffs.

GENERAL GEOLOGIC PATTERNS

The shore of the Pacific Ocean migrated westward during the Cenozoic Era, from an initial position close to the western edge of the Cascade Range Province to its present location. This means, of course, that the Coast Range has risen from the sea. Not all the uplift has been in the past few million years. By the middle of the Miocene Epoch, about 15 million years ago, most of the Coast Range region had emerged from the Pacific. Before that, subsidence rather than uplift had characterized the continental margin.

The stratigraphic section for the Eocene and Oligocene Epochs is particularly complete. Eocene strata consist of an exceptionally thick sequence of basaltic lavas, overlain by sandstones and shales. Fossils are quite abundant and indicate a relatively shallow-water marine environment for at least some of the beds. Some of the sediment came from the Klamath Mountains region, some

11-3 *The rain forest along the western flank of the Olympic Mountains, Washington. Dense vegetation like this is not conducive to detailed geologic studies of the region. (Photograph courtesy of Washington State Department of Commerce and Economic Development.)*

probably from British Columbia and the North Cascades as well. These sources were partly masked out beginning late in the Eocene Epoch by a large influx of volcanic debris, derived from the erosion of volcanoes within the Coast Range region and from cones that formed in what is now the Cascades. By Oligocene time, volcanism in the Coast Range had ceased, and most sediment came from the east. The region was still essentially a subsiding continental shelf and slope.

This pattern was destroyed in the Miocene Epoch by an orogeny as pronounced as the one currently under way. Formations throughout the Northwest were folded and faulted to a marked degree, and the Coast Range region essentially emerged from the sea for the last time. The past 20 million years have seen the accumulation of some shallow-water marine sediments in a few distinct embayments along the coastline and the uplift of our modern Coast Range. Pleistocene glaciation was restricted largely to the Olympic Mountains.

11-4 *A field party of the U.S. Geological Survey, employing a helicopter to provide access to the rugged interior of the Olympic Mountains. This is the only part of the Coast Range that is relatively free from dense vegetation. (Photograph courtesy of Richard Stewart.)*

THE ORIGIN OF CENOZOIC STRATA

Some general comments on the origin of Cenozoic sediments in the region of the "Columbia Embayment" are needed before commencing discussion of particular provinces. Considerable subsidence of the "embayment" must be accepted. This does not mean that most of its sediment was derived by erosion outside the region of pre-Tertiary rocks. Some of it was, especially in the Eocene Epoch. However, a much greater amount of sediment came from within. How could this be? It sounds like the law of mass conservation went on holiday, and matter was created somehow within the systems. In one sense, it was. The primary source of sediment was the erosion of Tertiary volcanic strata within the "embayment" itself. Cenozoic sections are markedly volcanic—a mixture of lava flows; volcanic mudflows; ash; and conglomerates, sandstones, and siltstones rich in volcanic rock fragments.

This is an important lesson about the history of the earth. Throughout geologic time, the earth has evolved from a relatively homogeneous sphere to a markedly layered one. The various layers or shells have distinct differences, both physically and chemically. In general, lighter elements and compounds have been concentrated near the outside of the earth and denser ones toward the center. (Thus the average specific gravity of the earth is 5.5, which means it is 5.5 times heavier than water. The average specific gravity of the earth's crust is only about 2.7 or 2.8.) The earth then has undergone a type of gravitational differentiation process through time.

Our principal interest here is with the crust. Chapter 2 introduced the three major rock types plus loose rock material at the surface, which together compose the crust. These were related through the concept of the rock cycle (Fig. 2-11), and there lies the potential trap. If the cycle is viewed as a closed system—one in which there is no input of new material but just recycling of older rock—then there is no possibility of generating additional crust. In actual fact, the system is not closed. Major input is provided by igneous processes, and throughout geologic time the earth's crust has been enlarging by the addition of magma or lava from the underlying mantle. Not all molten rock comes from the mantle, for some is produced by the melting of crustal rock. Much geologic research has gone into developing methods of analyzing igneous rock in order to determine if its parent magma was derived by melting of the crust or the mantle.

The Cenozoic history of the Northwest can be appreciated more fully by the reader who is aware of the tremendous amount of new material that reached the surface for the first time as a result of volcanic processes. This implies that the Cenozoic Era was different from preceding ones, and this is not really accurate. The role of Paleozoic and Mesozoic volcanism in the western part of the Cordilleran Geosyncline has been stressed again and again. So has the volcanic nature of much of the sedimentary strata interbedded with the lava. The biggest difference between the Cenozoic history of the Cordillera and that of the preceding eras was that most things have happened above sea level since the Cretaceous. The Coast Range is the only region in this part of the Cordillera where marine sedimentation was significant in the Cenozoic Era.

EOCENE SUBMARINE LAVA

More publicity is needed for the basaltic lava that erupted in the Eocene sea in the Coast Range region. In volume, it exceeded its younger but more famous relatives, the Columbia River Basalt and the Quaternary lava that built the Cascade volcanoes. Along the eastern flank of the Olympic Mountains, Eocene basalt accumulated to a total thickness of perhaps 10 miles, making it one of the thickest formations in the world. Yet relatively little is known about these Eocene eruptions.

Geologists have shied away from work in the marine basalts of the Coast Range for several reasons. Many of the flows have been altered to greenstone, due either to reaction of the lava with seawater or to incipient metamorphism under the weight of younger strata. Furthermore, there are few distinctive beds in this great lava pile—one part of it looks pretty much like another. This, combined with poor exposure (except in the rugged Olympics) and the presence of major faults and folds, makes stratigraphic study of the basalt most difficult.

11-5 *The southern part of the Olympic Mountains, looking west. Hood Canal is in the foreground, Lake Cushman is in the center, and the Pacific Ocean is in the far distance. All of the rock from Hood Canal west past the upper end of Lake Cushman consists of steeply dipping basalt of the Crescent Formation—here about 10 miles thick. (Photograph courtesy of Northwest Air Photos.)*

In western Washington, this early Eocene lava is known as the Crescent Formation, after exposures around Lake Crescent on the north end of the Olympic Peninsula. The same strata on Vancouver Island, across the Strait of Juan de Fuca, constitute the Metchosin Formation. Still other names are used in the Coast Range of Oregon—Tillamook Volcanic Series in the northern part of the range, Siletz River Volcanics in the central part, and Umpqua Formation toward the south end of the range. The submarine nature of these eruptions is proved by marine fossils contained in sedimentary rock interbedded with the lava. The widespread development of pillow structure in the flows (Figure 11-6) also supports a submarine origin. The base of the Eocene basaltic sequence is exposed only in the Olympic Mountains, not in the Willapa Hills or the Coast Range of Oregon. At some localities in Oregon, the total thickness of exposed flows is at least 15,000 feet, and it appears likely that the total volume of lava extruded in the Coast Range Province in the Eocene Epoch exceeded 100,000 cubic miles. (This is the equivalent of approximately 5,000 Mount Rainiers!)

How can we explain such a tremendous volcanic outpouring? The lava probably came from the mantle; chemical analyses of the basalts strongly resemble those from recent basaltic eruptions in oceanic areas where a subcrustal origin is certain. A comparison with Hawaiian eruptions has been suggested, and indeed this makes some sense. For one thing, the composition of the lava is comparable. Furthermore, the Hawaiian chain grew from the ocean floor by the building of flow upon flow of basaltic lava, until finally the pile built above sea level. A similar evolution for the Eocene lava accumulations is likely, for the flows near the top of the section appear to have been erupted above water. The nature and distribution of sediments interbedded with the lava suggests that they were deposited on or near volcanic islands.

If the Eocene lava was erupted to form oceanic islands, then subsequent spreading of the sea floor must have rafted them against the edge of the

11-6 *Pillow structure in basalt of the Crescent Formation, Olympic Mountains. Pillows form when lava erupts under water and are typically a few feet in diameter. Scale is provided by the geologist in the left foreground. (Photograph courtesy of Richard Stewart.)*

11-7 *Heceta Head on the central part of the Oregon coast. The bold, rocky headlands are composed of Eocene basaltic lava of the Tillamook Volcanic Series. (Copyrighted photograph courtesy of Delano Photographics.)*

continent. As an alternative, the lava might have extruded on the continental shelf or slope rather than in the deep ocean. The transfer of so much material from the mantle to the overlying crust should have produced marked subsidence of the original shelf (slope) surface, now the base of the lava pile. This subsidence could explain the great thickness of lava that was able to accumulate.

The nature of the rock beneath the Eocene volcanics holds the key to choosing between these two possible hypotheses. If it is continental crust, the analogy with the Hawaiian Islands would be invalid. If it is oceanic crust, the basalt could not have accumulated on the continental margin unless subduction wedged the oceanic plate between the basalt and the underlying continental plate. The Olympic Mountains are the only region of the Coast Range where the base of the Eocene lava section is exposed. Other topics should precede a discussion of the stratigraphic and structural relationship in the Olympics.

EARLY TERTIARY GEOGRAPHY

Perhaps the most distinctive formation in the southern part of the Oregon Coast Range is the Tyee Formation. It consists of light-colored beds of arkosic (that is, feldspar-rich) sandstone and siltstone that are typically 2 to 10 feet thick and has a maximum total thickness in excess of 10,000 feet. The few fossils that have been found in the formation indicate that it was deposited in the middle of the Eocene Epoch. Sedimentary structures suggest that the Tyee sediment came from the south. This is not surprising, for the most likely source rock that might have supplied arkosic sediment to the basin is the granitic rock of the Klamath Mountains. The Tyee beds are more coarse grained toward the south and consist of shallow-water conglomerates near the edge of the Klamath Mountains. Conversely, the nature of the Tyee fossil fauna in the central part of the Oregon Coast Range indicates sediment accumulation in water more than 5,000 feet deep.

By late Eocene time, the area that was to rise later as the Cascade Range was a region of very active volcanism. The climate was warm and subtropical. Immediately west of the line of volcanoes was a low, swampy coastal plain. The abundant plant material that accumulated in these swamps was transformed later by burial and compaction into coal. Eocene coal has been mined in the Bellingham, King County, and Centralia-Chehalis districts in Washington, and in the Coos Bay area on the Oregon coast. Sedimentary rocks interbedded with the coal consist primarily of arkosic sandstones and shales. However, in the Cascade foothills, andesitic volcanic rocks also occur. Some of the sandstones were probably deposited in deltas where rivers emptied into the ocean. Farther from shore, the sandstone grades into finer-grained (and deeper-water) marine siltstones and shales. Some basaltic volcanism also occurred offshore on the northern part of the Coast Range Province of Oregon (Goble Formation, Table 11-1).

Oligocene stratigraphy in the Coast Range is dominated by marine shales containing a very high proportion of volcanic ash. This ash was probably derived from volcanic vents in the ancestral Cascades and was probably carried to sea both by streams draining westward and by occasional east winds. Locally fossiliferous sandstone, conglomerate, and limestone are interbedded with the shale. Compared to the Eocene Epoch, the Oligocene was fairly stable, with little offshore volcanic activity, folding, or faulting. Mostly it was a time of relatively slow accumulation of fine-grained tuffaceous sediment, with coarser sands and gravels deposited nearshore.

The sedimentary basin or shelf had largely disappeared by the Miocene Epoch. This was probably due both to filling in by sediments and to uplift of the region. Early Miocene marine deposition occurred on the continental shelf (beyond the present coastline) and in a few shallow embayments inshore, most notably along the lower reaches of the modern Columbia River, in the Grays Harbor area in southern Washington, and at the northern edge of the Olympic Peninsula.

THE COLUMBIA RIVER BASALT

Flows of the Columbia River Basalt of Miocene age occur in the Coast Range from the Willapa Hills south to Newport, Oregon. The dark columnar lava forms many prominent capes and headlands along the northern coast of Oregon. In southwestern Washington, the best exposures occur along the north shore of the lower Columbia River, but poorly exposed outcrops are found also in many places in the Willapa Hills (Figure 11-9). Many eruptive vents have been identified in recent years, especially in the Oregon Coast Range. Some of the vents built volcanoes, but others were long fissures from which the fluid lava poured. Some of the eruptions occurred underwater, producing pillow basalt and layers of basaltic breccia. The interlayering of the Columbia River Basalt with both marine and nonmarine sedimentary strata indicates that the shoreline was in about the same position in the Miocene Epoch as it is today. Not all of the Miocene basaltic magma reached the surface. Sills up to 1,000 feet thick comprise much of Mount Hebo, 10 miles east of Pacific City, and Neahkahnie Mountain, immediately north of Nehalem.

THE OLYMPIC PUZZLE

The north side of the Olympic Peninsula is one of the best places to study Coast Range stratigraphy. The strata (Table 11-2) dip northward away from the core of the range and strike parallel to the Strait of Juan de Fuca. The major uplift of the Olympics apparently followed the deposition of the lower Miocene Clallam Formation. It has been

TABLE 11-1 Tertiary stratigraphy of the Oregon Coast Range. (From Baldwin, 1964.)*

AGE	NORTHERN	CENTRAL	SOUTHERN
Pliocene	Troutdale		Port Orford
			Empire
Miocene			
	COLUMBIA RIVER BASALT†	COLUMBIA RIVER BASALT	
	Astoria	Nye Mudstone	
		Yaquina	
	Scappoose		
Oligocene	Pittsburg Bluff	Toledo	Tunnel Point
	Keasey		Bastendorff
Eocene	GOBLE Cowlitz		Coaledo
	Yamhill	Tyee	Tyee
	SILETZ RIVER	SILETZ RIVER	Umpqua UMPQUA

*Green panels indicate unconformities.
†Formations in capital letters are primarily volcanic strata.

11-8 *Eocene sandstone of the Coaledo Formation at Shore Acres State Park, near Cape Arago, Oregon. Differential weathering and erosion have accentuated the uneven distribution of calcareous cement in the sandstone, as seen in the honeycombed surface near the top of the rock in the foreground and the brown, rounded concretions below. Note the uplifted wave-cut bench forming the flat surface to the left. (Photograph courtesy of Oregon State Highway Division.)*

folded and faulted to about the same degree as have the pre-Miocene formations (except those in the range core). Moreover, no major angular unconformities have been found in the Tertiary section.

A major problem in the Olympic Mountains has been interpreting the rocks in the core of the range. They occur beneath the Crescent Formation, which dips steeply away from the center of the range (Figure 11-10). The rocks consist of dark sandstones and siltstones, altered submarine basaltic rocks, and bedded chert. They have been markedly folded, fractured, and slightly metamorphosed. This assemblage, known as the Soleduck Formation, has long been assumed to be older than the Crescent Formation and probably of Mesozoic age. However, the few fossils found in the Soleduck suggest an Eocene or even Oligocene age for the formation.

Mapping in the remote interior of the Olympics has been done recently by the U.S. Geological Survey, university students, and several oil companies. It has shown that the Soleduck Formation is indeed Eocene and Oligocene in age rather than Mesozoic. Moreover, the unit appears to have been emplaced under the Crescent Formation along major east-dipping thrust faults.

This may be one of the clearest examples in the Cordillera of past underthrusting of the continent by the sea floor. The Soleduck Formation would represent oceanic crust plus its cover of sediment. The Crescent Formation could be either basalt that erupted on the edge of the continent or an oceanic island or seamount chain that was rafted against the continental margin. In either case, it was a part of the continent when it was underthrust by the Soleduck strata. The underthrusting was caused by spreading of the sea floor away from the Juan de Fuca Rise. The folding, faulting, and metamorphism of the Soleduck Formation was caused by the stresses inherent to the subduction process, but the rigidity of the massive Crescent Formation prevented much deformation of it and the overlying strata. Subsequent uplift of the Olympics and erosion have exposed the Soleduck Formation resting apparently under the normal Tertiary section. This hypothesis is compatible with our knowledge of the history of the Juan de Fuca Rise, but it still must be regarded as quite tentative.

The Cenozoic underthrusting suggested for the Olympics cannot be demonstrated in the Oregon Coast Range, although the greater uplift and erosion of the Olympics may have exposed oceanic material still concealed to the south. (In this con-

TABLE 11-2 Tertiary stratigraphy of the northern Olympic Peninsula. (After Brown, Gower, and Snavely, 1960.)

AGE	FORMATION	DESCRIPTION	THICKNESS, FEET
Middle Miocene	Clallam	Marine and nonmarine sandstone and siltstone; minor conglomerate, coal	>2,500
Late Eocene to early Miocene	Twin River	Marine sandstone, siltstone, and shale. Local thick conglomerate beds	6,000–12,000
Late Eocene	Lyre	Marine conglomerate and sandstone	<3,300
Middle Eocene	Aldwell	Marine siltstone and sandstone. Minor basalt and tuff near base	<2,900
Early to middle Eocene	Crescent	Marine basalt flows and flow breccia, with interbedded sandstone, siltstone, shale, and tuff beds. Very minor chert and limestone	7,000–>15,000

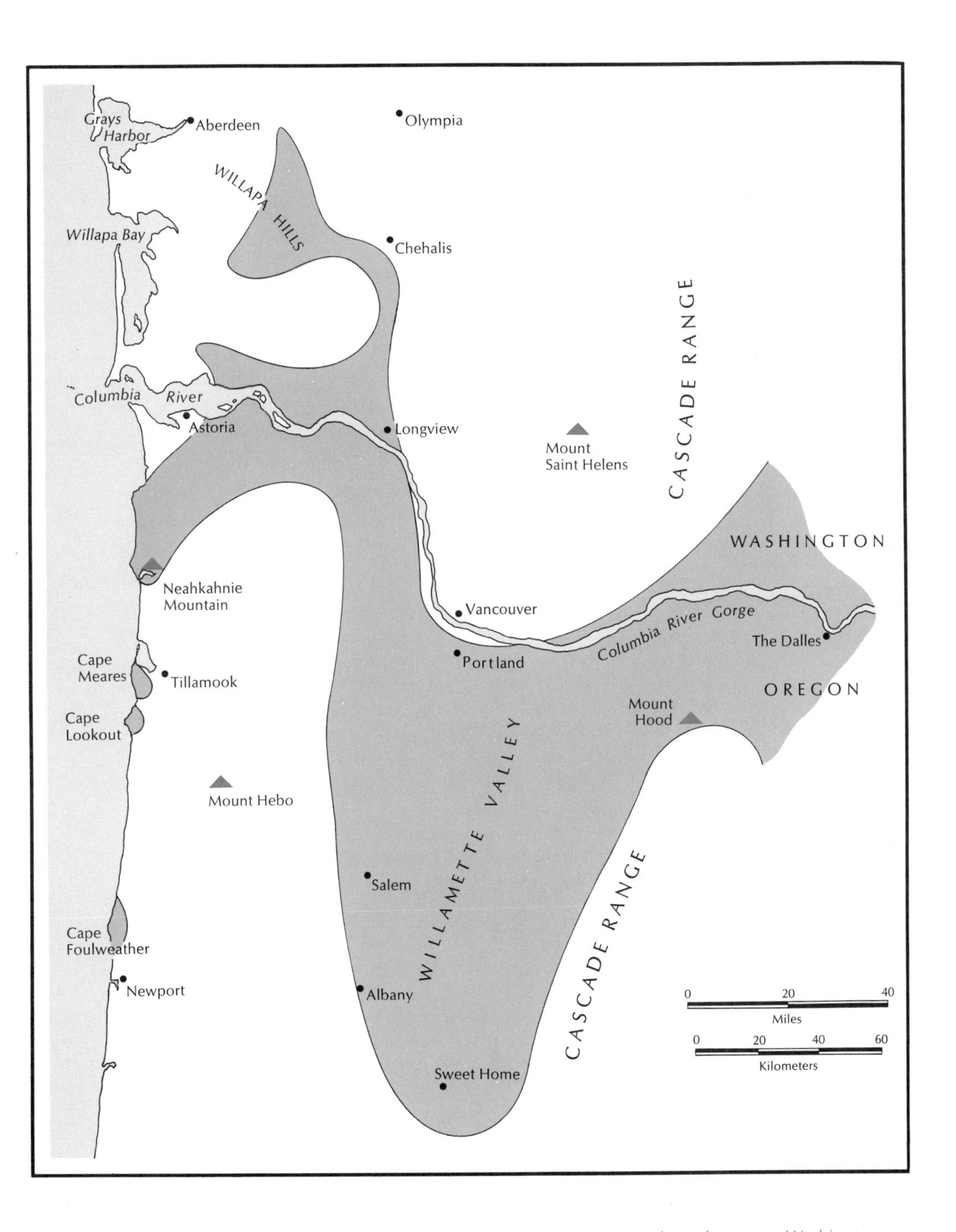

11-9 *The distribution of Columbia River Basalt in northwestern Oregon and southwestern Washington. (After Wells and Peck, 1961, and Huntting* et. al., *1961.)*

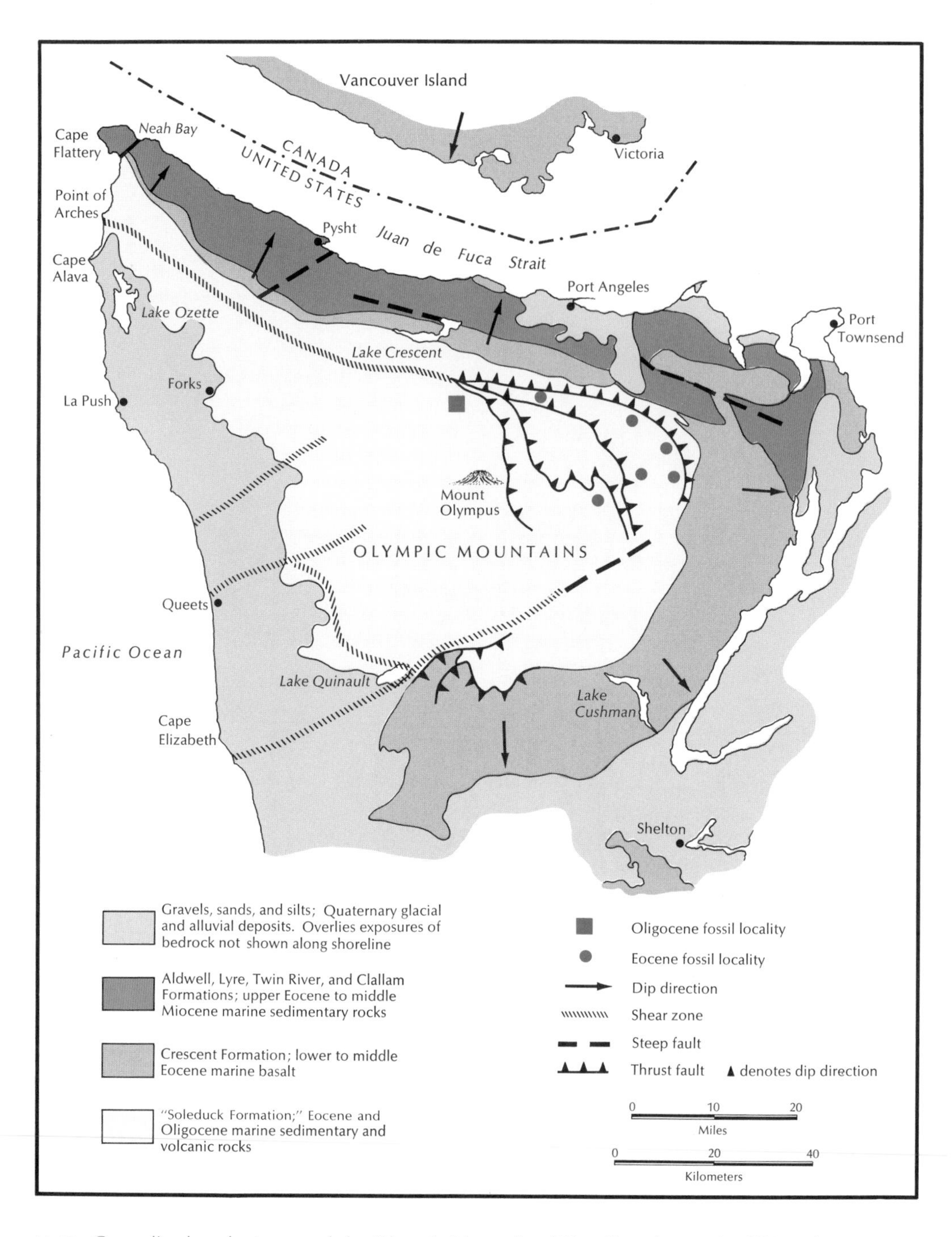

11-10 *Generalized geologic map of the Olympic Mountains. (After Huntting* et al., *1961, and unpublished map compiled by Richard Stewart.)*

nection, we might reemphasize that the base of the basaltic strata equivalent to the Crescent Formation is nowhere exposed in Oregon.)

THE SEARCH FOR PETROLEUM

Almost all of the world's petroleum occurs in marine sedimentary rocks. The Coast Ranges have seemed likely to contain petroleum pools and have received considerable attention from oil companies. Unfortunately, the many wells that were drilled there produced very little oil or gas, and company interest in the region waned. Then technological advances opened up the possibilities of offshore drilling; and, since the marine strata of the Coast Ranges were known to extend offshore, enthusiasm for the Northwest was renewed. This prompted considerable geologic and geophysical investigation, both on shore and offshore. Drilling rights for certain tracts on the shelf off Oregon and Washington were sold a few years ago at a total price of 35 million dollars. A few wells were drilled, but the results were disappointing.

Somewhere in the Northwest there must be a commercial accumulation of petroleum. As the world's reserve of oil and gas is depleted, the in-

11-11 *Point of Arches on the northern coast of Washington, with Cape Alava in the distance. (Photograph courtesy of Washington State Department of Commerce and Economic Development.)*

11-12 *The Washington coastline at La Push, looking north. Ozette Lake in the distance to the right. The Tertiary marine strata in the rocky islands and stacks continue offshore in the continental shelf and are potential source rocks for petroleum. (Copyrighted photograph courtesy of Delano Photographics.)*

11-13 *The Oregon Dunes area of the central Oregon coast. The mouth of the Siuslaw River is in the foreground, Heceta Head and Cape Perpetua in the distance. The white dunes are active; the others have been stabilized by vegetation. (Copyrighted photograph courtesy of Delano Photographics.)*

centive for finding it will increase. However, the specter of an oil spill is sufficiently fresh in people's minds to discourage some companies from expending more time and money for offshore prospecting (at least for now). Even if the presence of a major oil pool was strongly suspected, drilling for it might not be greeted by much public enthusiasm. Unfortunately, there is no method devised to prove the presence of oil underground other than drilling a "wildcat" well.

LATE CENOZOIC TERRACES

The Late Cenozoic record of the Coast Ranges is not extensive. Near the Washington coast, the Pliocene Montesano and Quinault Formations consist of shallow-water marine and nonmarine sandstones, siltstones, and conglomerates. These strata suggest that the Pliocene shoreline was not far removed from its present position. In Oregon, Pliocene marine strata are found only in the southern part of the Coast Range. The Empire Formation in the Coos Bay area consists of 3,000 feet of sandstone. It contains the bones of whales, sea lions, and seals, as well as abundant clam and snail shells.

The complete history of marine terrace development is perhaps the most interesting Pleistocene feature of the Coast Range. Wave-cut terraces occur as high as 1,600 feet above present sea level. The highest are the oldest, some dating back perhaps to the late Pliocene Epoch. Their position so far above sea level suggests considerable Pleistocene uplift of the Coast Range. The uplift has not been uniform. The most prominent terrace, late Pleistocene in age, lies as much as several hundred feet above sea level in some areas, but elsewhere it is below sea level. This demonstrates warping of the coastline, for the terrace was formed by wave erosion at sea level. The interpretation of these terraces is difficult because of Pleistocene sea-level changes brought about by widespread continental glaciation. At times of maximum glaciation, sea level was as much as 400 feet below its present position. Conversely, during some of the interglacial intervals there was much less glacial ice on land than today, and sea level may have been higher, up to 100 feet. Sea level has been rising during the past 15,000 years, causing a partial "drowning" of the Oregon coast. The lower reaches of river valleys have been turned into estuaries and tidal lagoons. Wave erosion and longshore currents have been effective in straightening the Oregon coast. Steep, rocky cliffs alternate with beaches and spits of sand that have been deposited by longshore currents across the mouths of rivers. Where sand is plentiful and the hills inshore are neither too close nor too high, the prevailing onshore wind produces various types of sand dunes that migrate inland. Sand dunes are particularly prominent in the Oregon Dunes area in the central part of the Oregon Coast Range (Figure 11-13).

GLACIERS IN THE OLYMPIC MOUNTAINS

The Olympics are the only part of the Coast Range Province that has glaciers—in fact, it is the only part of the region that receives much snowfall. Today's glaciers are concentrated in the high peaks around Mount Olympus; and the longest one, the Hoh Glacier, is only 3 miles long. These are midgets compared with the great Olympic glaciers of the past. Possibly four times in the Pleistocene Epoch an ice cap covered the center of the range, burying all but a few of the highest summits. Alpine glaciers flowed down and shaped all of the major valleys. Some of these reached as far as the present coastline near Taholah, Queets, and the mouth of the Hoh River. When these glaciers wasted away, they left behind sheets of till mantling the ground and some distinct recessional moraines (such as the one that dams Quinault

Lake). Valley glaciers also reached the Puget Lowland and the shore along the Strait of Juan de Fuca, but the evidence beyond the front of the range was destroyed by the large Puget and Juan de Fuca Lobes of lowland ice (Chapter 18).

For many years, glaciologists have been studying some of the glaciers around Mount Olympus. These have not been casual observations. They have involved permanent camps and the use of the most advanced techniques of glacial research and have yielded much significant data on the physics of ice and the mechanics of glacial flow. Thus the few glaciers remaining in the Olympic Mountains, puny as they may seem when compared with their ancestors, are important laboratories for the study of glacial processes.

11-14 *A small glacier on the south flank of Mount Olympus, highest peak in the Olympic Mountains. (Photograph courtesy of Richard Stewart.)*

12 the cascade range of oregon and southern washington

■ Certainly, the most renowned geologic features of the Cascade Range are the many young volcanoes. In the high eastern half of the Oregon Cascades, Quaternary lava constitutes a major part of the range; but elsewhere it is vastly subordinate to older Cenozoic rocks. These too are volcanic, but few of the pre-Quaternary volcanoes survive as distinct physiographic features. In this chapter, we will deal with the Tertiary events that really built the range. The story of the illustrious Quaternary volcanoes of the Cascades deserves separate treatment (Chapter 13).

REGIONAL PHYSIOGRAPHY

The Cascades connect in the south with the Sierra Nevada near Mount Lassen and in the north they merge with the Coast Mountains of British Columbia. The Cascades vary in width from a minimum of about 30 miles in southern Oregon to a maximum of about 80 miles in northern Washington. They have an average elevation of about 5,000 feet above sea level but rise somewhat above this at each end. The most rugged topography occurs in the North Cascades of Washington and southern British Columbia (Chapter 7).

The Cascades are breached only by the Columbia and Klamath Rivers. Numerous rivers drain the west flank of the range; these are tributary to the Rogue or Willamette Rivers in Oregon, The Cowlitz or Chehalis Rivers in southern Washington, or Puget Sound in northern Washington. The east side of the Cascades is drained by rivers that are tributary to the Columbia, except for streams at the south end that flow out into the arid Basin and Range Province. There they generally disappear.

The best look "into" the Cascades is in the Columbia River Gorge. The various roads over the range in Washington and Oregon generally provide only isolated rock exposures. High amounts of precipitation are dropped on the western flank of the range. This has produced a heavy vegetative cover, which makes movement on foot away from highways and secondary roads exceedingly difficult. This is one explanation for why the Cascades have not been mapped in detail.

AN OVERVIEW OF CENOZOIC HISTORY

The Cascades provide some of the thickest sections of Tertiary strata found anywhere in the West. The base of the Tertiary sequence is exposed at the edge of the Cascades where the range merges with the Klamath Mountains and the Sierra Nevada. It is also seen in a few localities along the southern boundary of the North Cascades. Elsewhere, the base of the Tertiary sequence has yet to be uncovered by erosion, so that the nature of the underlying strata is unknown.

The Tertiary history of the Cascade Range is very complex. A great number of individual formations have been named locally, especially along the western flank of the range. Correlating from one area to another is very difficult, because the characteristics of units change over short distances,

TABLE 12-1 Cenozoic stratigraphy of the western Cascades, Oregon. (After Peck *et al.*, 1964.)

AGE	FORMATION	DESCRIPTION	THICKNESS, FEET
Pliocene to Quaternary	Cascade Andesite	Andesite and basalt flows, breccias. Minor pyroclastic beds. High Cascades suite	>3,000
Pliocene	Troutdale	Conglomerate, sandstone, and siltstone. Stream deposits. Poorly consolidated	<400
Middle to late Miocene	Sardine	Andesite flows, tuff breccia, and tuff	<10,000 (3,000 av.)
Middle Miocene	Columbia River Basalt (Group)	Basalt flows, minor andesite	<1,500
Oligocene to early Miocene	Little Butte Volcanic Series	Dacitic and andesitic tuff and less abundant flows and breccias	3,000–15,000
Late Eocene	Colestin	Andesitic tuff, conglomerate, sandstone. Minor andesite flows and breccias	<3,000 (1,500 av.)

▬▬ Major unconformity

12-1 *The southern part of the Cascade Range in Washington, looking south toward Mount Adams and, in the distance, Mount Hood. The Goat Rocks area in the foreground is part of a deeply eroded volcano of Miocene or Pliocene age. (Photograph courtesy of Rayner Studio.)*

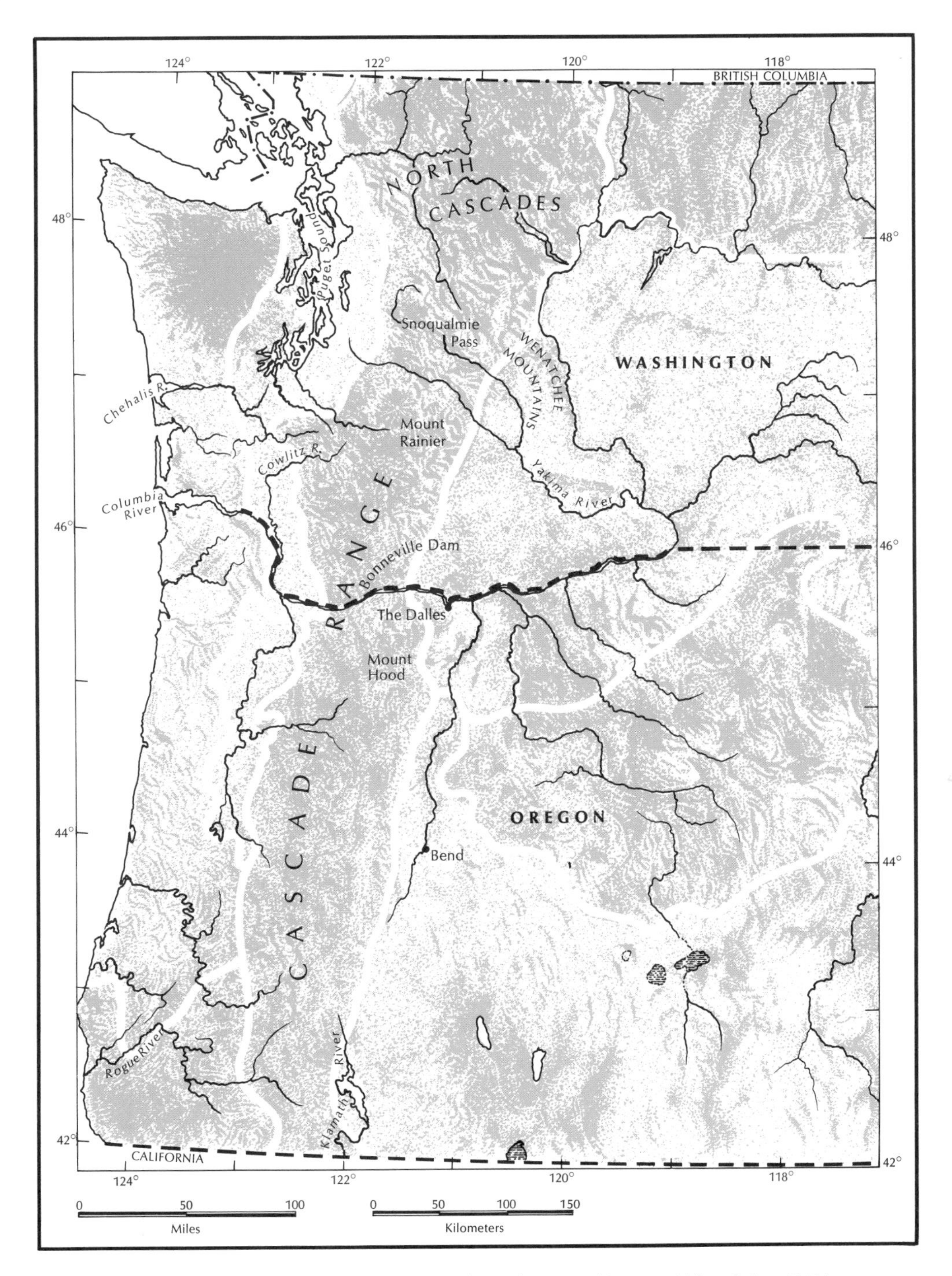

12-2 *Physiography of the Cascades of Oregon and southern Washington. (After Raisz, 1941.)*

and fossils are scarce. Moreover, most fossils that are found are leaves, which do not permit very reliable age dating. Fortunately, the general characteristics if not the details of the stratigraphic record of the range are much the same from central Washington to California.

Volcanic activity has dominated the entire Cenozoic history of the Cascades. It has produced a tremendously varied assemblage of rocks, especially lava flows, volcanic ash beds, mudflow deposits, and sedimentary rocks consisting mostly of angular volcanic debris. The volcanic rocks vary in composition from dark, iron-rich basalt to light-colored, silicic dacite and rhyolite. In places, quite diverse types are interbedded.

The major events in the Cenozoic history of the Cascades have been:

1. Deposition between late Eocene and early Miocene time of nonmarine sedimentary and volcanic strata to a total thickness of 3 to 6 miles.
2. Folding, faulting, and some intrusion of igneous batholiths and stocks toward the middle of the Miocene Epoch. Mountains produced by this orogeny were largely leveled by later Miocene erosion.
3. Partial burial of the range beneath fluid lava flows of the Columbia River Basalt.
4. Late Tertiary volcanism and initiation of range uplift.
5. Quaternary formation of the Cascade volcanoes. Continued uplift of the range. Several episodes of alpine glaciation, especially on the volcanoes and in the higher parts of the Cascades.

A very generalized cross section through the range in southern Washington (Figure 12-3) summarizes major stratigraphic and structural relationships found throughout the region.

LOWER TERTIARY STRATA

The early history of the Cascades is known primarily from studies in three areas—the western flank of the range in northern Oregon and the Columbia River Gorge, Mount Rainier National Park, and the area south of Interstate Highway 90 and Snoqualmie Pass in central Washington. Two distinct series of strata have been recognized. The oldest is generally late Eocene in age and contains significant amounts of conglomerate, sandstone, and siltstone interbedded with volcanic rocks. This sequence can usually be distinguished from the overlying one because it has been folded somewhat more severely and has undergone very slight metamorphic alteration. The metamorphism has given the strata a distinctive reddish or greenish color not found in younger rocks. This older series is more than 10,000 feet thick in the Mount Rainier area, where it is called the Ohanapecosh Formation (Table 12-2).

The younger sequence rests unconformably on the folded and eroded strata of Eocene age. In

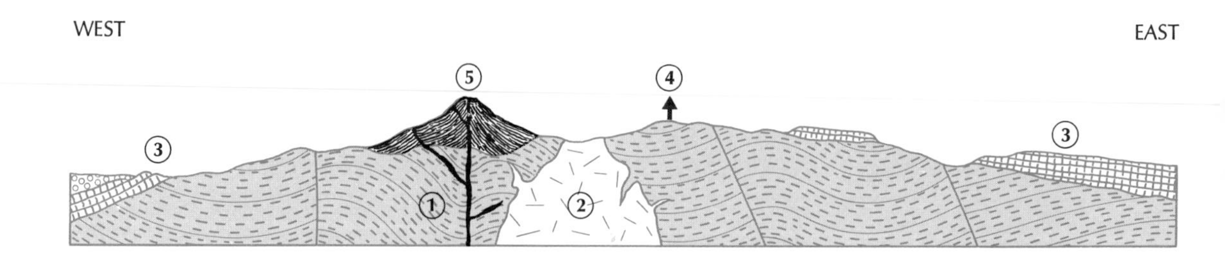

12-3 *Generalized cross section through the Cascade Range in southern Washington. The sequence of events is shown by the numbers as follows: (1) Eocene to early Miocene Epochs—deposition of volcanic strata (Keechelus, and so on); (2) Middle Miocene folding, faulting, and granitic intrusions (Tatoosh, and so on); (3) Middle to late Miocene eruption of Columbia River Basalt; (4) Plio-Pleistocene uplift of Cascade Arch and consequent erosion; (5) Pleistocene-Holocene buildup of Cascade volcanoes, uplift, and erosion.*

Oregon, it is known generally as the Little Butte Volcanic Series and in Washington as the Keechelus Volcanic Group. Its age is considered to be Oligocene to perhaps early Miocene. Characteristic rock types include beds of light-colored volcanic ash (tuff), massive flows of andesite, and layers of andesitic breccias, many of them deposited by mudflows. Basaltic and dacitic flows and breccias are not uncommon; neither are sedimentary interbeds. This younger sequence is more than 5,000 feet thick in the Mount Rainier area but attains a maximum thickness of 15,000 feet along the west flank of the range in Oregon and twice that figure in the Cascades of central Washington north and east of Mount Rainier.

Six miles of strata, much of it volcanic, would seem to suggest a very inhospitable environment. But does it really? Remember that perhaps 10 million years or more of time is represented by those beds, meaning that the average rate of accumulation may not have exceeded 3 feet in a thousand years. Naturally, an "average" rate has little significance when many of the layers are congealed lava flows 50 to 100 feet thick. These would have been deposited in a few hours, but thousands of years might have elapsed between successive eruptions.

TYPES OF VOLCANIC ERUPTIONS

A volcano is a conical hill or mountain built up around one or more vents in the ground from which erupt some variety of volcanic product. The material expelled may range from solid pieces of congealed lava to exceedingly fluid lava flows. The shape of a volcano is determined largely by the type of volcanic product erupted, the distribution of vents and fissures in the cone, and surface processes that are eroding or otherwise modifying the slopes. Volcanologists classify volcanoes according to their morphology and their composition. Four major types of cones are distinguished. The first is the *shield volcano,* which has a very low profile. Slope angles seldom exceed about 10° and may be only a few degrees near the summit. The most impressive feature of shield volcanoes is their size. Newberry Crater in central Oregon (Chapter 15) is typical—its base diameter is about 20 miles. The "big island," Hawaii, consists of five overlapping shield volcanoes, including the active ones, Mauna Loa and Kilauea. The island rises almost 3 miles above sea level, but the gentle volcanic slopes descend more than 3 miles underwater to the sea floor, making this the tallest mountain mass in the world. The width of this shield volcano complex at its base (on the sea floor) approaches 1,000 miles!

The shield volcano is composed, invariably, of basalt. Basaltic lava is relatively fluid and hot, and the shape of the volcano reflects the ability of the lava to spread rapidly over great distances before it solidifies. The eruption of lava from long rifts or fissures that extend away from the summit crater also tends to produce a broad cone. (The extreme

TABLE 12-2 Early to Middle Tertiary stratigraphy, Mount Rainier National Park. (After Fiske, Hopson, and Waters, 1963.)

AGE	FORMATION	DESCRIPTION	THICKNESS, FEET
Oligocene to early Miocene (?)	Fifes Peak	Basalt and andesite flows; minor pyroclastic and clastic interbeds	>2,400
Oligocene to early Miocene (?)	Stevens Ridge	Rhyodacite ash flows; minor pyroclastic and clastic interbeds	450–3,000
Late (?) Eocene (base not exposed)	Ohanapecosh	Volcanic breccia, sandstone, siltstone; pyroclastic beds; mudflow breccias; local andesite and rhyolite flows	>10,000

(heavy line) Major unconformity

was reached in the so-called fissure eruptions that built the Columbia Plateau. In eruptions of this type, the basaltic lava was even more fluid than that of the shield volcano. The lava issues entirely from long rifts rather than central vents and as a result no cone was built at all.)

The *dome cone* is at the other extreme (built of the least fluid or most viscous lava). Here the lava erupts in a nearly solid state, so stiff that it can flow down only the steepest slopes. Often it issues forth as hot, jagged blocks, slabs, or spires. The temperature of the lava is lower than in basaltic eruptions, explaining some of the difference in fluidity. More important, the lava is of quite a different composition. It consists of silica-rich dacite or rhyolite rather than basalt; silica content has an important control on the viscosity of lava. Dome cones are normally very rough, irregular cones consisting largely of a great pile of loose debris with some jagged outcrops of solid rock projecting through. They hardly look like volcanoes at all. The most famous dome cone in the Cascades is Mount Lassen at the south end of the range, but there are many smaller dacite domes in Oregon and Washington. Some of them have a more symmetrical, bulbous form than Mount Lassen. They were formed by a single eruptive pulse of viscous dacite.

Volcanoes that erupt viscous lava are apt to be the more violent ones. All volcanoes give off considerable amounts of steam, produced by boiling off of surface water, groundwater, or water that was contained in the magma at depth. The pressure built up by the steam is a very important source of the energy of a volcanic eruption. If the lava is fluid, like basaltic lava, the steam can bub-

TABLE 12-3 Early to Middle Tertiary stratigraphy, central Cascades, Washington. (After Hammond, 1961.)

AGE	FORMATION		DESCRIPTION	THICKNESS, FEET
Early (?) Miocene	Cougar Mountain (unconformity)		Andesite flows and breccias; subordinate boulder conglomerate and volcanic sandstone	3,210–5,300
Late Eocene to early Oligocene	Keechelus Volcanic Group	Snow Creek	Pyroclastic breccias; minor andesite flows, sediments, and tuff	3,600–4,800
		Stampede Tuff	Dacite tuff; local pumice beds and tuff breccia	40–2,820
		Eagle Gorge Andesite	Andesite flows, flow breccia, and minor tuff	730–4,150
		Huckleberry Mountain	Tuff-breccia with minor volcanic sedimentary rocks, tuff, and andesite flows	2,200–11,100
		Enumclaw	Andesite flows and flow breccias; minor volcanic sediments and tuff. (Equivalent to Fifes Peak Formation)	3,600–9,600
	Mount Catherine Tuff		Rhyolitic to dacitic tuff, with minor volcanic sediments	1,420
Middle to late Eocene (?)	Guye		Sandstone, shale, chert, minor conglomerate. Nonmarine	>4,400

12-4 *Stonewall Ridge, south of Mount Rainier, looking west. The valley of the Cowlitz River is in the distance. The strata exposed in the ridges consist of Lower Tertiary volcanic rock typical of much of the range. (Copyrighted photograph courtesy of Delano Photographics.)*

12-5 *A volcanic bomb on the flank of a cinder cone in the Craters of the Moon National Monument, Idaho. The pattern on the surface of the bomb resulted from shrinkage of the crust during cooling. The bomb may have split on impact or at a later time. Note pen beyond bomb for scale. (Photograph by Bates McKee.)*

ble off without producing anything more spectacular than a lava fountain. If the lava is viscous, however, escape is not easy. An analogy would be trying to remove the cork from a bottle of champagne by somehow increasing the gas pressure inside the bottle. You might be able to slowly push the cork out in this manner or drain the gas pressure off by slow leakage and thus prevent violence. The result would probably be more spectacular, however. The cork might be blown out explosively or the neck of the bottle shattered, producing an "eruption" of champagne, gas, broken glass, and bits of cork. Dacitic volcanoes sometimes act like this, with violent eruptions that blow out hot, fiery mixtures of rock from the volcano (the glass), shattered fragments of the lava congealed in the volcano's throat (the cork), globs or pieces of new, hot lava (the champagne), and gas. Some of the Tertiary volcanoes in the Cascades had just this type of eruption. The extensive sheets of siliceous ash in, for instance, the Keechelus Volcanic Series were formed by eruptions of this type, which are called *glowing avalanches*.

A third type of cone, the *cinder cone,* is formed by a less cataclysmic type of eruption. Cinder cones are seldom more than 1,000 feet high. They have typically a rather symmetrical shape with smoother slopes of about 30° inclination (Figure 8-13). They are produced by the ejection from a central vent of small solid fragments of lava rock or globules of molten lava that solidify in flight. The solid ejecta from volcanoes is referred to as *pyroclastic* material or debris; and it is classified in order of decreasing size into blocks, cinders, lapilli, and ash. Large molten clots of lava cool in flight to form volcanic bombs, which have a characteristic rounded, streamlined form (Figure 12-5). The coarser blocks and bombs fall back into the vent or around its rim, while the finer material may be transported a considerable distance if the wind is blowing during the eruption. The cone enlarges primarily by a combination of upbuilding and downslope sliding of the ejecta. The characteristic 30° slope angle is essentially the angle of repose of this type of material; as more falls around the vent it must slide or roll down the flanks. (An analogy would be building a cone by pouring loose sand from a spout.)

Quaternary cinder cones are common throughout the Oregon Cascades and on the surface of the High Lava Plains to the east. Most of these cones are of basaltic composition.

The last major type of cone, the *stratovolcano,* or composite cone, is the one that best fits most people's idea of how a volcano should look. Its slopes are steepest near the summit and decrease uniformly toward the lower flanks of the mountain. Stratovolcanoes consist of alternating layers of lava (typically andesite) and pyroclastic material, dipping away from the summit crater. The classic form is produced when the eruptions feed from the same conduit throughout the history of the volcano. If the vent shifts, the result is an asymmetrical cone or two or more overlapping cones.

THE CASCADE LANDSCAPE 30 MILLION YEARS AGO

Geologists cannot study directly Oligocene volcanoes, for those cones largely disappeared long ago, the victims of erosion. Yet they can reconstruct the Oligocene landscape with some assurance, for the stratigraphic record of the Cascade Range has preserved the variety of volcanic products that were erupted. Moreover, mapping in some areas has located the lava congealed deep in the vents of some Oligocene volcanoes (Figure 12-6). Probably all of the major types of volcanic cones were present in the region, although not perhaps all active at the same time. The large amount of ash in parts of the section suggests periods of explosive volcanism. Some was deposited directly from glowing avalanches or fell cold from ash clouds. Much of the ash, however, was reworked by running water and came to rest finally in stream or lake beds. Single lava flows may have covered large areas—especially the more fluid basaltic flows, some of which probably issued from long fissures. Presumably, the more siliceous flows never spread far from their vent; they built stratovolcanoes or dome cones.

The sediment on the ground at any time consisted almost entirely of particles and fragments either blown or eroded from nearby cones, just as it does today in the High Cascades of Oregon. In fact, a trip today along the crest of the Oregon Cascades (Figure 12-7) might give quite an accurate feeling for what the region was like 30 million years ago. Then it was probably somewhat lower, the climate considerably warmer (the fossil leaves indicate this), and the animals very different. However, the variety of volcanic forms found today in this part of the Cascades also certainly existed in the Oligocene. The scarcity of fossils in Tertiary strata in the Cascades is probably due more to the unfavorable conditions for fossil preservation in these kinds of beds than to a meager flora or fauna. Certainly, many animals roamed the plains to the east in what is now part of the Blue Mountains Province (Chapter 14), and a considerable variety of marine life inhabited the continental shelf to the west.

MIOCENE OROGENY

Earlier in the chapter, we noted that a period of folding, faulting, and widespread erosion occurred late in the Eocene Epoch, between the time of deposition of the two major Cascade stratigraphic sequences. Following this, the region remained relatively stable for perhaps 20 million years. This is suggested by the absence of major unconformi-

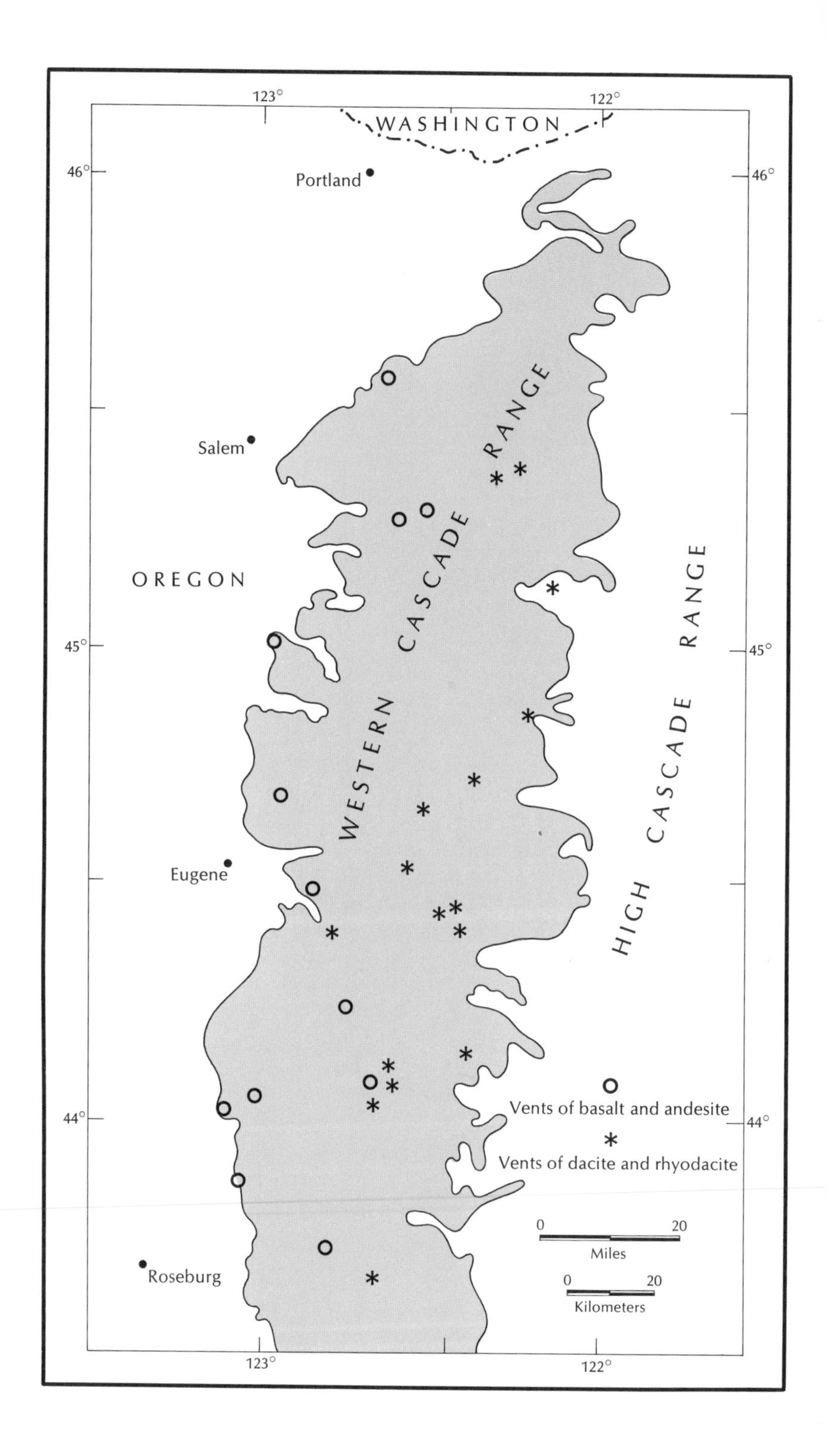

12-6 *The distribution of known eruptive vents of the Little Butte Volcanic Series, Cascade Range, Oregon. (After Peck et al., 1964, p. 21.)*

12-7 *The High Cascades of central Oregon, perhaps not unlike the scene here 30 million years ago. Mount Washington in the foreground, Three-fingered Jack in the middle distance, Mount Jefferson beyond. Glacial erosion has modified all three volcanoes, and this would not have occurred with Oligocene volcanoes. (Copyrighted photograph courtesy of Delano Photographics.)*

ties in the Oligocene and early Miocene stratigraphic record. Then, in about the middle of the Miocene Epoch, there began one of the major orogenies of the Cordillera—an orogeny that in some ways is continuing today. It began before the great eruptions of the Columbia River Basalt; and, in a few million years' time, it had produced great folds in the strata thousands of feet high and fault displacements of comparable magnitude. The actual effect of this deformation on the topography of the region was less pronounced, for erosion partly offset uplift. Perhaps no major ranges were produced—at least much of the Cascade region had been worn down to rolling hills by the time of the first outpourings of basalt.

The cause of the Miocene orogeny is unclear. The trends of the folds and faults were not everywhere the same, for although a general northerly trend characterized these structures in Oregon, those in Washington strike northwest, diagonal to the present range. The orogeny also deformed strata in other regions, including the Coast Range, the Blue Mountains, and central Idaho. Whatever the cause, clearly it acted over a large area.

The emplacement of granitic batholiths and stocks roughly coincided with the folding and faulting. The North Cascades had witnessed essentially continuous crystallization of intrusions since the Cretaceous Period, but intrusions in the range to the south are primarily of middle Miocene age. The largest of these are in Washington, principally the Snoqualmie Batholith (which lies mostly in the North Cascades), the Tatoosh Pluton under Mount Rainier (Figure 12-8), and the Bumping Lake Pluton 25 miles east of Mount Rainier. Smaller stocks are exposed locally along the entire western flank of the Cascade Range. Radiometric age dating of the Snoqualmie and Tatoosh intrusions indicates emplacement about 14 million to 18 million years ago. Probably these were shallow intrusions that crystallized near the surface. (The andesite lava at The Palisades in Mount Rainier National Park appears to have come from fissures that connect directly with the Tatoosh Pluton.) The intrusions may have been formed by the rise of magma from deep chambers that had supplied some lava to the Oligocene and early Miocene volcanoes. This possibility is suggested by the chemical similarity between the intrusive granodiorite and the andesite of the intruded Tertiary section. Whatever the origin of the magma, its generation at depth has continued since the Miocene, for andesitic lava has continued to erupt in the Cascades to the present day.

THE RISE OF THE CASCADE RANGE

The modern Cascade Range has been built by Late Cenozoic uplift along a north-south axis. Whatever the reason, this uplift is only a part of a truly grand orogeny (the Cascade Orogeny) that has produced much of the present physiography of the North American Cordillera. The nature of the uplift is most clearly seen in southern Washington and northern Oregon, where the Columbia River Basalt has not been stripped from the range by erosion. The basalt is discussed in detail in Chapter 17; all we need to know here is that the formation consists of many separate lava flows, each of which was exceptionally fluid and spread almost like water. Consequently, the top surface of any flow was essentially horizontal after flowage had ceased; any slope now observed on the top of a flow reflects post-Miocene deformation.

The Columbia River Basalt is exposed along both walls of the Columbia River Gorge all the way through the Cascade Range. The section, which is about 2,100 feet thick, climbs gradually westward out of the Columbia Plateau to a crest near Bonneville Dam. From there it descends until finally it disappears under younger strata at the west end of the gorge. The total structural relief of the arch in this part of the range is about 2,800 feet. In other words, the axis of the Cascades has been raised (relative to the adjacent lowlands) this amount since late Miocene time. Undoubtedly, this is close to a minimum figure for Cascade uplift,

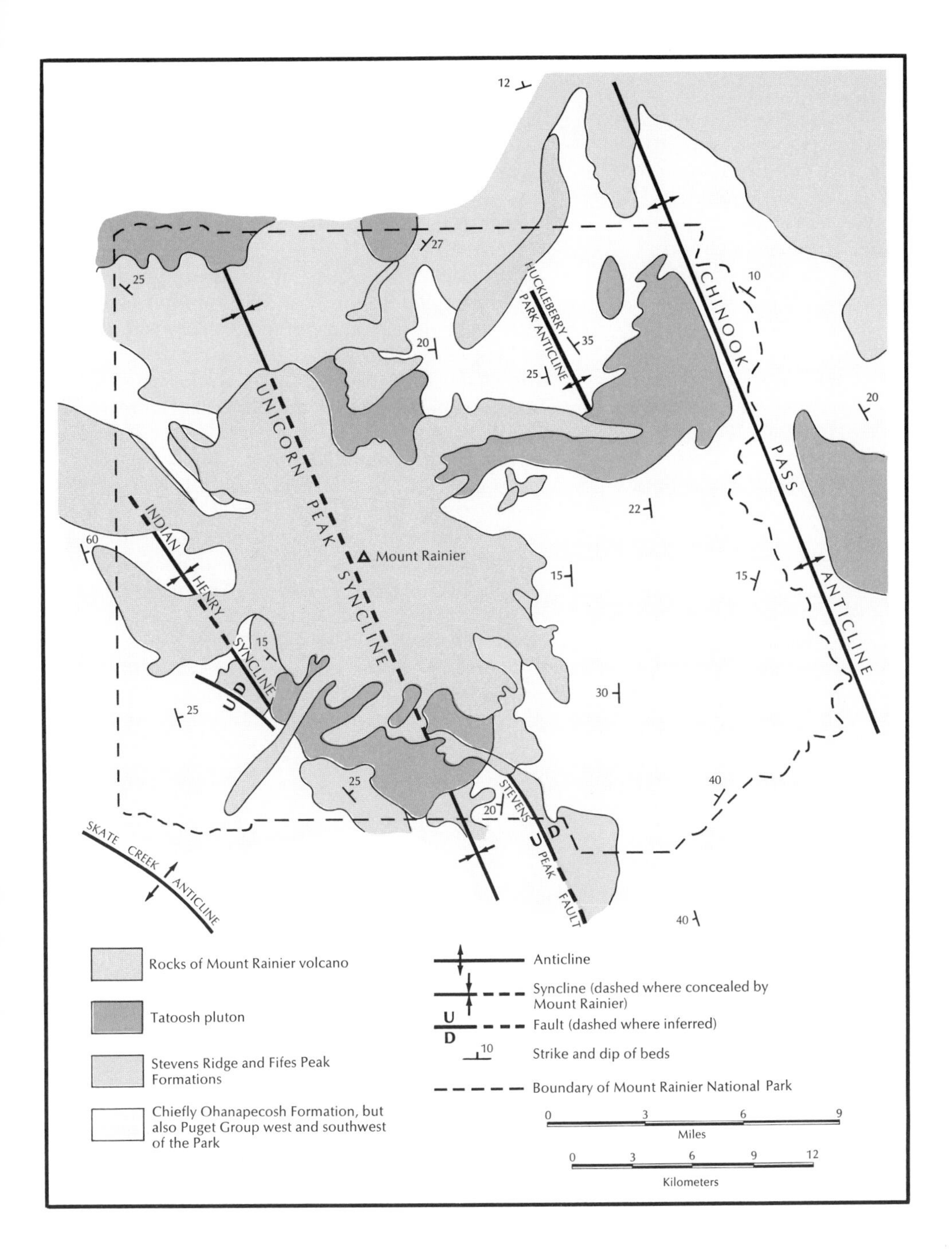

12-8 *Generalized geologic map of Mount Rainier area. (After Fiske, Hopson, and Waters, 1963, p. 38.)*

however, since the region around the Columbia River Gorge is the lowest part of the range. To illustrate this, basalt flows in the Wenatchee Mountains of central Washington rise westward into the Cascades at about 8° to an elevation of 7,000 feet, about a mile higher than the nearby Columbia Plateau. Furthermore, the Wenatchee Mountains are far to the east of the axis of the arch. Unfortunately, erosion has stripped the Columbia River Basalt from the axial region of the Washington Cascades.

South of Mount Hood, evidence for the Cascade Arch is less satisfactory. Common sense would suggest that since the Cascade Range is a moderately uniform physiographic feature, its origin should remain similar along its length. Unfortunately, the High Cascades of central and southern Oregon contain such a thick section of Pliocene and Quaternary volcanic rock that the Columbia River Basalt is concealed. Perhaps it never did reach farther south than the vicinity of Bend. At any rate, at least some of the southern part of the Cascade Range owes its elevation to the accumulation of Late Cenozoic lava and pyroclastic debris rather than uplift.

The Columbia River Basalt in the gorge also demonstrates that the post-Miocene deformation of the region has involved more than a simple uparching of the Cascades. Between Hood River and The Dalles four very large folds in the basalt trend northeast across the Columbia River (Figure 12-9). Pliocene beds are also folded. Along the western margin of the Columbia Plateau, similar folds trend east-west or northwest, essentially at right angles to the Cascade Arch. These trends are the same as those of the Miocene orogeny. They indicate that the stress field of the Miocene, whatever its origin, continued to operate for the remainder of the Tertiary Period, and probably is still active today (Chapter 17). The Late Cenozoic deformational history of the Cascades must ultimately explain the superimposition of the Cascade Arch across this older stress field.

12-9 *Flows of Columbia River Basalt, dipping off the flank of a fold that trends northeast, diagonal to the Cascade Arch and the general course of the Columbia River. View looking northeast toward Lyle, Washington. (Photograph courtesy of John Whetten.)*

THE PLIOCENE RECORD

The geography of the Cascade Region in the Pliocene Epoch is surprisingly difficult to interpret. The Cascade Arch probably began its rise at this time, for by the middle of the epoch considerable quantities of volcanic sand and gravel were being washed from the range and deposited along its flanks. These deposits are especially prominent east of the Cascades. They constitute the Ellensburg Formation of south-central Washington (Figure 12-10), the Dalles Formation at the east end of the Columbia River Gorge, and the Deschutes Formation of north-central Oregon. Each consists of a thousand feet or more of stream deposits. Two pebble types predominate—andesite and white pumice. Sedimentary structures demonstrate clearly that at least the Ellensburg and the Deschutes Formations were deposited by east-flowing streams, presumably draining the east flank of the Cascades. The composition of the sediment suggests considerable contemporaneous volcanic activity in the range, especially pumice eruptions. Ironically, were it not for the preservation of these strata marginal to the Cascades, Pliocene volcanism would be much more difficult to prove. In the Cascade Range of southern Washington (Figure 12-3), few strata younger than the Miocene intrusions and Columbia River Basalt have been identified, except for the Quaternary cones like Mount Rainier and Mount Saint Helens and the young lava field around Mount Adams. Similarly in the Oregon Cascades, formations positively identified as being of Pliocene age are rare, although the older lavas of the Cascade Andesite Formation, which covers much of the High Cascades, may be as old as Pliocene (Figure 12-11). Even if most of the Pliocene cones were worn away by erosion, the throats of the volcanoes should be still recognizable. Beacon Rock, on the north side of the Columbia River Gorge 5 miles west of Bonneville Dam, has been interpreted as the eroded core of a volcano that helped to produce the Cascade Ande-

12-10 *Tilted river beds of the Pliocene Ellensburg Formation, overlain by Pleistocene gravel. The light color of some of the layers is due to abundant pumice fragments, derived from volcanoes in the Cascades to the west. The tilting reflects continued folding along Miocene trends. View is to the southeast, down the Wenas Valley toward Yakima. (Photograph by Bates McKee.)*

site. Wind and Shellrock Mountains, 15 miles to the east, are thought to have had a similar origin. Undoubtedly, future work in the Cascades will locate additional vents of possible Pliocene volcanism.

THE ANCESTRAL COLUMBIA RIVER

How old is the Columbia River? Northwest geologists have been wrestling with this question for many years, but as yet have not come up with a wholly satisfactory history for the river. Rivers comparable to the Columbia must have existed throughout much of the Cenozoic Era, for drainage from the interior of the continent must have passed through the volcanic Cascade region to reach the Pacific. This is suggested by the nature of some of the Tertiary strata of the Coast Range, especially metamorphic and granitic pebbles in formations like the Astoria around the mouth of the modern Columbia River. Such pebbles had to have come from a region like the Blue Mountains or Columbia Mountains, far to the east of the Cascade Range Province.

The Cascade Arch grew across the paths of these west-flowing rivers—a situation identical to the rise of the Coast Mountains in British Columbia (Chapter 8). The rivers that could erode downward at a rate equal to the rate of uplift survived;

12-11 *Jointed flows of the Cascade Andesite Formation along U.S. Highway 20 near the crest of the Cascade Range in northern Oregon. (Photograph by Bates McKee.)*

12-12 *The Columbia River Gorge, looking downstream toward Stevenson, Washington (by the distant bend in the river). Wind and Shellrock Mountains, probable Pliocene volcano remnants, flank the river in the middle distance. The Columbia River cut the gorge as the Cascade Range rose. (Copyrighted photograph courtesy of Delano Photographics.)*

less vigorous rivers were blocked and were diverted elsewhere. The Columbia River apparently survived, while other rivers crossing what is now the Cascade Range did not. (Its success was no doubt aided by the fact that the Cascade Arch is lower here than it is to the north or south.)

The antecedent nature of the Columbia relative to the Cascades is suggested also by the stratigraphy. The Pliocene Dalles Formation contains, in addition to volcanic pebbles, stones of granitic and metamorphic rock types that must have been transported across the Columbia Plateau Province. Similar crystalline pebbles are common in the Troutdale Formation (Table 12-1). It consists of stream gravels of Pliocene age and is thickest at the west end of the Columbia River Gorge. Thus on both sides of the arch there is good stratigraphic evidence of an ancestral Columbia River in the approximate location of the present river by the middle of the Pliocene Epoch. How long it had been there is not clear.

We will reenter the Columbia River Gorge in Chapter 17, following the course of gigantic torrents of water that swept across the Columbia Plateau and through the gorge in the Pleistocene Epoch. Now it is time to look at the great volcanoes that have been built on top of the Cascade Range during the Quaternary Period.

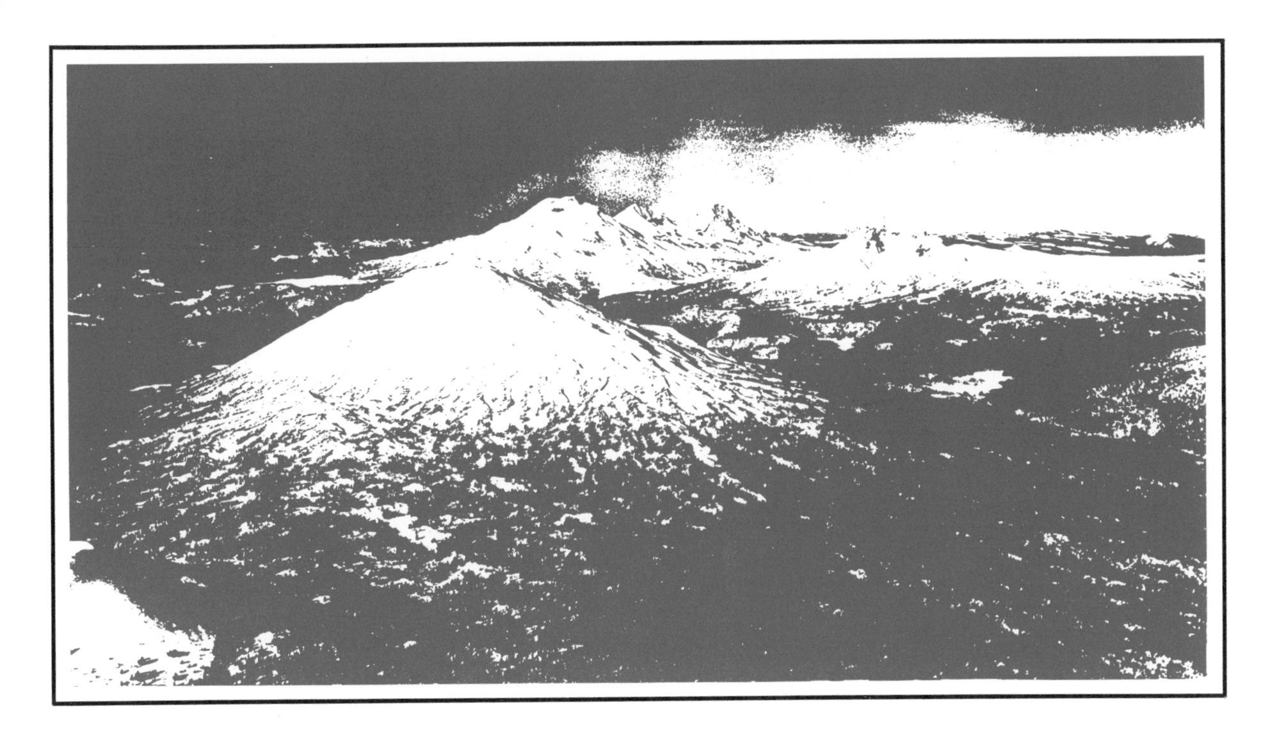

13 the cascade volcanoes

■ The young volcanoes of the Cascade Range are part of a nearly continuous ring of volcanoes that surround the Pacific Ocean. Rather close parallels exist between the Quaternary histories of such distant regions as the Andes, the Cascades, the Aleutians, Japan, and some of the island chains of the southwest Pacific. The volcanoes of these regions are of much the same type, and knowledge gained by the study of volcanism in part of the Pacific margin is more or less applicable to other parts.

Some regions of the Pacific rim have been studied in considerable detail—others are virtually unexplored. Our knowledge of the Cascade volcanoes is likewise spotty.

It is impossible to talk about the "typical" Cascade volcano for each one is unique. On the other hand, the history of every cone could fill many volumes, so this chapter is something of a compromise. It starts with generalizations that apply to most of the cones, then looks at the history of some particular ones. In this way, the chapter will build a composite picture of a Cascade volcano, drawing on features of particular interest found on individual cones.

SOME GENERAL CHARACTERISTICS

The elevations of the principal volcanoes of the Cascades are shown in Figure 13-1. The cones rise generally about a mile above the average elevation of the surrounding range. In the southern part of the Oregon Cascades, where deeper erosion or (especially in the case of Mount Mazama or Crater Lake volcano) eruptive violence has reduced the original size of the volcano, the relief is somewhat less. In general, the volcanoes have the classic form of a stratovolcano with rather steep upper flanks grading into more moderate lower slopes. This shape has been more or less modified by glacial erosion and by some shifting in position of eruptive vents. The higher cones still nurture alpine glaciers; and most of the lower ones have been glaciated in the past, presumably at times of maximum Pleistocene glaciation.

Geologists are still somewhat uncertain as to the age of the volcanoes. They have been regarded as Quaternary, which means they formed during the past several million years. Stratigraphic evidence for this is not very good. The lavas of some cones rest unconformably on rocks as young as middle Pliocene and fill deep canyons that had been cut into a mature Cascade Range. Since the Cascade Range itself is a young uplift, and since canyons thousands of feet deep had been cut into this range before formation of the volcanoes, the latter are probably Quaternary. As radiometric age measurement techniques are improved, this uncertainty will disappear. Unfortunately, at the present time the least reliable age determinations come from rocks that are 50,000 to 1,000,000 years old. Consequently, geologists seldom try to date rocks suspected to fall within this age range.

The time scale for magnetic reversals (Chapter 4) offers a new method for dating the Cascade volcanoes. The earth's magnetic polarity has generally been "normal" (that is, with the north magnetic pole near the north geographic pole) for the past 700,000 years. Before that, back to about 2.5 million years ago, the polarity was mostly "reversed." The position of the earth's polarity is fixed in a volcanic rock as it cools past a certain temperature. This position can be measured in the rock, either in the laboratory or in the field. Thus if a volcanic rock from a Cascade volcano has a normal magnetic polarity it cooled, in all probability, less than 700,000 years ago, while a reversed polarity would suggest an older age. This technique has not been applied systematically to many of the Cascade volcanoes, but the polarity of those flows studied has been normal.

The question of the age of the most recent eruptions on the Cascade volcanoes is answered more easily, for the radiocarbon method allows accurate dating of organic material back to approximately 50,000 years ago. Unfortunately, organic material usually is destroyed in a hot lava flow, but volcanic ash beds commonly bury carbonaceous material. The age of a lava flow can often be bracketed by dates obtained from sedimentary or ash beds above and below the flow. Radiocarbon dating, combined with detailed stratigraphic studies, has produced a well-documented history of recent events on a few volcanoes, most notably Mount Rainier.

The relationship of the volcanoes to the last major episode of glaciation also has provided an approximate measure of recent volcanic activity on some of the cones. In general, most of the volcanoes were built primarily before the last major glacial advance, approximately 15,000 years ago. Valleys that begin high on the cones show evidence of glacial shaping and erosion from their upper ends (where glaciers may exist today) many miles down the flank of the volcano. Presumably this

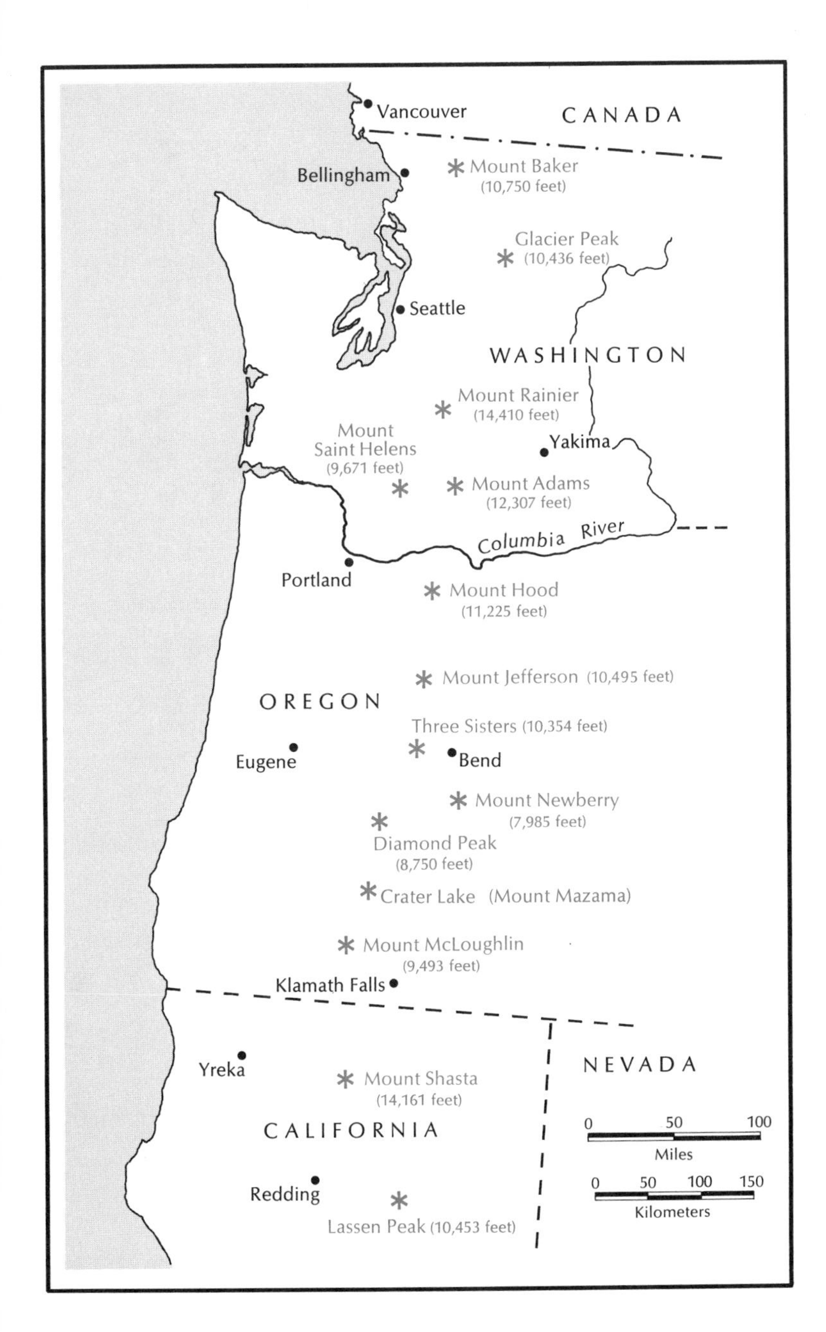

13-1 *The major Quaternary volcanoes of the Cascade Range.*

occurred during the last glacial epoch. Some cones, such as Mount Washington (Figure 12-7), have little evidence that suggests eruption since that time. Other cones, such as Rainier and Hood, have not been entirely quiescent since the last glacial maximum but have had only relatively minor eruptions since then. Finally, several cones, most notably Mount Mazama (Crater Lake) and Mount Saint Helens, owe most of their present form to postglacial volcanism.

Volcanologists like to classify volcanoes in various ways. Chapter 12 introduced a morphological classification. Another scheme classifies volcanoes according to whether they are active, dormant, or extinct. An active volcano erupts fairly frequently. Clearly, none of the Cascade cones qualify as active volcanoes, since the last Cascade eruption was in 1915 when Mount Lassen put on a rather violent and spectacular display. At the other extreme, an extinct volcano is one that has not erupted for a long time and is not expected to erupt again. This is a somewhat subjective determination, and sometimes an "extinct" volcano erupts unexpectedly. The basis for predicting whether a particular cone is dead or only asleep (that is, dormant) is a knowledge of its past history. Only if we know that a volcano has not erupted for tens of thousands (or better yet, millions) of years can we confidently rule out future activity. The five major volcanoes of Washington and most of the cones of Oregon are best categorized as dormant. This assessment can be extended eastward to include the many young volcanic centers of southeastern Oregon (Chapter 15) and the Craters of the Moon area of Idaho (Chapter 16). After we have looked at the characteristics of some past eruptions in the

13-2 *Mount Adams, a typical stratovolcano in the southern part of the Cascade Range of Washington. Note the modification of the volcano's classic shape by glacial erosion. (Photograph courtesy of the Washington State Department of Commerce and Economic Development.)*

13-3 *The Three Sisters, central Oregon, viewed from the northwest. North Sister, on the left, has been deeply scarred by glaciation and has not erupted for a long time. The same applies to Broken Top (in the distance at left) and Middle Sister (in line with taller peak of South Sister in middle distance). South Sister, though moderately eroded by glaciers, has a young summit crater, and Bachelor Butte (in the distance between North and South Sisters) has not been glaciated at all. Note the fresh lateral moraines of Collier Glacier west of North Sister, and breached young cinder cone in the left foreground. (Copyrighted photograph courtesy of Delano Photographics.)*

Cascades, we will be in a better position to discuss the possible hazards posed by these sleeping giants.

Most of the Cascade volcanoes have been built by alternating eruptions of andesitic lava and pyroclastic debris. The andesite is typically medium gray and may contain large gray crystals of the mineral plagioclase. The overall composition of the andesite is similar to that of the most prominent types of igneous intrusions we find in the Northwest—diorite, quartz diorite, and granodiorite.

The degree of compositional variety of lavas varies markedly from one Cascade volcano to the next. For example, andesite eruptions have characterized the history of Mount Rainier from eruption of the earliest flows to the most recent ones. Nearby Mount Saint Helens, on the other hand, has erupted a complex mixture of basalt, andesite, and dacite. Other cones consist primarily of basalt or of dacite, although some andesite is found in all the larger Cascade volcanoes.

Where erosion has cut valleys into the flanks of the volcanoes, the internal layering of the cone is quite obvious. The strata dip away from the vent at angles up to about 30°. They consist of alternating layers of lava and pyroclastic debris. From a distance, the pyroclastic layers look quite dark compared to the lava flows (Figure 13-4). On closer inspection, the most pronounced difference is the even columnar structure or jointing of many of the flow layers. This contrasts markedly with the irregular fractures seen in the more massive pyroclastic layers.

13-4 *Alternating layers of dark pyroclastic debris and light-colored andesite in a glaciated valley wall, Mount Mazama. Southeast rim of Crater Lake is just out of sight to the right. (Photograph by Bates McKee.)*

13-5 *Table Mountain, underlain by massive andesite flows from Mount Baker which rest unconformably on crystalline rocks of the Shuksan Metamorphic Suite, exposed in the foreground. (Photograph courtesy of Howard Coombs.)*

ORIGIN OF THE LAVA

The origin of the lava that has built the great volcanoes around the Pacific rim is a complex subject that is not well understood. For many years, the leading theory on lava (magma) genesis called for a "parent" magma of basaltic composition. This magma was derived by melting either of the earth's mantle, as happens beneath Hawaii, or of the deeper part of the crust. More silicic lava types (andesite, dacite, and rhyolite) were considered to have separated from the basaltic parent magma by various processes of magmatic fractionation. An obvious objection to this theory arises in a region like the Cascades, where the total quantity of lava of intermediate and silicic composition exceeds that of the basaltic parent. This problem is even more acute for the great volume of older Cascade intrusions of intermediate composition.

One alternative theory called for contamination of the basaltic magma by the melting and incorporation of surrounding rocks as the magma rose to the surface. Another suggested that the magma forms by partial or complete melting of the crust in orogenic regions. Both of these theories are reasonable, and undoubtedly these processes do occur. Serious problems arise, however, when they are invoked to explain the origin of the Cascade lavas. One is the relative uniformity of lava types from one volcano to the next, despite major apparent differences in the thickness and composition of the underlying crust. For example, the lavas of Mount Baker and Glacier Peak rose through a great thickness of pre-Tertiary metamorphic and granitic rocks. Lava in Cascade cones to the south passed through a crust that consisted primarily of Tertiary volcanic rocks. Yet these differences in the nature of the bedrock beneath the volcanoes are not reflected in the chemistry of the erupting lava, which is essentially the same.

Geologists in recent years have considered seriously the possibility of a parental magma of andesitic composition, produced by some degree of melting of the mantle beneath continental margins. In this theory, both the basaltic and the dacitic lavas of some stratovolcanoes would be products of the fractionation of this andesitic parent magma. The problems of the relative volumes of the lava types and the apparent lack of bedrock control on lava composition would be explained. Several models involving the underthrusting of continents by oceanic crust have been proposed to produce the andesitic magma.

This is a "hot" subject at present. With all the brain cells now at work, right answers will surely be forthcoming.

MOUNT BAKER

Mount Baker rises about 5,000 feet above the surrounding peaks of the North Cascades to an elevation of 10,750 feet above sea level. The early flows were andesites. They filled deep canyons cut into various Paleozoic, Mesozoic, and Lower Cenozoic sedimentary and volcanic formations (Figure 13-6). Subsequent eruptions varied little in composition, and the cone is a rather ordinary stratovolcano in most respects. Perhaps the most significant event in Mount Baker's history was the shift of the eruptive vent several miles to the east. The older lavas and pyroclastic layers dip symmetrically away from Black Butte, a jagged, deeply eroded plug of andesite that marks the original vent. The eruption of new lavas to the east built a cone that now overshadows the original one (Figure 13-6). It contains all the more recent flows. The age of this new vent is not known, but judging from the deep glacial erosion of the modern cone it probably preceded the last major glacial episode, approximately 15,000 years ago. No proof of significant eruptions in the past few thousand years has been found. The last known eruption of this volcano was in the early 1840s when it emitted some ash and considerable steam. Minor amounts of steam issue today from the summit crater. Mount Baker is not one of our more thoroughly studied volcanoes. It may prove to be more interesting than it now appears.

13-6 *Mount Baker from the east. Crags to the left of the main cone are erosional remnants of the original cone. (Photograph courtesy of Washington State Department of Commerce and Economic Development.)*

LONELY GLACIER PEAK

This volcano, one of the least accessible cones in the entire Cascade Range, has been studied only in the past few years. The cone lies immediately west of the crest of the range. It has been more thoroughly eroded by glaciers than any of the other volcanoes in Washington.

Several aspects of Glacier Peak's history are unique. First, it was built on a high ridge, so that the thickness of the main cone is only a few thousand feet. Second, the eruptions consisted almost entirely of dacitic lava, without much pyroclastic debris or flows of basaltic or andesitic composition. Finally, Glacier Peak erupted with considerable force approximately 12,000 years ago. An immense blanket of volcanic ash drifted to the east and northeast during this eruption. It was second in volume only to the younger Mazama Ash from Crater Lake. Other aspects of the volcano are also noteworthy. This is one of the Cascade volcanoes that sits on top (partially) of a relatively young intrusion, the Cloudy Pass Batholith of Miocene age. Several Pliocene eruptions of lava bridge the time gap between the batholith and the birth of Glacier Peak in the Quaternary. Recent studies suggest a possible direct relationship between these various magmatic events, perhaps even a common magma source at depth.

The oldest flows from Glacier Peak have a normal magnetic polarity, indicating that they are younger than 700,000 years old. The cone building was largely finished by the last glaciation. Since then, a younger eruption of viscous dacite high on

13-7 *The ice-mantled east face of Glacier Peak. (Photograph courtesy of Stephen Porter.)*

the south flank built the satellitic dome known as Disappointment Peak, and a similar eruption on the east flank (presumably buried now under Chocolate Glacier) generated large mudflows. These built a great debris fan that fills the upper reaches of the Suiattle River. This probably preceded the eruption of the Glacier Peak Ash, which occurred approximately 12,000 years ago. This ash has been found as far east as central Montana and as far northeast as central Alberta. Surprisingly no trace of it is found west of the volcano except for layers of ash and pumice that washed off Glacier Peak and filled the upper part of the Whitechuck River. One can only conclude that the volcano put out a great amount of ash in a short period of time, or else the wind direction was constant (from the west and southwest) for a long time. Unlike Mount Mazama, Glacier Peak was not destroyed by the cataclysmic eruption of ash and associated volcanism, for the mountain does not have any great summit basin or *caldera* such as Crater Lake. Glacier Peak's principal crater has been breached by glaciers, especially on the east side of the cone, and is scarcely recognizable. The eruption of Glacier Peak Ash was apparently the last major eruption from the cone, although minor eruptions have occurred on and around Glacier Peak since. Of all of Washington's young volcanoes, Glacier Peak has had probably the least activity in recent times. It might be considered extinct, but we really do not know how long one of these Cascade cones can lie dormant between eruptions. Even 12,000 years of inactivity is probably not enough time to prove extinction.

MOUNT RAINIER—KING OF THE RANGE

Only a modest amount of prejudice enters in my assessment of Mount Rainier as one of the most magnificent mountains in the world. Its summit elevation, 14,410 feet above sea level, puts it far down the list of tall mountains. Few peaks so completely dominate a region, however. The cone lies west of the Cascade crest; and, although a few neighboring peaks rise to almost 7,000 feet, they are overshadowed completely by Rainier (Figure 1-4). The volcano's upper regions are mantled by glaciers, which cover 37 square miles of the mountain. When viewed from afar, the glaciers appear to form a gigantic white cloud resting on an immense dark base. What makes Mount Rainier so awesome is that it rises almost 3 miles above the Puget Lowland. Moreover, local weather conditions are such that the mountain is not always "out"; and, when it is, Rainier is a conversation piece. (They say you can predict the weather by Mount Rainier. When you can see the mountain, it is going to rain. When you cannot see it, it is raining.)

Mount Rainier is one of our more studied volcanoes; we have an especially thorough history for the past 10,000 years. The time of the oldest eruptions is unknown, but it was certainly many tens of thousands and perhaps hundreds of thousands of years ago. These early flows of Rainier, like those of Baker and Glacier Peak, had to fill canyons several thousands of feet deep. Some of the first flows traveled as far as 15 miles from the erupting vent, but later flows were generally more viscous and spread only a few miles. The bedrock under the volcano consists of Lower to Middle Tertiary volcanic and sedimentary rocks, intruded by Miocene granodiorite of the Tatoosh Pluton (Chapter 12). The fact that some of the early canyon-filling flows now cap ridges suggests that considerable uplift has occurred since the volcano's birth. This reversal of topography resulted from a greater resistance to erosion of the solidified flow than of the bedrock. Because of this, ridges of older bedrock were eroded below the level of adjacent valleys, which were protected by their fill of lava (Figure 13-8). Many Cascade volcanoes have these types of complex stratigraphic relationships brought about by the interplay of erosion and eruption. Geologists do not have detailed information about the internal structure and stratigraphic sequence of any of the large Cascade stratovolcanoes.

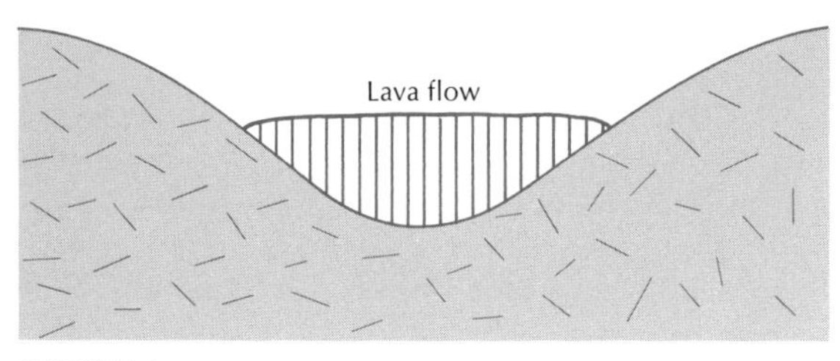

SECTION A

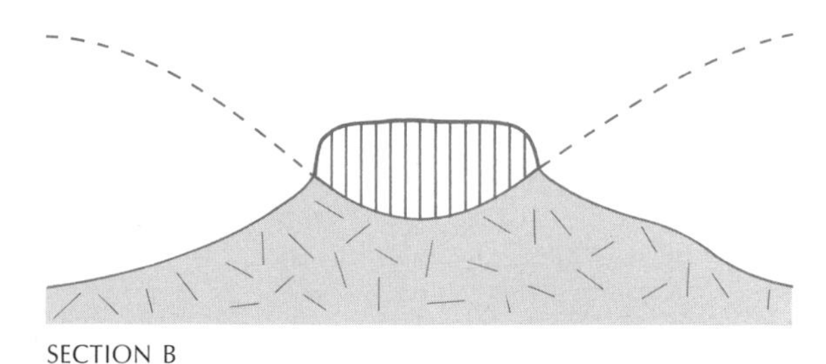

SECTION B

13-8 *Cross-sectional views that illustrate reversal of topography. In view A a valley has been partly filled by a lava flow. After considerable erosion (view B) the lava caps a ridge because the softer walls of the original valley have been worn down faster than the lava flow.*

Mount Rainier consists largely of andesite. Little apparent difference in composition is found from the earliest eruptions to the latest. The cone is a typical stratovolcano, with alternating layers of lava and pyroclastic debris dipping steeply away from the central vent. The layering is particularly prominent in the steep headwalls of several of the glacially scoured amphitheaters on the north side and in the flanks of Little Tahoma Peak, a remnant of the original cone preserved as a "*cleaver*" between the Emmons and Ingraham Glaciers 2 miles east of the summit. Because glacial erosion has removed so much of the original cone, geologists presume that the major eruptions largely preceded the last glaciation.

The modern summit of Mount Rainier is several thousand feet lower than was the original summit. Yet there is a well-defined summit cone (actually, several overlapping cones; Figure 13-9). These young cones have been built within a rather large depression in the main cone—a basin that resulted from some combination of eruptive violence, collapse, and glacial erosion. This collapse and partial refilling of the depression explains the more rounded and broad-shouldered profile of Mount Rainier compared with many of the other Cascade volcanoes.

Studies by geologists of the United States Geological Survey during the past several decades have made Mount Rainier one of the foremost laboratories in the world for examination of the environment of a volcano. Eruptions during the past 10,000 years have been infrequent. They have consisted primarily of pumice and ash falls and have not been voluminous compared to the eruptions that produced the Glacier Peak Ash or the Mazama Ash. The interplay between eruptions and erosive processes has been documented particularly well. Several aspects of the young history of Rainier are noteworthy. One is the frequency of very large mudflows. Another is the detailed stratigraphy that has been established by careful study of the many ash layers that blanket different flanks of the volcano. The source vent for most of these ash eruptions has been identified. Many, of course, erupted from Mount Rainier, but some came from Mount Saint Helens, and one, the Mazama Ash, came from the Crater Lake area. Finally, the mountain's history of recent alpine glaciation has been investigated. Certainly, Rainier is not the only volcano in the world that is suitable for these types of investigations. However, many of the particular techniques of study have been perfected on Rainier.

Numerous mudflows, some of tremendous volume, have swept down the flanks of Mount Rainier. Volcanoes are natural environments for the generation of large mudflows. They are especially common on cones that are active and covered partially by ice and snow. In this situation, eruptions produce, by melting, large quantities of water in a short time. Furthermore, torrential rains are commonplace around erupting volcanoes, since large amounts of steam rise, cool, and condense. The solid debris of the mudflow is provided also by the volcano—either directly by pyroclastic eruption or indirectly as the rock debris mantling the cones is eroded by water runoff. Vegetation, which tends to retard runoff and slow erosion, is often destroyed by killing ash falls or by fires begun by eruption. (The effectiveness of veg-

13-9 *The summit of Mount Rainier from the southeast. Note the several young cones contained with a depression in the summit of the original cone. (Photograph courtesy of U.S. Geological Survey.)*

etation in retarding erosion is demonstrated dramatically in the hills of southern California. Winter mudflows follow summer fires with rather dreadful predictability.)

The best-studied mudflow from Mount Rainier occurred approximately 5,000 years ago. Termed the Osceola Mudflow, it originated high on the northeast slope of the cone and poured down the valleys of both the White River and its west fork to their intersection. Reunited, the mudflow swept on beyond the mountain front at Enumclaw, approximately 30 miles northeast of the summit. It then spread as a great apron of mud across the lowland, burying the present sites of Enumclaw, Buckley, Kent, Auburn, Sumner, and Puyallup. The total volume of mud was perhaps 2.5 billion cubic yards; the thickness of the mudflow deposit in the lowland is as much as 70 feet. The mud of the Osceola flow may have been produced primarily by the alteration and subsequent ejection of the upper part of Mount Rainier. This theory supposes that the hot rock material of the volcano slowly changed to clay as the volcano "stewed in its juices." This neatly explains the missing top of Mount Rainier. If the relative volumes do not quite match, there were other large mudflows at about the same time as the Osceola. One covered the Paradise Park area on the south flank and swept on down the Nisqually River valley for at least 18 miles.

Mudflows probably have coursed down all the valleys that originate on the mountain, including some certainly younger than the Osceola and Paradise flows. Approximately 600 years ago, the Electron Mudflow traveled westward down the Puyallup River and buried the present site of Orting, 25 miles northwest of Mount Rainier. Not all of Rainier's mudflows have been generated by volcanic eruptions. Some were initiated by rockfalls and avalanches, and others by the sudden outburst of meltwater trapped within glaciers. An example of the latter kind occurred in 1947 in the valley of Kautz Creek, on the southwest side of the mountain.

Rockfalls and rock avalanches represent smaller but no less significant processes at work wearing down Mount Rainier. A dramatic example occurred during December of 1963. A series of seven rockfalls, with a total volume of approximately 14 million cubic yards, cascaded off the north face of Little Tahoma Peak onto the Emmons Glacier, several thousand feet below. The falls were not observed, but their short-lived histories were reconstructed by a careful study of the blanket of rock debris that covered the glacier and the head of the White River valley beyond. The falling rock and snow were propelled downvalley as a rock avalanche by the slope of the Emmons Glacier. It formed a very buoyant cloud of debris that traveled more than 4 miles and contained rocks weighing hundreds and even several thousand tons. An estimate of the velocity was made by noting the height the debris reached as the rock avalanche banked off obstructing walls. The speed attained was greater than 60 miles per hour—perhaps as much as 100 miles per hour. These

TABLE 13-1 Pumice and ash beds in Mount Rainier National Park. (After Crandell, 1969, p. 10.)

PUMICE LAYER	MAXIMUM THICKNESS, IN.	COLOR	SOURCE	APPROXIMATE AGE, YEARS
X	1	Lt. olive gray	Mt. Rainier	100–150
W	3	White	Mt. St. Helens	450
C	8	Brown	Mt. Rainier	2,150–2,500
Y	20	Yellow	Mt. St. Helens	3,300–4,000
D	6	Brown	Mt. Rainier	5,800–6,600
L	8	Brown	Mt. Rainier	5,800–6,600
O	3	Yellow orange	Mt. Mazama	6,600
R	5	Red brown	Mt. Rainier	8,750–11,000 (?)

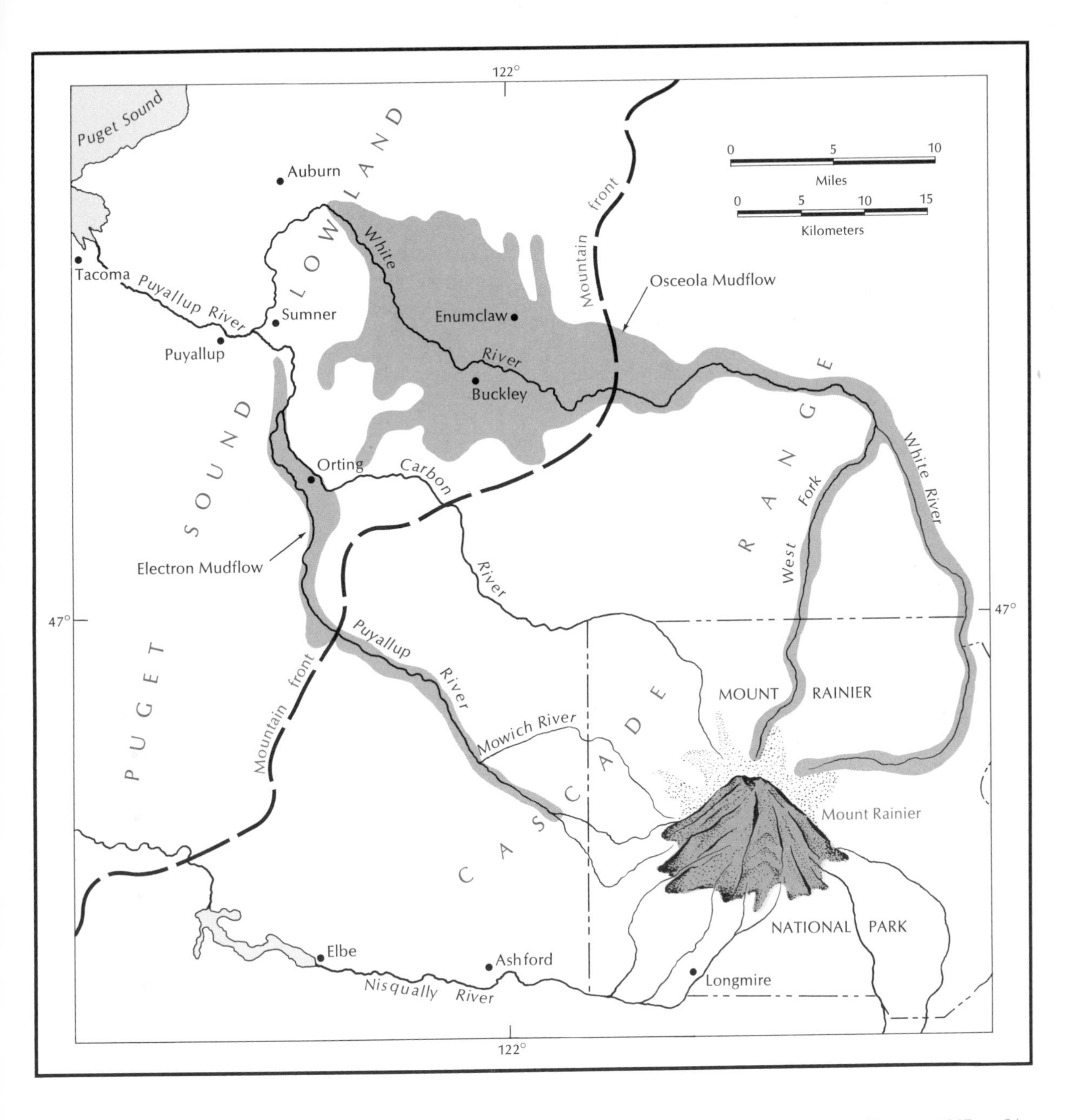

13-10 *Distribution of the Osceola and Electron Mudflows from Mount Rainier. (After Crandell and Mullineaux, 1967, p. 2.)*

were not mudflows—in fact, water had little to do with them. Instead, they were turbid clouds of rock debris, flowing downhill on a cushion of air trapped and compressed beneath the rapidly moving front. Similar rock avalanches have occurred recently in the Andes and in the earthquake-generated Madison Canyon slide near Yellowstone National Park in 1959.

When was the last eruption of Mount Rainier? This question has received much attention; and, when we contemplate the probable consequences of renewed activity, we are naturally curious about the date of the last eruption. Fourteen reports of eruptions were published during the nineteenth century, but all these were based on distant sightings of supposed eruptive clouds. Certainly, some of these were dust clouds from rockfalls or simply atmospheric clouds. None of the lava flows appear to be much younger than a thousand years or so, and it has been estimated that the summit cone was built approximately 2,000 years ago. Younger ash beds occur in the soils, however. One, known as ash layer *X*, was deposited on the east flank between 1820 and 1854 and probably represents the most recent eruption. The dating was incidental to a study of recent glacial events. Trees living on young glacial moraines were cored and dated by

13-11 *Surface of the debris deposited by rockfall avalanches in the White River Valley near Mount Rainier during December, 1963. All of the rock came from the face of Little Tahoma Peak, almost 4 miles away. (Photograph by Bates McKee.)*

counting of their annual growth rings. In theory, the moraine would be a few years older than the oldest tree growing on it. The study showed that the glaciers on Mount Rainier reached farthest down the mountain (in the past several thousand years) between the fourteenth and nineteenth centuries, different glaciers peaking at different times. The dating of the various recessional moraines then allowed dating of layer *X*, for the ash was presumed to be younger than an 1820 moraine on which it settled but older than an adjacent 1854 moraine that has no ash layer on it.

Retreat of the front of glaciers on Rainier was rapid during the late nineteenth and early twentieth centuries. This general recession was halted around 1950, however, and since then the fronts of some glaciers (such as the Nisqually and Emmons) have advanced as much as 1,000 feet. We have no way of knowing if this recent advance portends a significant change or is simply a short-lived reversal. Some of Rainier's glaciers are apparently still receding.

What about future eruptions from Mount Rainier? We can be fairly confident that they will occur, but when? We have no very good way of predicting them, except that a future eruption probably will be preceded by many local earthquakes that will begin a few days or weeks before eruption. If an eruption occurred next week we should not be surprised; but, conversely, if none of us lives to see a Rainier eruption, this should not astound us either. It could happen anytime. What can we do in preparation for such a possibility? Unfortunately, very little, for if an eruption should occur it could represent forces so immense that our best efforts would make little difference. Actually the potential hazard to the general region is not too great. Lava would not flow farther than a few miles from the erupting vent, if past history can be relied on; and ash falls would be more of a nuisance than a threat to life. The greatest hazards would be mudflows. Clearly an Osceola-type event would be disastrous for all life and property within the affected valley or valleys and upon the lowland region at the mouth of the valley. The potential volume of mud would make damming or diversion of such a mudflow an impossible task. Loss of life could be minimized, however, if the mudflow was associated with an eruption that was preceded by seismic foreshocks that provided some warning. A perfectly prudent man would stay far away from Mount Rainier; but, from an actuarial point of view, even active volcanoes are infinately safer than are highways.

YOUNG MOUNT SAINT HELENS

Geologists have suspected for many years that Mount Saint Helens is younger than the other major cones of the Cascades. They base this conclusion on the fact that it has a more perfect profile, having been eroded less intensely than have the others. The glaciers on modern Saint Helens have never been much larger than they are today. This suggests that the cone postdates the last major glaciation.

Mount Saint Helens is not tall by Cascade standards, rising to only 9,677 feet above sea level; but the volcano lies well west of the range crest and rests on hills no more than 2,000 feet high. This means it rises about as high above the surrounding region as do its biggest brothers, and it is a favorite of many "volcano collectors."

Mount Saint Helens is compositionally more interesting than the other volcanoes of Washington. Its history is not known in detail, but apparently there have been two major volcanic cones on the site—the one we see today, which is made up of basalt, andesite, and dacite, and an older cone that consisted primarily of basalt and basaltic andesite. The eroded base of the older cone underlies the present volcano and is visible up to an elevation of nearly 5,000 feet. Its girth suggests it was once much higher and must have been greatly reduced by erosion before the new cone started to grow. Deposits from glaciers that extended many miles down the flanks of ancient Saint Helens have been dated by radiocarbon methods as approximately 20,000 years old.

13-12 *The west flank of Mount Saint Helens. Mount Adams in the distance at left. Note the relatively smooth flanks of this cone, suggesting its youth. (Copyrighted photograph courtesy of Delano Photographics; also see frontispiece.)*

Mudflows have played a major role in the history of the mountain. Several large mudflows partially fill the valley of the Lewis River, which flows westward past the south side of Mount Saint Helens. Others occupy the valleys of the Kalama River and the North and South Forks of the Toutle River, north and west of the mountain. Some of these mudflows traveled all the way to the Cowlitz River at Castle Rock, a distance of approximately 50 miles. One dammed the North Fork of the Toutle River at the foot of the cone, producing scenic Spirit Lake (Figure 13-13). Also noteworthy are the several prominent dacite domes low on the flank of Mount Saint Helens, especially on the south side. A 2,000-year-old basalt flow, also on the south flank and readily accessible by road, contains well-preserved tree casts and several nearly intact lava tunnels, including one (Ape Cave) 2 miles long. (Tree casts and lava tunnels are discussed in Chapter 15.) In short, Mount Saint Helens has much of interest. It lacks only the volume of tourists found around its more famous neighbors, Mount Rainier and Mount Hood.

13-13 *Mount Saint Helens from the north. Spirit Lake in the foreground was created when a large mudflow from the volcano dammed the North Fork of the Toutle River. (Copyrighted photograph courtesy of Delano Photographics.)*

THE DEBRIS FAN ON MOUNT HOOD

Mount Hood, 11,235 feet high, sits atop one of the lowest parts of the Cascade Range and dominates the Portland area as Rainier dominates the Puget Lowland. The cone was built upon the eroded root of a Pliocene volcano. The composition of the lava and interbedded pyroclastic debris, while not completely uniform, consists largely of andesite.

The earliest flows filled north-sloping canyons tributary to the Columbia River. By the same process of topographic reversal discussed in the section of Mount Rainier, they now cap ridges. Some are as much as 500 feet thick and 8 miles long. The main cone, a typical stratovolcano, was built upon this earlier platform. It apparently reached its maximum height (500 to 1,000 feet higher than today) before the last glacial maximum, during which time accelerated glacial erosion cut deeply into the cone. On the north side, the main vent is apparently exposed in the cliff immediately below the summit. Chaotic structures in lava flows high on the mountain suggest that many of them flowed over snowfields and perhaps glaciers. These flows presumably generated some of the mudflows that are found in many of the river valleys that originate on Mount Hood.

A gigantic debris fan, which forms the southwest flank of the volcano, is the most noteworthy feature of Mount Hood. It heads at Crater Rock, just south of the present summit, and dips uniformly to the base of the cone at Government Camp. (Timberline Lodge is built near the eastern edge of the fan. Most of the area's ski runs are down its relatively smooth surface.) The best exposures of the fan are in the walls of Zigzag Canyon, a young ravine that has been cut through the fan and into the interbedded flows and pyroclastic beds of the main cone beneath. The fan consists of crudely layered and poorly sorted sand, which contains blocks of lava as large as 20 feet in diameter. The blocks are andesitic and contain numerous crystals of the mineral hornblende. This same distinctive andesite makes up Crater Rock at the apex of the fan. The Crater Rock Flow erupted late in the volcano's history as a pasty plug, apparently in a crater that had been blown out of the southwest flank about 1,000 feet below the summit. The eruption occurred over some period of time, for melting ice and snow must have contributed large volumes of water while the growing dome contributed great amounts of andesite to explain the size and layering of the fan. This volcanic slurry buried numerous trees, some of which have been radiocarbon dated. The debris fan was deposited about 2,000 years ago. Fracture patterns in some of the boulders suggest that the material was still hot at the time of deposition. This eruption was apparently the last major emission of lava from Mount Hood, although other eruptions, especially of ash, may have occurred since. The Crater Rock Flow and its resultant debris fan are of particular interest, for they constitute a well-documented example of the effects of lava extrusion on a snow- and ice-clad volcano. They illustrate the process that probably produced the Suiattle River fan immediately east of Glacier Peak and many other Cascade debris flows. They also show once again the impossibility of controlling such forces. If such a flow happens on some cone in the future, Man had best get out of the way.

CRATER LAKE AND MOUNT MAZAMA

Crater Lake occupies the core of a volcano that has largely disappeared. The name Crater Lake is inappropriate, for volcanic craters are the circular pits immediately above a volcano's conduit. They are almost never more than a mile across, while Crater Lake partly fills a depression some 5 to 6 miles wide. Crater Lake is a classic example of a geologic feature known as a *caldera*—in fact, the concept of a caldera was conceived in large measure as a consequence of a study of Crater Lake. A caldera forms by a combination of collapse and eruption. The foundering of the top of a volcano is a fairly logical consequence of continued erup-

13-14 *The west side of Mount Hood. The smooth right-hand slope of the cone is underlain by a thick debris fan that originated near Crater Rock, the spire in the summit amphitheater. (Photograph courtesy of the Oregon State Highway Division.)*

13-15 *Crater Lake from the northwest, with Wizard Island on the right. Note the U-shaped profile of glaciated valleys hanging above the far rim of the lake. (Copyrighted photograph courtesy of Delano Photographics.)*

tion, especially if the cone is underlain at some relatively shallow depth by a chamber of molten magma. If the magma is expelled as lava or pyroclastic debris, this material loads the ground surface. Eruption also reduces the subsurface pressure as magma is withdrawn. A common result is the collapse of the central part of the volcano. The collapse is along steep faults peripheral to the summit or in grabens that cross the cone. Subsequent rise of magma may be along these faults with a close interplay between subsidence and eruption. This was well illustrated during recent eruptions in the summit caldera of Kilauea Volcano on Hawaii. The origin of the Crater Lake caldera is somewhat more complex, however.

The original volcano, Mount Mazama, was not a single cone. Rather, it consisted of a series of overlapping cones built as the erupting vent shifted from place to place. For this reason, geologists cannot simply project the remaining slopes upward to determine the original maximum height of the mountain. Mount Mazama must have had a relatively flatter and less regular profile than did Mount Rainier. The original cone probably rose another 2,000 to 3,000 feet above the general level of the present caldera rim (which is approximately 7,000 to 8,000 feet above sea level). Mount Mazama was built upon a complex base of Pliocene basaltic shield volcanoes and older lavas that are part of the Cascade Andesite Formation (Chapter 12). The major cone-building phase of Mount Mazama consisted of numerous eruptions of andesitic lava without much emission of pyroclastic debris. Toward the end of this stage, however, more viscous dacite flows issued from subsidiary vents on the south and east flanks of the volcano, and dacitic pumice spewed from the summit crater. Mazama must have supported many glaciers because glacial deposits (tills) are interbedded with flows in the caldera walls. By the time of the last glacial maximum, individual glaciers flowed long distances down the flanks of the volcano. They smoothed and shaped the valley walls, producing the characteristic U-shaped profile clearly seen in Kerr (Sand Creek), Sun, and Munson Valleys on the south flank. The ice in these valleys was a thousand feet thick or more; one glacier extended a maximum distance of 17 miles from the summit.

Arcuate fractures, perhaps representing incipient collapse of the cone, then opened on the north flank of Mount Mazama. These positioned new vents, which erupted viscous flows of andesite and dacite that are now exposed in the north wall of Crater Lake. Dacitic ash was erupted from the summit, and small andesitic and dacitic dome cones formed low on the east slope of Mount Mazama. Following this activity, more violent eruptions began. At first, these were of dacitic ash that drifted to the east. Then an immense pyroclastic cloud, which contained coarse pumice fragments, spread to the north and east. The resulting sheet of ash (the Mazama Ash) blanketed at least 350,000 square miles of northern Nevada, Idaho, Oregon, Washington, southern British Columbia, southwestern Alberta, and western Montana. The pumice sheet is 10 feet thick 25 miles northeast of Crater Lake. Not all the pyroclastic debris rose high above the volcano. Some of it bubbled over the crater rim and formed glowing avalanches that roared down the flanks of the volcano. Some traveled 25 miles to the east and northeast, carrying blocks of pumice as large as 14 feet in diameter. One that coursed down the valley of Sand Creek is particularly well exposed at the Pinnacles, 5 miles south of the crater rim (Figure 13-16).

Charred wood from within and beneath these pumiceous sheets has been radiocarbon dated. The eruptions occurred about 7,000 years ago. This date is extremely significant, for wherever the Mazama Ash is found it provides a dated reference surface in the recent stratigraphic record. Geologists are thus able to determine such things as the average rate of sedimentation for the past 7,000 years in Lake Washington or in Cascadia Basin (Chapter 5). They can also tell which of the many young glacial moraines in the Cascades preceded and which are younger than the eruption of Mount Mazama.

Mount Mazama had largely disappeared after the climax of ash eruption. Where it had been, there was instead the Crater Lake caldera. Probably about 15 cubic miles of the volcano had disappeared. The total volume of solid rock and lava in the Mazama Ash and the various pumice flows is only about 10 cubic miles. If these estimates are correct, approximately one-third of the missing volume would have to be explained by collapse of Mount Mazama, presumably because magma inside the volcano somehow moved down and away from the summit.

Initially the caldera was dry, but rain and melting snow gradually filled it to the present level of Crater Lake. This level is probably relatively stable, representing a balance between water added by precipitation and runoff and that lost through evaporation and seepage outward through the lower flanks of the volcano. Wizard Island, close to the west shore of the lake, is a relatively young cinder cone. It was built by andesitic eruption through the caldera floor, with several eruptions of blocky lava constructing a wide base outward from the lower flank of the cone. (I once heard Wizard Island identified as the original top of Mount Mazama that was blown into the air during the climax of eruption and feel back intact into the newly formed caldera. While imaginative, this theory is not correct!)

A recent bathymetric survey of Crater Lake has proved it to be the deepest lake (1,932 feet deep)

13-16 *Deposits of pumice debris from a glowing avalanche that originated on Mount Mazama approximately 7,000 years ago. The Pinnacles, 5 miles south of the rim of Crater Lake. (Photograph by Bates McKee.)*

in the United States. This study also discovered another cinder cone (Merriam Cone) that rises in nearly perfect symmetry from the lake floor near the north rim to within about 500 feet of the lake surface. The ages of Merriam Cone and Wizard Island are not known. Their morphology suggests however, that they probably were not built under water and may be almost 7,000 years old.

OTHER CASCADE VOLCANOES

Many fine volcanoes in the Cascades have not been mentioned—Mount Adams, Mount Jefferson, Mount Washington, and the Three Sisters, to name a few. Some are better known geologically than others, but all are interesting. They have been omitted here for reasons of space. The reader who understands the principles and processes mentioned in this chapter's discussion of six of the Cascade volcanoes should be able to strike out on his own and recognize major elements in the history of other cones in the range. Furthermore, some of the more accessible volcanic fields in the Oregon Cascades are the subject of excellent field guides cited in the reference list for this chapter. A few days spent exploring central Oregon with these volumes in hand will provide a most rewarding experience as well as the best possible feeling for Cascade volcanism.

13-17 *Wizard Island from the south rim of Crater Lake. (Photograph by Bates McKee.)*

14 the blue mountains province

■ The Blue Mountains Province is probably the best all-around region in the Northwest for ease of viewing a great variety of geologic structures and rock types. The Blue Mountains have enough topographic relief to guarantee good exposure, but not so much to severely limit access. Furthermore, they have the right amount of vegetation to provide a pleasing landscape without covering up all of the geologic history, which has been long and interesting.

The Blue Mountains have proved a paradise for "rockhounds," and the region's popularity with the rock clubs will probably go on and on. The prime interest has been in agates, "thunder eggs," and petrified wood, for which the region has a worldwide reputation. If this is not your interest, the Blue Mountains have excellent mammal fossil beds, some of them set aside in state parks established specifically for collecting by the public. Finally, the lover of old mines and ghost towns can take many trips to the Blue Mountains before he has exhausted the possibilities there. In summary, the region really has something for anyone with an interest in geology.

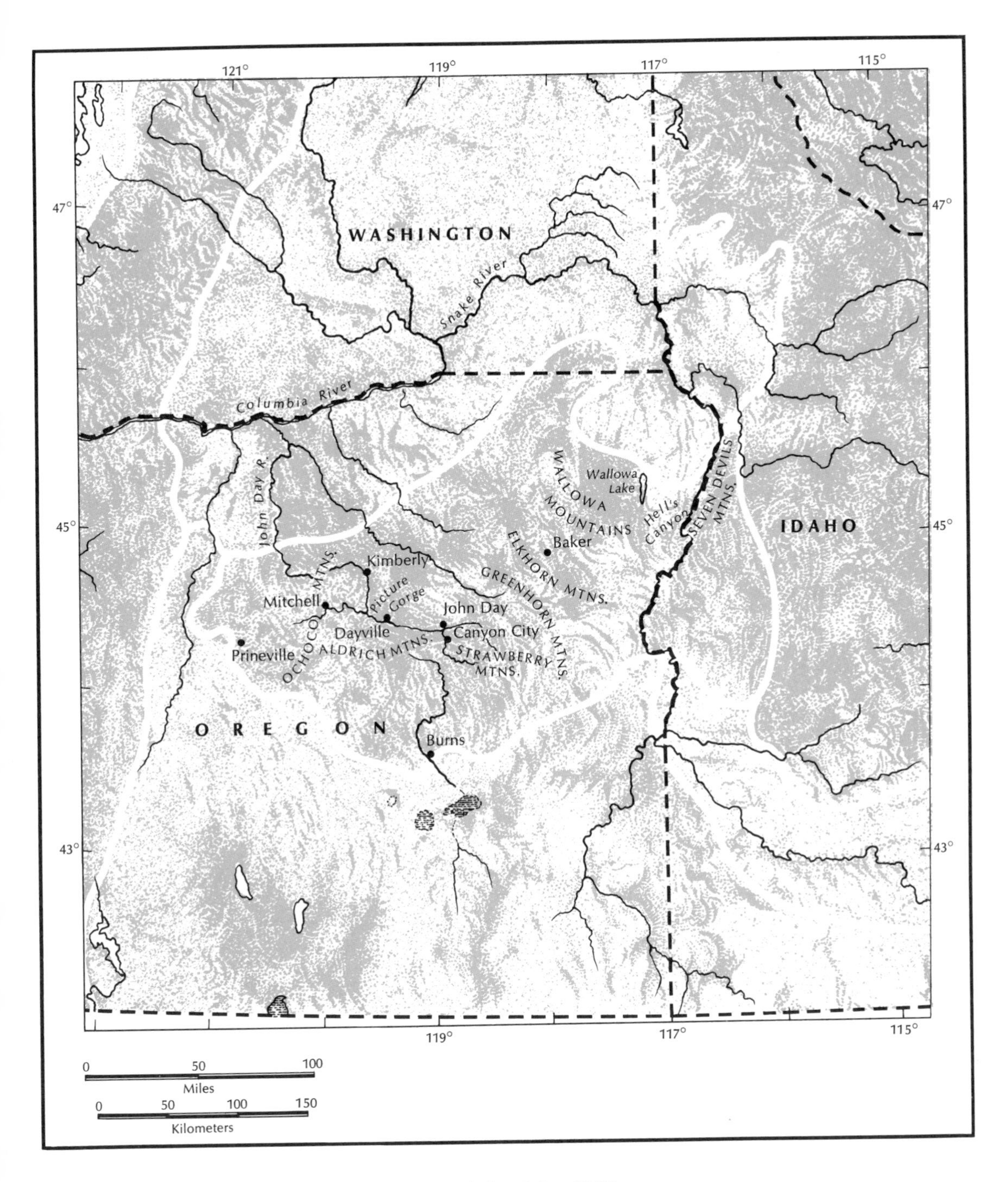

14-1 *Physiography of the Blue Mountains Province. (After Raisz, 1941.)*

GEOGRAPHY AND STRUCTURE

The Blue Mountains Province trends east from near Prineville in north-central Oregon into western Idaho. This uplift contains many individual ranges, most notably (from west to east) the Ochoco, Aldrich, Strawberry, Greenhorn, Elkhorn, Wallowa, and Seven Devils Mountains (Figure 14-1). The greatest relief is found at the east end. There the highest peaks in the Wallowas rise as much as 8,000 feet above the floor of the Snake River, which has cut the spectacular Hells Canyon through the Blue Mountains. The topography is complex, reflecting the varied lithology and structure of the Blue Mountains.

The largest river within the province is the John Day, which with its many branches and tributaries carries much of the drainage of the western half of the Blue Mountains north into the Columbia. Major rivers in the eastern part include the Grande Ronde, Imnaha, and Powder Rivers, all tributary to the Snake. Shorter intermittent streams that drain the south flank of the range generally end in depressions in the Basin and Range Province.

The Blue Mountains separate the Columbia Plateau Province on the north from the Basin and Range Province to the south. Both of these neighboring regions contain thick sections of Upper Cenozoic volcanic strata. The Blue Mountains contain these formations too, but "windows" eroded through them there provide exposures of older strata not seen in the neighboring provinces. The overall structure of the Blue Mountains is a large, asymmetric anticline with a steep north flank and a gentle south flank. As a result, the Blue Mountains rise sharply up from the Columbia Plateau, but merge gradually into the high plateau country to the south.

THE HISTORY IN BRIEF

The pre-Tertiary rocks in the Blue Mountains look familiar—Upper Paleozoic and Lower Mesozoic eugeosynclinal sedimentary and volcanic strata, cut by granitic intrusions of Cretaceous age. Triassic intrusions of peridotite, gabbro, and serpentine also are found. They are similar to the Triassic intrusive plutons of the Klamath Mountains. The entire Paleozoic and Mesozoic section was folded and faulted before the Eocene Epoch. Metamorphism was intense only in the eastern part of the Blue Mountains and occurred primarily adjacent to the granitic batholiths. Late Cenozoic uplift of the Blue Mountains and subsequent erosional stripping have been in just the proper proportions to provide good exposures of both the older rocks and the Cenozoic strata above.

The Cenozoic history of this region was dominated by volcanism, and extensive volcanic units erupted during every epoch. Not all of the units are lava flows, however. There are considerable thicknesses of sedimentary rocks between the flow units; most of the sediment was produced by the erosion of nearby volcanic rocks. The Late Cenozoic history of the Blue Mountains was dominated by folding, faulting, and uplift of the range. Alpine glaciers formed in the Wallowa Mountains in the Pleistocene Epoch but have since disappeared.

THE CORDILLERAN GEOSYNCLINE IN THE BLUE MOUNTAINS

The greatest exposures of Paleozoic strata are near the west and east ends of the province. Originally, the rocks were marine sandstones, shales, and andesitic to basaltic lava, with lesser amounts of limestone and chert. The strata were deformed by several orogenies, and those at the east end of the range were metamorphosed to schists, slates, and greenstones. The least deformed and recrystallized Paleozoic section is near the head of the Crooked River at the west end of the Blue Mountains. It contains the oldest fossils (Middle Devonian) found in Oregon, plus Late Mississipian, Pennsylvanian, and Early Permian fossils. Only a few Permian fossils have been found in the eastern part of the Blue Mountains. The Paleozoic strata there

14-2 *Hells Canyon, the deepest canyon in North America, formed where the Snake River cuts through the Seven Devils Mountains at the eastern end of the Blue Mountains Province. (Photograph courtesy of Oregon State Highway Division.)*

are at least several miles thick, however. Like the Chilliwack Group of the North Cascades or the Cache Creek Assemblage of south-central British Columbia, they may have been deposited over a major part of the Late Paleozoic Era.

One formation in the Wallowa Mountains, the Clover Creek Greenstone, consists of 5,000 feet of marine andesite and basalt. These are interbedded with minor sandstone and shale units which contain a few fossils. Those in the lower part of the section indicate an Early to Middle Permian age for this formation; those higher up a Late Triassic age. Consequently, there must be a major unconformity within the Clover Creek greenstone, for no strata of Late Permian or Early and Middle Triassic age are found. This break is found in this part of the record throughout the Northwest. It must represent a major interval of uplift and erosion, but clearly conditions in the Cordillera were much the same after Middle Triassic time as they had been in the Permian Period. Upper Triassic formations that overlie the Clover Creek Greenstone consist of marine sandstones, shales, volcanic rocks, and some rather thick beds of limestone.

An extraordinarily thick sequence of Triassic and Jurassic strata is exposed around the small communities of Izee and Suplee on the south flank of the Aldrich Mountains. The total thickness of the Mesozoic strata in these mountains approaches 50,000 feet. The Triassic rocks (Table 14-1) are marine sedimentary and volcanic types, much the same as equivalent strata in the Wallowa Mountains but without the limestone. The Jurassic formations rest unconformably on the Triassic beds. They consist of marine sandstones and shales.

14-3 *Hells Canyon Dam on the Snake River, looking north (downstream). Most of the massive rock in the canyon walls belongs to the Seven Devils Volcanics, a greenstone formation that correlates with the Clover Creek Greenstone in the Wallowa Mountains to the southwest. The peak in the distance is capped by the Columbia River Basalt. (Photograph by Bates McKee.)*

There are no lava flows in the section, but andesitic tuff beds are numerous, as well as andesitic sediment eroded from nearby volcanoes.

MESOZOIC INTRUSIONS

Two principal episodes of deformation, accompanied by the emplacement of large, igneous intrusions, occurred during the Mesozoic Era in the Blue Mountain Province. The first episode was in the middle part of the Triassic Period; it accounts for the aforementioned unconformity. This orogeny involved the folding, faulting, and mild metamorphism of Paleozoic formations, plus the emplacement of the serpentinized masses of gabbro and peridotite found around Canyon City.

The second major period of orogeny occurred early in the Cretaceous Period. It was punctuated by intrusion, most notably the Wallowa and Bald Mountain Batholiths in the eastern part of the province. These may have been offshoots of the larger Idaho Batholith to the northeast, as the intrusions are similar in composition (granodiorite) and have the same radiometric age (approximately 100 million years).

The major ore deposits of the Blue Mountains are closely associated with these Cretaceous intrusions. Gold and silver have been the principal metals produced. Since 1880, mines in the Blue Mountains (especially in Baker and Grant Counties) have provided 70 percent of the state's gold and 74 percent of its silver production, the rest having come from the Klamath Mountains. Gold and silver minerals occur in quartz veins around the margins of intrusion and also in placer deposits, both in modern stream sediments and in Tertiary sandstones and conglomerates.

Other metals that have been obtained from mines in the Blue Mountains include copper, cobalt, mercury, manganese, antimony, tungsten, and chromium. Limestone has led the long list of nonmetallic resources mined in the province.

The Pacific Ocean retreated reluctantly from the

TABLE 14-1 Mesozoic stratigraphy of the Suplee-Izee area. (After Dickinson and Vigrass, 1965.)

AGE	FORMATION	DESCRIPTION	THICKNESS, FEET
Early to Late Cretaceous	Bernard	Pebbly sandstone	1,500
Late Jurassic	Lonesome	Volcanic sandstone and mudstone	10,000
Late Jurassic	Trowbridge	Black mudstone and volcanic sandstone	3,000
Early to Late Jurassic	Snowshoe	Interbedded andesitic flows, breccias, and volcanic sandstone and shale	3,000–4,000
Early Jurassic	Mowich (Group)	Andesitic tuff and sandstone, with minor shale, limestone, conglomerate	1,500
Early Jurassic	Graylock	Dark siltstone, with minor limestone	500
Late Triassic	Rail Cabin Argillite	Black to green siltstone, mudstone, tuff	1,000
Early Late Triassic	Brisbois	Black, gray, and green mudstone, sandstone, limestone; basaltic to andesitic flows and tuffs	5,000
Early Late (?) Triassic	Begg	Mudstone, sandstone, conglomerate. Minor lava and limestone	7,500

(Heavy rule) Major unconformity

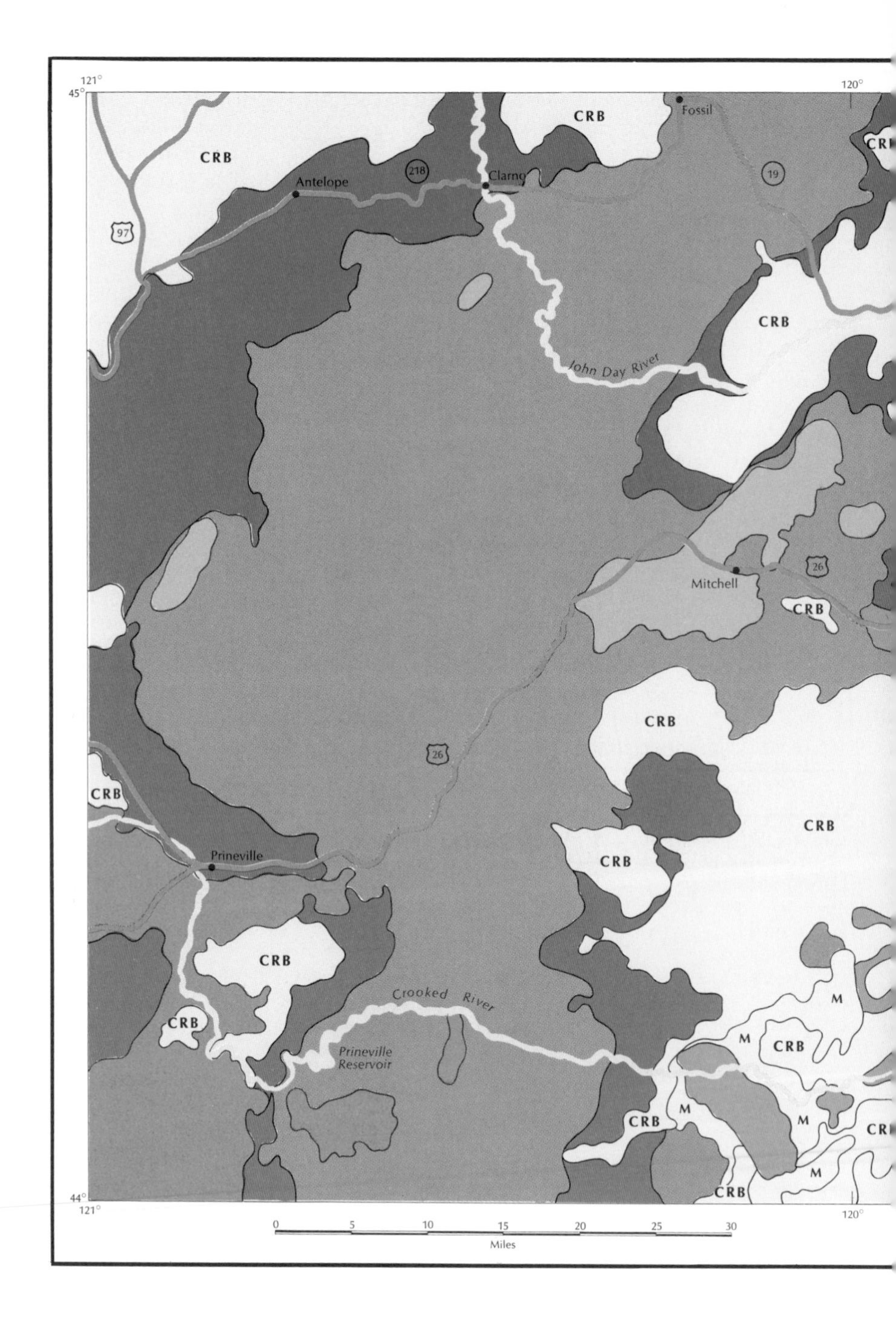

14-4 *Generalized geologic map of the western end of the Blue Mountains Province. (After Brown and Thayer, 1966, and Swanson, 1969.)*

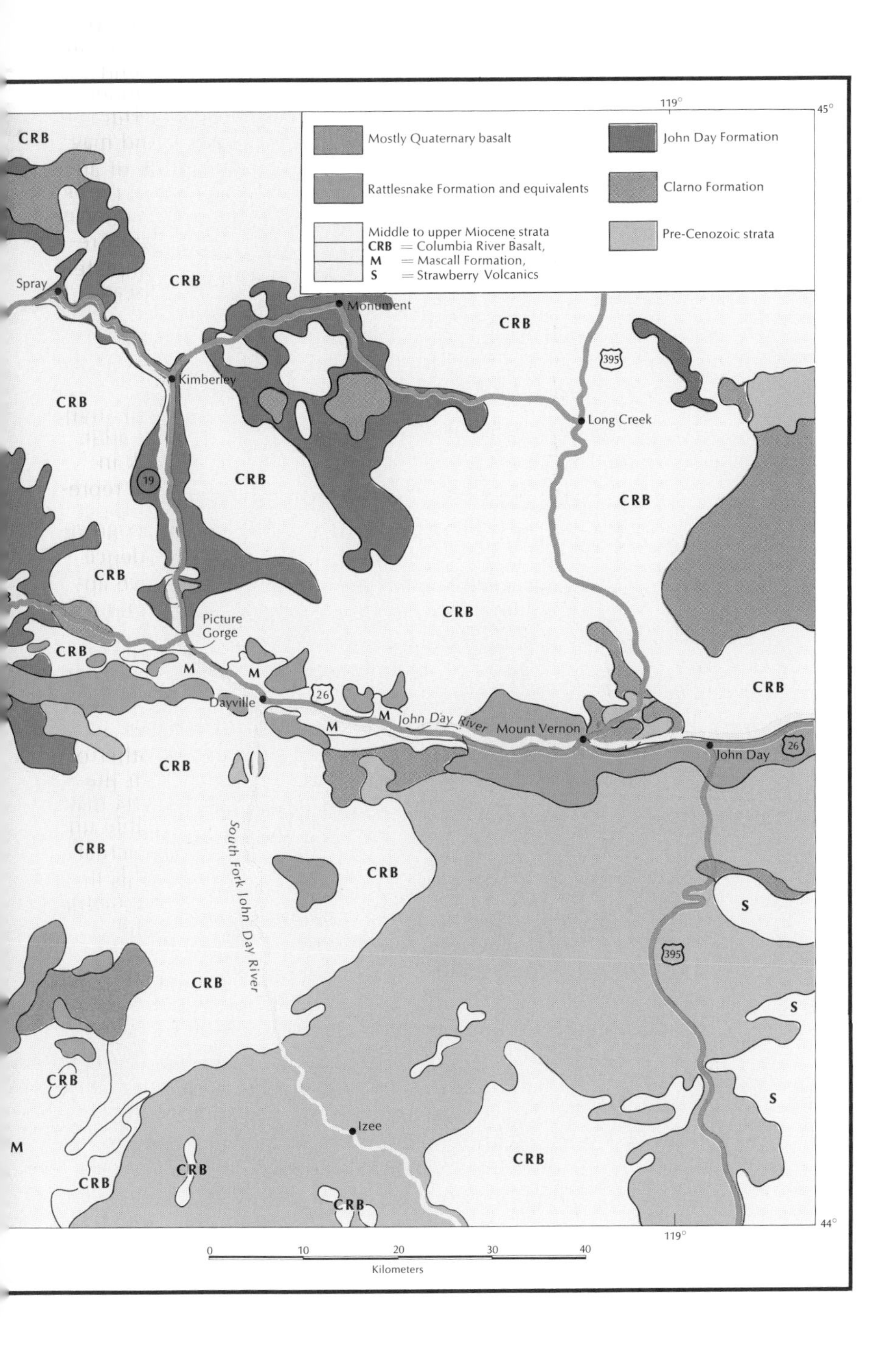

Mostly Quaternary basalt
Rattlesnake Formation and equivalents
Middle to upper Miocene strata
CRB = Columbia River Basalt,
M = Mascall Formation,
S = Strawberry Volcanics
John Day Formation
Clarno Formation
Pre-Cenozoic strata
119°
45°
44°
Spray
Monument
Kimberley
Long Creek
Picture Gorge
Dayville
John Day River
Mount Vernon
John Day
South Fork John Day River
Izee
CRB
M
S
395
26
19
0
10
20
30
40
Kilometers

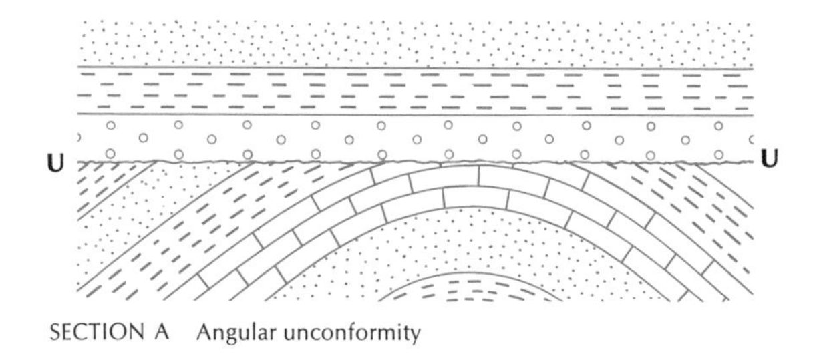

SECTION A Angular unconformity

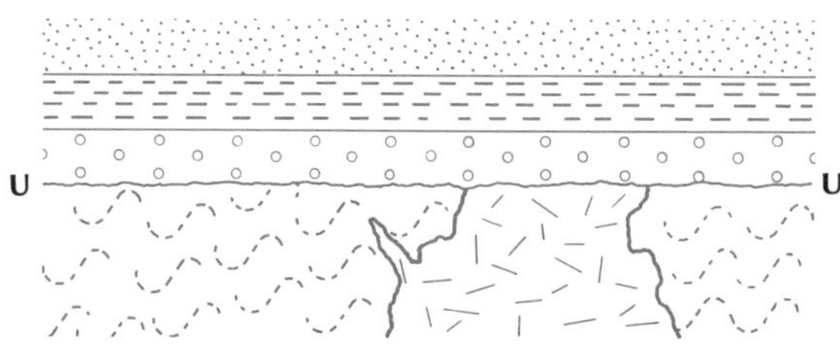

SECTION B Nonconformity

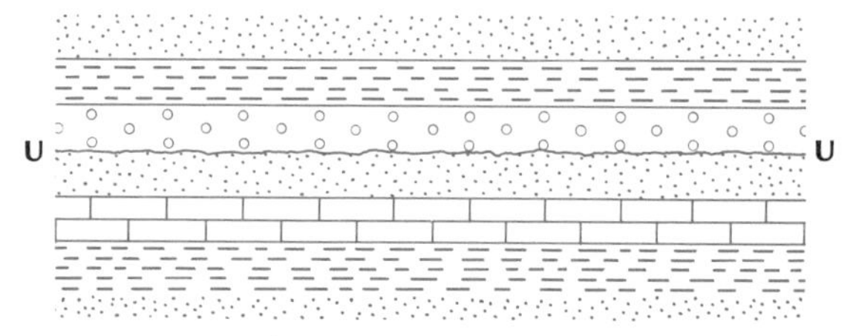

SECTION C Disconformity

14-5 *The three major types of unconformities.*

region, as it did elsewhere in the Cordillera. Approximately 9,000 feet of Cretaceous conglomerate, sandstone, and shale beds are exposed around Mitchell in the Ochoco Mountains, and thinner sections of similar strata occur elsewhere. The conglomerate beds are particularly thick and may have been deposited along the south flank of an emerging landmass. Most of the strata were probably deposited in shallow water, some of the gravel perhaps on land at the water's edge. By Late Cretaceous time, the seas had disappeared completely from the Cordillera east of the modern Cascades.

THE MEANING OF UNCONFORMITIES

Unconformities are such a significant part of stratigraphic interpretation that they merit some additional comment. An unconformity is a break in time within a stratigraphic sequence. This is represented by a surface within the rocks which separates strata of different ages. In order to recognize an unconformity, a geologist must find evidence that the rocks on two sides of a surface have appreciably different ages, and this is not always obvious. What type of evidence is useful?

First the geologist must prove that the relationship between two rock units is actually a stratigraphic one—in other words one unit was deposited on top of the other. (Other possibilities would be that one unit had intruded the other or the contact between the two was a fault.) If the contact is stratigraphic, many types of criteria may be helpful to the evaluation of it as a possible unconformity. First of all, information on the actual age of both units is obviously significant. This information may come from fossils, from radiometric age measurements, or from correlation of the strata with formations elsewhere whose age is known. Thus when Middle Permian fossils were found in the lower part of the Clover Creek Greenstone and Late Triassic fossils were found higher in the same unit, this required the existence of an unconformity within this formation. (Sometimes geologists "discover" an unconformity in this way before they see the surface actually exposed in an outcrop.)

Unconformities can be recognized without knowledge of the true age of the adjacent strata. In one case, that of an *angular unconformity* (Figure 14-5, Section A), the older strata were folded and eroded before the next formation was deposited. The position of one unit across the bevelled edge of the other is a clear indication of a time break. In the *nonconformity* (Figure 14-5, Section B), strata have been laid down on a surface composed of intrusive igneous rock or metamorphic rock. Often this represents the most amount of lost time, for a very considerable amount of uplift and erosion must have occurred in order for crystalline rock, formed deep underground, to be exposed at the surface ready to receive new strata.

The third major type of unconformity is the *disconformity* (Figure 14-5, Section C). In this case, the strata above and below the time break have parallel layering, and there is no obvious structure to indicate the disconformity's existence. Perhaps the surface itself, if studied in detail, would provide significant clues. More likely, the geologist would have to know from other evidence that the strata on opposite sides of the surface had different ages before he could recognize it as an unconformity.

Unconformities are of great help in dating deformational events. For instance, if Lower Cretaceous rocks are separated from Upper Triassic rocks by an angular unconformity, folding (and/or faulting) and erosion must have occurred in the Jurassic Period. If they are separated by a nonconformity (in other words, the Triassic strata have been metamorphosed but not the Cretaceous rocks), then metamorphism and considerable erosion took place in the Jurassic. But what about the disconformity—the case where neither metamorphism nor folding occurred? Geologists formerly referred to such unconformities as surfaces that represented periods of erosion or of nondeposition, the latter implying that there could be parts of the earth's surface which were experiencing neither erosion nor deposition. Such regions have not been discovered, and all unconformities are now thought to represent a buried erosion surface.

This leads, finally, to one other important conclusion. If all unconformities are erosion surfaces, then some strata must have been removed by erosion before deposition began again. If Lower Cretaceous strata rest unconformably on rocks of latest Triassic age, there must have been some deposition in the Jurassic Period. Otherwise, the Triassic beds would have been partly removed by erosion. The unconformity could be explained by deposition in the first part of the Jurassic Period, followed by erosion (down to Triassic strata), and then resumption of sedimentation in the Cretaceous Period. Thus the time missing in the record is always longer than the actual interval of erosion. To illustrate this point, right now in the upper part of the Crooked River, recent stream gravels are washing across outcrops of Devonian strata. If this area were to be buried somehow and this relationship preserved, a geologist of the future might discover an angular unconformity between Devonian beds and Holocene conglomerate. But clearly this region has not been exposed to erosion for 350 million years, even though that much of the record is missing. Instead, the unconformity results from uplift and erosion during the past few million years. There have been other periods of erosion in this region since the Devonian, but there has been much deposition also.

THE CLARNO AND JOHN DAY FORMATIONS

Volcanism and sedimentation resumed in the Blue Mountains region in the Eocene Epoch, approximately 35 million to 40 million years ago. By this time the region was perhaps one of grassy plains and low hills that exposed all the various older formations. Volcanoes were erupting in the Cascade Range Province, and this activity spread to the east into the Blue Mountains region. Within the next 15 million years, two very extensive formations were deposited around and on top of the volcanoes of the Blue Mountains. These were the Clarno and the John Day Formations.

The type area for the Clarno Formation is Clarno, a former ferry crossing on the lower reaches of the John Day River. The formation consists of lava flows, mudflows, volcanic breccias, and beds of volcanic ash. Basalt, andesite, dacite, and rhyolite flows are included, suggesting that quite a variety of volcanoes were developed. Sandstones and siltstones, deposited in rivers and lakes, are interbedded with the volcanic rocks. They contain fossils that prove a late Eocene age for most of the beds. The remains of leaves, wood, and nuts are most common. Bones and teeth from large vertebrate animals have also been found, proving that the Eocene landscape was populated by such exotic creatures as crocodiles, tapirs, and rhinos. The climate was apparently warm and moist. Remnants of the Clarno Formation are found everywhere in the Blue Mountains except in the northeastern part of the range, where they may have been stripped off by erosion.

The John Day Formation consists of several thousand feet of nearly pure volcanic ash. Much of this ash washed into depressions and small lakes from nearby hills. The formation also contains several extensive layers of welded tuff, which is a compact ash deposited by "glowing avalanches." The John Day beds have yielded a great diversity of vertebrate remains, which has made it one of the most famous fossiliferous units in the Northwest.

14-6 *Massive volcanic mudflow breccia of the Clarno Formation, near Clarno. The cliff is nearly 200 feet high. (Photograph courtesy of Oregon State Highway Division.)*

The John Day Formation is brightly colored. The lower beds are bright brownish-red, the middle beds are green, and the upper beds are white.

The origin of the bright colors is of some interest. The red is due to the presence of the iron oxide mineral hematite (an important constituent of rust). The hematite may have come from red soils developed in the Oligocene Epoch on the volcanic rocks of the Clarno Formation. A more likely explanation is that the hematite developed in the John Day ash sometime after deposition. This would have been a later alteration of the ash due to an increase in temperature and pressure after burial. The green color of the middle member definitely indicates later alteration of the ash, since it is caused by the presence of a low-grade metamorphic mineral known as clinoptilolite. The buff beds near the top of the John Day Formation probably retain their original color.

The John Day fossil fauna is extremely varied. The richest fossil horizons are in the upper part of the formation and finds are very rare in the lower redbeds. More than one hundred species of mammals have been unearthed. They include the ancestors of the modern cat, dog, pig, rabbit, opossum, rodent, horse, camel, tapir, and rhinoceros. One of the most common fossils is the oreodon, and extinct herbivorous mammal that is sometimes referred to as a ruminant (that is, cud-chewing) hog.

The flora and the fauna preserved in the John Day beds tell quite a bit about the environment. They suggest a climate that was both warmer and moister than that of today. Undoubtedly, the ash falls must have drastically reduced the population on occasion, but the landscape was quickly repopulated. Most of the bones and teeth that are found are disarticulated, and whole skeletons are rare. This results from the ash having been reworked by running water. A few leaves found in the redbeds suggest a late Oligocene age for the lower part of the John Day Formation. The vertebrate fossil remains in the green and white members indicate that they were deposited early in the Miocene Epoch.

14-7 *Dikes of Clarno basalt, filling fissures that fed some of the Clarno lava flows. The dikes have been accentuated by differential weathering. Near Burnt Ranch, south of Clarno. (Photograph by Bates McKee.)*

14-8 *The John Day Formation at Painted Hills State Park near Mitchell. Hill in the middle distance is capped by a welded tuff unit. Far cliffs consist of many lava flows of the Columbia River Basalt. (Photograph courtesy of Oregon State Highway Division.)*

The welded tuffs form prominent cliffs in the John Day Formation. One toward the middle of the formation is particularly thick and widespread. Its pale orange-brown color contrasts markedly with the green tint of the enclosing ash beds. These glowing-avalanche units are difficult to comprehend, even though it was a deposit like this that, in 1902, devastated the city of Saint Pierre on the island of Martinique. The fiery cloud contains ash, pumice, gas, clots of lava, and pieces of solid rock. It moves as a type of density current, somewhat like the turbidity current in water (Chapter 5) or the rockfall avalanche (Chapter 13). Its velocity may be as high as 100 miles per hour, due initially perhaps to a violent explosion out the flank of a volcano but maintained subsequently due to flowage down low slopes. When it finally comes to rest, the flow may harden (weld) almost like a lava flow—especially if it still retains much heat. Compaction markedly flattens pumice near the base of a thick bed of welded tuff, but most of the pumice "floats" to the top before final solidification of the unit.

A welded tuff is a very useful marker bed in the stratigraphy of a region, because it is almost a perfect "time plane." This means that the unit is everywhere the same age (or very nearly so), which is not the case for many geologic formations.

The welded tuffs in the John Day Formation become thicker and contain larger lava fragments toward the western end of the Blue Mountains; their source was probably a volcano located somewhere west of Mitchell. Several geologists have cited Smith Rocks, 10 miles north of Redmond, as the possible vent from which one or more of the John Day tuffs came. Others feel that Smith Rocks are older than Miocene, but possibly produced welded tuffs found in the Clarno Formation.

Some of the most accessible outcrops of John Day strata lie along the John Day River between Picture Gorge and Kimberly, and in Painted Hills State Park northwest of Mitchell. Because each rain washes down some new material, the outcrops are never picked clean of fossils. The collector is most likely to find rodent or other small vertebrate remains; but larger fossils, especially teeth, are not uncommon. Collecting takes patience, but a walk into an area of John Day strata is worth while, if only because this brightly colored "badland" scenery is not often found in the Northwest.

TABLE 14-2 Tertiary stratigraphy in the Picture Gorge area. (After Baldwin, 1964.)

AGE	FORMATION	DESCRIPTION	THICKNESS, FEET
Pliocene	Rattlesnake	Gravel, sand, and silt, with vertebrate fossils. Prominent welded tuff	<800
Late Miocene	Mascall	Light-colored tuffaceous sandstone and siltstone; tuff beds. Vertebrate and leaf fossils	<1,000
Middle to late Miocene	Picture Gorge Basalt	Multiple basalt flows; minor sedimentary interbeds	>2,000
Oligocene to early Miocene	John Day	Fine-grained tuffaceous sandstone and siltstone; welded tuffs; basaltic flows near base. Vertebrate fossils	1,500
Late Eocene to early (?) Oligocene	Clarno	Varied volcanic flows, tuffs, breccias. Clastic interbeds, including massive mudflow breccias. Vertebrate and leaf fossils	3,000

▬▬ Major unconformity

14-9 *The upper part of the John Day Formation along the John Day River, 1 mile north of Picture Gorge. Sheep Rock in the center is topped by an erosional remnant of Picture Gorge (Columbia River) Basalt. The dark, massive unit capping the lower cliff is a welded tuff. (Photograph courtesy of Oregon State Highway Division.)*

LATE TERTIARY VOLCANISM

The character of the volcanic eruptions in the Blue Mountains Province changed markedly in the Middle Miocene after the silicic ash eruptions of the John Day Formation had ceased. Great floods of rapidly moving, fluid basalt covered the low hills of the region, and produced a totally new scene. Individual flows spread many tens of miles in a few hours, and the lava pooled to depths of 50 feet or more. There were many such outpourings, resulting in a great pile of basalt more than 2,000 feet thick. These lavas constitute part of the Columbia River Basalt Group.

The stratigraphy of the group has been studied more thoroughly in Washington than in Oregon. One of the named formations, however, the Pic-

14-10 *Smith Rocks near Redmond at the west end of the Blue Mountains Province. The strata are massive welded tuffs, possibly in or close to vents from which came one or more of the tuff units in the John Day or Clarno Formations. A Quaternary basalt flow underlies the plateau surface to the left. Three Sisters volcanoes in the distance to the right. (Photograph courtesy of Oregon State Highway Division.)*

14-11 *Picture Gorge and the John Day River, looking north (downstream). The canyon walls in the right foreground offer excellent exposure of numerous flows of the Columbia River Basalt. Sheep Rock is to the right of the big meander in the John Day River to the left of center, in front of Middle Mountain. Light-colored strata in the valley beneath Sheep Rock are tuffaceous sediments of the John Day Formation. The Mascall Formation underlies the grassy slopes in the foreground. (Copyrighted photograph courtesy of Delano Photographics.)*

ture Gorge Basalt, has as its type area the steep-walled canyon west of Dayville in the Blue Mountains. This is one of the best places in the state to see the Columbia River Basalt.

The basaltic lava did not erupt from volcanoes. Instead, it issued quietly from many long cracks or fissures that broke through the underlying rock. Following each eruption some lava cooled in the fissures and formed dikes. These are well exposed in the Monument area, 20 miles northeast of Picture Gorge, where the dark basalt of the dike contrasts strongly with the light-color ash beds of the enclosing John Day Formation. Other feeder dikes for Columbia River Basalt are found in the Wallowa Mountains. Flows of Columbia River Basalt cap many peaks in the Blue Mountains, especially along the north side. At one time, the basalt may have covered most of the region.

Water-laid ash beds very similar to the John Day Formation overlie the Picture Gorge Basalt near Picture Gorge. These strata constitute the Mascall Formation. Conglomerate lenses in the formation indicate that the material was derived largely by the erosion of volcanoes that lay a few miles to the south and east. Vertebrate and plant fossils indicate a late Miocene age for the Mascall Formation.

Volcanoes 30 miles east of Picture Gorge were erupting at about this same time. They produced the Strawberry Volcanics, which cover most of the high country in the eastern part of the Strawberry Mountains. They consist of a complex assemblage of basaltic to rhyolitic flows, breccias, and tuffs, with a total thickness of at least 6,500 feet.

The only extensive Pliocene formation in the Blue Mountains Province is the Rattlesnake Formation. It occurs principally in the upper part of the John Day River drainage. The Rattlesnake strata consist of at least 700 feet of poorly sorted stream gravel plus a welded tuff layer. The latter forms a very prominent bench from Picture Gorge east past the town of John Day. A radiometric determination of the tuff indicates that it was erupted 6.4 million years ago. The Rattlesnake gravels rest unconformably on the tilted and eroded edges of the Columbia River Basalt and the Mascall Formation and locally on rocks as old as Paleozoic. A rather varied mammalian fossil fauna has been col-

14-12 *Cross section through Middle Mountain, Sheep Rock, and Picture Gorge.*

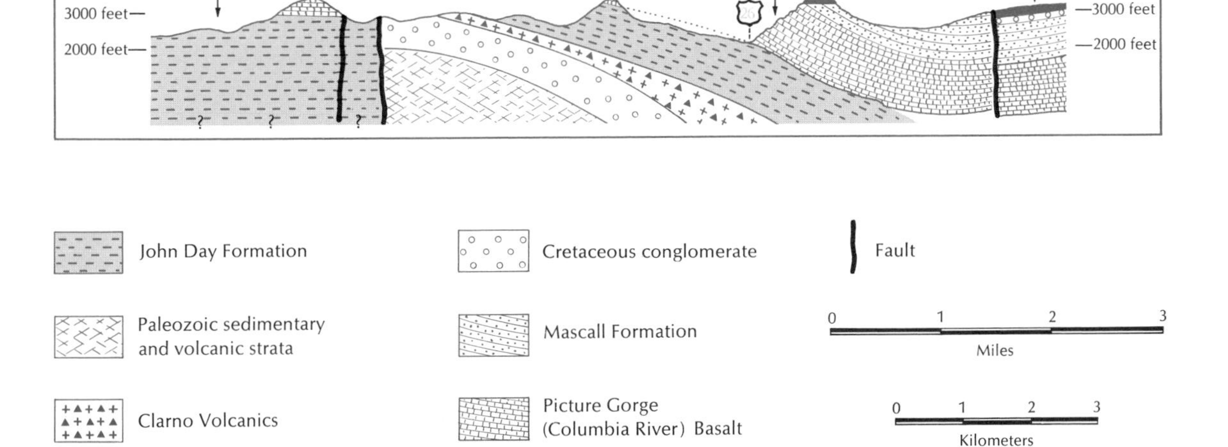

lected from the Rattlesnake gravels. It suggests that the middle Pliocene environment in the Blue Mountains Province was that of a semiarid grassland region.

BLUE MOUNTAINS STRUCTURE

The overall structure of the Blue Mountains Province, like that of the Cascade Range to the west, consists of one major structural pattern superimposed upon another. The east-west arch that defines the Late Cenozoic uplift of the mountains has been noted. A second structural trend is seen in a north to northwest orientation for folds and nearly vertical faults that cut across the axis of the Blue Mountains uplift. This second trend is essentially a continuation through the range of a structural pattern developed more clearly in the Basin and Range Province to the south. Both trends are seen most clearly in the offset by faulting and the warping of the layered Columbia River Basalt.

Deformation along one or the other of these directions has been going on during much of the Cenozoic Era. Although strata from every major epoch of the Tertiary Period are found in the Blue Mountains, there are numerous gaps in the record; most of these are angular unconformities. Each formation buried or partly buried hills eroded in older formations. Consequently, each unit is thickest where it was deposited in valleys, and thinnest where it buried hills. Geologists can gain some idea of the topography buried by a particular formation from information on the unit's distribution and thickness.

The John Day Formation appears to be thickest near the center of northwest-trending synclines. The folds are, in part, younger than the Columbia River Basalt, as that unit has been warped by the folding. The thickness data from the John Day strata suggest that the same folds existed in the Oligocene and that deformation along northwest-

14-13 *Entrance to Picture Gorge from the east. Mesa beyond the gorge is underlain by the welded tuff of the Rattlesnake Formation, dipping south at about 10°. The saddle in the left distance has been eroded in the Mascall Formation. (Photograph courtesy of Oregon State Highway Division.)*

trending lines has been going on for at least 30 million years. (This relationship is analogous to that found in the southern Cascade Range in Washington. There northwest-trending folds and faults that deformed the Lower Tertiary strata during the Miocene orogeny have remained active to the present day.)

A more obvious example of the longevity of structural trends in the Blue Mountains is found along the John Day Valley east from Picture Gorge. The Blue Mountains uplift is not actually a simple arch, for it contains some east-west folds and faults that parallel the major fold. The largest of these are the John Day Syncline and the John Day Fault. Both structures lie along the John Day Valley, the fault south of the axis of the fold. Thus the Mesozoic strata in the Aldrich Mountains to the south have been uplifted both by folding and by movement on the John Day Fault. The Columbia River (Picture Gorge) Basalt dips south under the valley along its north side (Figure 14-13) and then rises steeply along the south side of the valley. Moreover, the John Day Valley is also the area where the Pliocene Rattlesnake Formation is the thickest. It too has been folded and faulted in the John Day Syncline; but, whereas the Miocene basalt flows dip into the syncline at 30° or more, the Rattlesnake strata dip at only about 10°. Thus the folding began well before the middle of the Pliocene Epoch and has continued since then.

PLEISTOCENE GLACIATION IN THE WALLOWAS

The colder climatic conditions that prevailed during parts of the Ice Ages created some rather large glaciers in the Wallowa Mountains and small ones in the Elkhorn, Greenhorn, and Strawberry Mountains. In the Wallowas, they formed on high peaks toward the center of the range, some of which reach an elevation of about 10,000 feet, and radiated out from this central cap like spokes on a wheel. The largest glacier was 20 miles long and at least 2,000 feet thick, so these were not just little snowfields. The most striking effects of the glaciation are seen along the northeast flank of the mountains. Wallowa Lake, just south of Joseph, is situated at the lower end of a classic U-shaped valley. It is confined by a series of end moraines that loop from one valley wall out onto the flat plain toward Joseph and then back to tie into the other side of the valley. The highest of these glacial moraines stands about 2,000 feet above the floor of the valley.

The Wallowa Mountains were probably glaciated at least three times in the Pleistocene Epoch. Establishing the glacial sequence in such an area is not easy. For one thing, if the most recent glaciation was quite extensive (and in the Cordillera generally it was), much of the evidence for previous glaciations may be destroyed by glacial erosion or concealed beneath sediment deposited during the last glaciation. Furthermore, a glacier recedes spasmodically rather than at a steady rate. Its terminus may stand in one position for a relatively long time. When this happens, a substantial moraine may be deposited, so a receding glacier can produce a number of separate moraines. (These are called *recessional moraines*). Thus three moraines do not necessarily indicate three separate periods of glaciation.

Glacial geologists have developed some special methods for establishing glacial sequences. Sometimes radiocarbon dating of organic material in the glacial sediment helps, but this method is useless for material older than 50,000 years. Likewise, detailed stratigraphic studies making use of volcanic ash beds of known age (Chapter 13) has proved useful only for the most recent glaciation. Most studies of older events are based on the principle of superposition plus observation of the effects of postglacial weathering on the deposits. Granitic rocks tend to decompose and disintegrate more readily than do fine-grained or glassy volcanic rocks. The granitic boulders that mantle the surface of the most recent moraines in the Wallowa Mountains are still hard and relatively unweath-

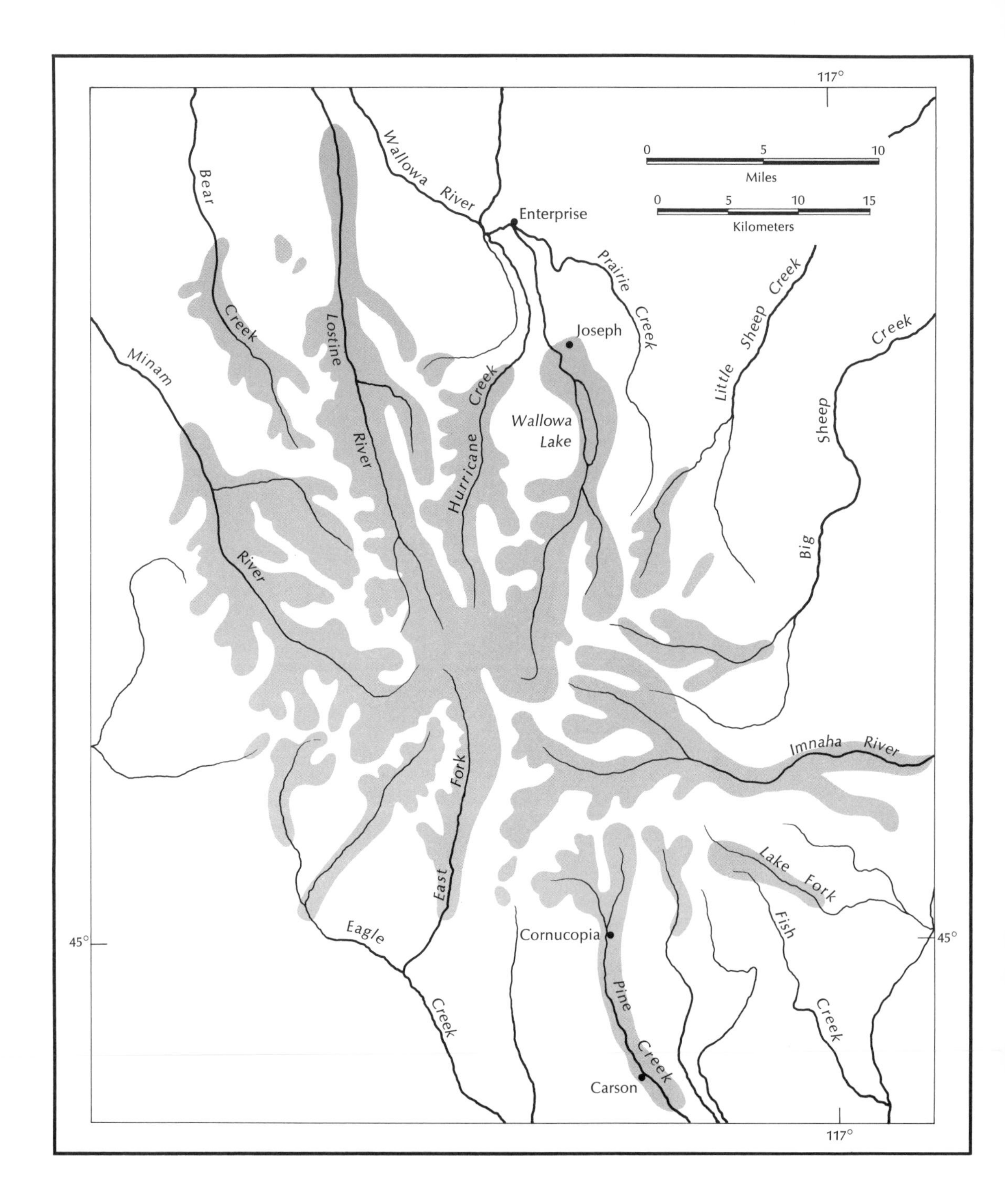

14-14 *Distribution of Pleistocene glaciers in the Wallowa Mountains. (After D. R. Crandell, Glacial History of western Washington and Oregon, in H. E. Wright, Jr. and David G. Frey (eds.),* The Quaternary of the United States, *copyright © 1965 by Princeton University Press, fig. 4, p. 350. Reprinted by permission.)*

ered. Those on older moraines crumble more easily, and many have disintegrated away to sand. Similarly, the thickness of the soil developed on a moraine is some indication of its age. Even hard, volcanic pebbles or boulders are useful, for the thickness of the discolored weathering rind on such a stone increases with time. Thus rind thickness measurements on pebbles of, say, the Columbia River Basalt might help to distinguish moraines produced during different glacial periods. None of these methods yield absolute ages—they only indicate the relative ages of two or more moraines.

Counting the number of granitic boulders per acre or breaking 100 pebbles and measuring the thickness of the weathering rind does not sound like very exciting field work. Most glacial geologists would agree, but these basic methods are among those that permit geologists to say with confidence, "The Wallowa Mountains were probably glaciated at least three times in the Pleistocene Epoch."

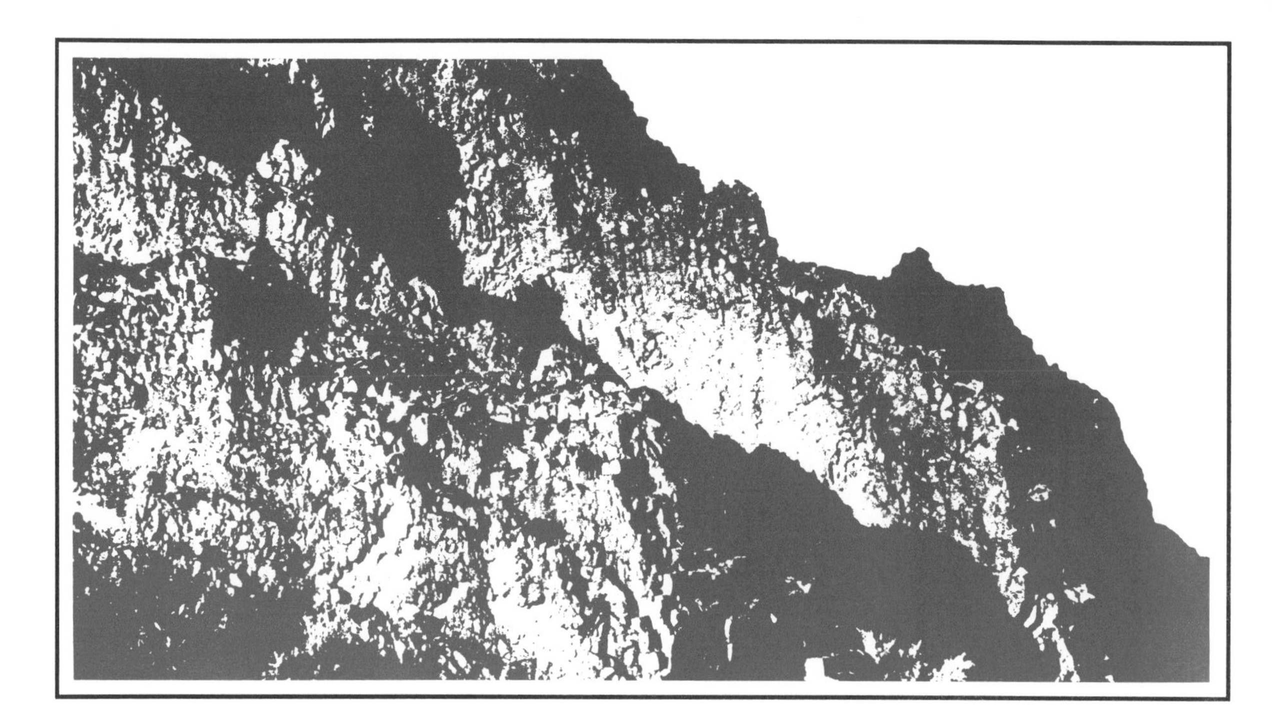

15 southeastern oregon— a volcanic highland

■ The desolate southeastern quarter of Oregon, that part of the state lying south of the Blue Mountains and east of the Cascades, has one of the world's best displays of volcanic landforms. The High Lava Plains offer the most variety of recent volcanoes, while the Basin and Range Province and the Owyhee Uplands provide excellent exposures of older fissure basalt flows. The aridity of southeastern Oregon prevents more than a grass-sage-juniper flora from growing (except close to the Cascades), but the sparse vegetation means generally good rock exposure. In some ways, this country seems almost timeless—a place where geologic processes might work very slowly, and changes in the landscape might occur at almost imperceptible rates. Yet here is a land that only yesterday (geologically) was dotted with many large lakes—where the highest mountains even contained a glacier or two.

PHYSIOGRAPHIC PATTERNS

Southeastern Oregon is characterized by north-trending mountain ranges and intervening flat valleys, which are blanketed with alluvium or recent lava flows. Most of the region lies at an elevation of more than 4,000 feet. The highest point, almost 10,000 feet above sea level, is in the Steens Mountains. The drainage is largely interior, with short, intermittent streams flowing into low lakes and playas that do not have outlets. Rivers that exist from the region and drain parts of the margin are the Owyhee River (a tributary of the Snake River), the Deschutes River (a tributary of the Columbia), and the Klamath River. The Klamath flows from Klamath Lake to the Pacific Ocean via a very tortuous route through the Cascades and the Klamath Mountains of northern California.

The topography of southeastern Oregon strongly reflects the regional stratigraphy and structure. The rocks consist primarily of extensive sheets of solidified lava. Much of the lava is basalt, but some widespread silicic ash flows and tuffs are present as well. The volcanic strata are interbedded with nonmarine sandstone, shale, and conglomerate beds.

Movement on steeply dipping faults has tilted many of the sections. The faults trend generally north-south and define many of the mountain fronts. Some of the ranges have been raised (relative to the adjoining basins) by faults on both sides; others are tilted fault-block mountains raised on one side only. Some faulting has occurred recently, for many of the fault scarps are fresh, and drainage patterns have been disturbed.

The evidence of Pleistocene volcanism throughout the region is impressive. Most of the eruptions issued from vents and produced volcanoes. These are mostly small cinder cones that literally dot the surface of parts of the High Lava Plains. The vents were localized to some extent by faulting, and

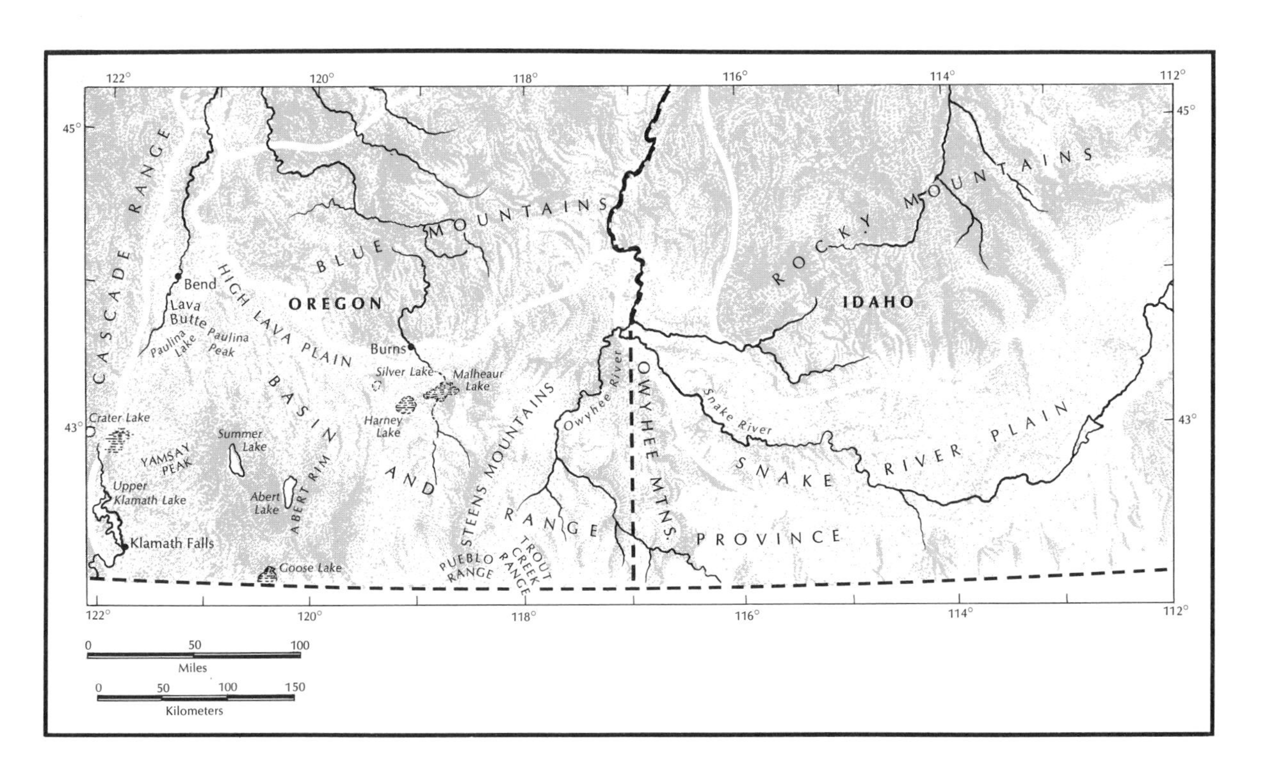

15-1 *Physiography of southeastern Oregon. (After Raisz, 1941.)*

15-2 *Alvord Desert, looking east from near the top of Steens Mountain. The topography of most of this Basin and Range Province is controlled by steep north-south faults, such as the one along the base of this range. The strata in the foreground are fissure basalt flows that were erupted in the Miocene Epoch before Steens Mountain was raised up by faulting. (Photograph courtesy of Oregon State Highway Division.)*

definite lines of cones have developed along some of the fractures. The flat plateau surface beneath much of the High Lava Plains is formed by one or more extensive fissure flows of basalt.

PRE-QUATERNARY STRATIGRAPHY

The oldest rocks in southeastern Oregon are found in the Pueblo and Trout Creek ranges, south of Steens Mountain. They consist of nonfossiliferous sedimentary and volcanic rocks that have been somewhat metamorphosed. The strata were cut by diorite and granodiorite intrusions, probably in the Cretaceous Period. The sedimentary and volcanic rocks may be related to some of the Triassic formations of the Blue Mountains Province. Comparable pre-Cenozoic rocks probably underlie the entire region, but this is the only place where the older basement is exposed. (Geologists use the term "basement" in a relatively informal manner. Sometimes it refers to the oldest rocks found in an area, especially if they are crystalline in type and rest unconformably under all of the other formations. In this sense, the "basement" is Mesozoic in southeastern Oregon, pre-Devonian in the North Cascades, and Early Precambrian in the Canadian Shield. Some glacial geologists sometimes refer jokingly to all rock beneath Pleistocene glacial deposits as "basement." "Hard rock" geologists counter by loosely designating all unmetamorphosed volcanic and sedimentary strata that overlie the units that interest them as "overburden.")

Upper Oligocene and lower Miocene strata occur in many ranges but are not well exposed. They consist of a wide variety of volcanic flows, pyroclastic units, and minor interbeds of nonmarine sedimentary rocks. The rock types plus some meager fossil evidence suggest a correlation with the John Day Formation of the Blue Mountains.

The Steens Basalt is probably the most famous formation of southeastern Oregon. It is made up of more than 100 separate flows, with a thickness locally of more than 4,000 feet. The formation covers an area of several thousand square miles east of Abert Rim. The flows issued from numerous long fissures, some of which are exposed as dikes along the eastern flank of Steens Mountain. The Steens Basalt is late Miocene in age.

The Owyhee Basalt is the most prominent unit in the Owyhee Upland. It probably correlates with the Steens Basalt, and both of these formations erupted at approximately the same time as the Columbia River Basalt. Each of these three great piles of upper Miocene basalt has some unique characteristics, but they could have come from different parts of one immense subcrustal magma chamber.

The Pliocene stratigraphic record of southeastern Oregon consists mostly of pumiceous sandstones, shales, and claystones that were deposited in lakes and along streambeds. Thin sheets of basalt and of welded tuff are interbedded with the sedimentary rocks. These Pliocene strata compare directly with the Deschutes and Dalles Formations of north-central Oregon, the Rattlesnake Formation of the western Blue Mountains, the Ellensburg Formation of south-central Washington, and the Idaho Group of southern Idaho.

In summary, the pre-Quaternary stratigraphy of southeastern Oregon contains no real surprises. Each of the major formations has its counterpart in one or more nearby provinces, where generally more intensive study has provided geologists with a clearer picture of the ancient environments.

QUATERNARY VOLCANOES

The variety of young volcanic features found in southeastern Oregon can be illustrated by four volcanoes—Newberry Crater, Lava Butte, Hole-in-the-Ground, and Fort Rock. All these occur within an area about 30 miles long and 10 miles wide in the western part of the High Lava Plains.

Newberry Crater, 25 miles south of Bend, is quite a spectacular volcano. The original cone (Newberry Volcano) was a broad, gently sloping shield volcano, 20 miles across and rising only about 4,000 feet above the surrounding plain. Like Mount Mazama (Crater Lake), the summit region disappeared, partly by explosion and partly by collapse, leaving a large caldera. Eruption continued after the formation of the caldera. This built two overlapping cinder cones in the center of the caldera floor; these now separate East and Paulina Lakes. In addition, a large, viscous flow of glassy obsidian issued from a vent along the south rim of the caldera. It flowed slowly toward the center, producing a very fresh-looking, wrinkled surface that contrasts markedly with anything else in the area. This flow is the source of much of the obsidian that is used for decorative stone in central Oregon. Undoubtedly, the obsidian was used long ago by Indians for arrowheads and other implements. (A walk of only a few yards across the surface of this flow is a very convincing demonstration of how sharp obsidian is.)

An excellent view of the entire caldera is gained by the road to the lookout on Paulina Peak. This, the highest part of the caldera rim, is composed in part of dikes that fed siliceous flows on the vol-

15-3 *Kiger Gorge toward the north end of Steens Mountain, looking south. The strata exposed in the canyon walls are flows of the Steens Basalt. Note the gentle dip to the east of the strata, produced by tilting of the range during fault uplift. (Copyrighted photograph courtesy of Delano Photographics.)*

15-4 *Newberry Crater from the west. This large volcanic caldera is occupied by two lakes, Paulina (in the foreground) and East Lakes, separated by two overlapping cinder cones and lava flows which issued from the caldera rim. The toe of a fresh obsidian flow is seen along the right edge of the photograph, near the south shore of Paulina Lake. (Copyrighted photograph courtesy of Delano Photographics.)*

cano's southwest flank. The views to the south and west are equally impressive. The many small cones that dot the surface of the High Lava Plains are visible to the south, and the many overlapping volcanoes that cap the Cascade Range are seen to the west.

Lava Butte, located 10 miles south of Bend along U.S. Highway 97, is a readily accessible cinder cone. The summit of the cone, only a few hundred feet high, is reached by a good road. A small display at the summit explains the history of the cone and points out the features that are seen along a short trail around the crater. The surface of the plain immediately south and west of Lava Butte is covered by a very blocky basalt flow that issued from a vent at the southern edge of the cone. The flow looks very young, but it is probably 5,000 to 6,000 years old. Lava Butte is the youngest of a series of similar cinder cones (with basaltic flank eruptions) located along a fracture zone that extends southeast up the flank of Newberry Volcano.

Hole-in-the-Ground, 20 miles south of Newberry Crater, is a fine example of a volcanic feature known as an explosion crater. The crater is several thousand feet in diameter and has a rim that is almost a perfect circle. The rim rises several hundred feet above the surrounding plain, while the flat crater floor is a little below the level of the plain. At first glance, one might attribute the origin of the crater to the fall of a meteorite. The crater was blown out by volcanic explosion, however, for the rim consists of pyroclastic debris rather than fragments from the rock beneath the surface of the plain. The similarity in form between explosion and impact craters explains why some geologists have suggested that certain of the moon's craters might have had a volcanic rather than an impact origin.

A few miles southeast of Hole-in-the-Ground, there is a tuff ring or tuff cone known as Fort Rock. It also was built by explosions, but the eruption lasted longer and ejected more volcanic debris than did the Hole-in-the-Ground eruption. In both volcanoes, the explosions were caused by steam, which was produced, presumably, when hot lava rising toward the surface encountered

15-5 *Lava Butte, a cinder cone that formed on the High Lava Plains 5,000 to 6,000 years ago. The jagged rock in front of the cone is a blocky basalt flow that issued from the base of Lava Butte. (Photograph by Bates McKee.)*

15-6 *Hole-in-the-Ground south of Newberry Volcano. This is a classic example of an explosion crater, although its form resembles a meteorite crater. (Copyrighted photograph courtesy of Delano Photographics.)*

groundwater. Steam is the driving mechanism for virtually all eruptions that produce cinder cones. The differences in morphology between many of the cones of southeastern Oregon are explained in part by how much lava and what type of pyroclastic debris were ejected by the steam.

LAVA CAVES AND TREE CASTS

Southeastern Oregon is a good place to go underground in a lava flow, both figuratively speaking and actually. Basaltic lava flows that issue from volcanoes and flow down a slope sometimes develop a feature known as a *lava cave* or a *lava tube*. This happens in flows that have ceased moving and have partly solidified. At this point, the top, sides, bottom, and front of the flow have solidified; but the interior of the flow is still molten. This molten lava may burst through the front of the flow or (on a moderate slope) through the top surface and flow farther downhill. Its departure leaves a hollow cavern inside the flow, into which the top of the flow may or may not collapse. Some lava caves are quite long, such as Ape Cave on the south flank of Mount Saint Helens, which has a total length of more than 2 miles. Lava River Cave, located just southeast of Lava Butte close to U.S. Highway 97, is a readily accessible cave. It contains an interesting feature found in many other lava tubes. This is a series of "bathtub rings," made in the congealed basalt plastered on the walls of the cave. Each ring presumably repre-

15-7 *Fort Rock, rising more than 300 feet above the High Lava Plains, is the remnant of a tuff cone. Much of the erosion was caused by wave action in a Pleistocene lake that covered the valley floor. (Photograph courtesy of Oregon State Highway Division.)*

sents a former short-lived stand of the lava level as the interior drained away. (The reader can appreciate the impossibility of observing such processes in action.)

The *tree cast* is another feature common to basalt flows found in the High Lava Plains. This is formed when lava engulfs a tree (either upright or downed). The tree is charred and ultimately disappears, either by burning or by rot. The shape of the tree is cast in the flow as a tube, whose walls preserve as ridges the cracks of the charred tree. Many excellent tree casts can be seen in the Lava Cast Forest, several miles east of Lava Butte.

15-8 *Tree cast near the top of a blocky flow of basaltic lava in the Oregon Cascades. (Photograph courtesy of Oregon State Highway Division.)*

AGATES AND PETRIFIED FOSSILS

The underground cavity in a lava flow that ultimately becomes an agate, a geode, or a "thunder egg" is the one that interests the most people. These features form in many types of volcanic rocks, including pyroclastic deposits; and they are found in almost all regions, especially arid ones, in which many volcanic strata are exposed. This includes a large part of eastern Oregon.

Agates, geodes, and thunder eggs are lovely objects that form in an unglamorous manner. Each consists primarily of some silica (silicon dioxide) mineral, either quartz, chalcedony, or opal. The silica is precipitated in a cavity by the action of groundwater, in much the same way as dripstone (calcium carbonate) is deposited in some caverns and caves. The process is a very slow one. It requires the passage through the cavity of a considerable volume of groundwater, for even very "hard" water does not contain much silica. Thus the cavities in a lava flow or ash bed may slowly be filled by silica, and some millions of years later weathering and erosion may produce good hunting for the rockhound.

The petrifaction of fossils is a process very similar to the formation of agates, but there is an important difference. Both processes involve the gradual precipitation of silica from groundwater. To form an agate, a cavity is filled; to petrify a fossil, the original organic material of the bone (or shell or tree, or whatever) is *replaced*. The re-

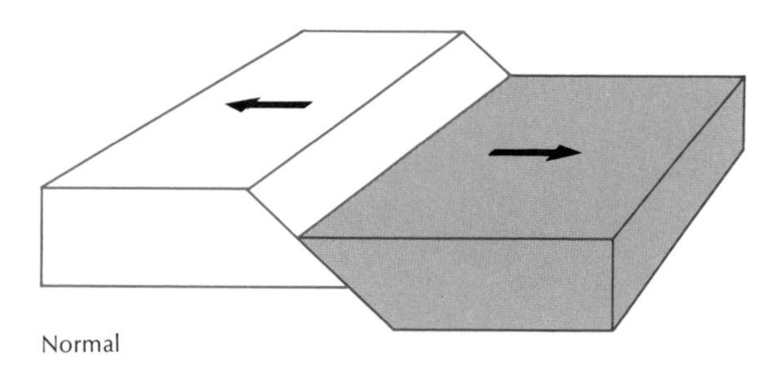

Normal

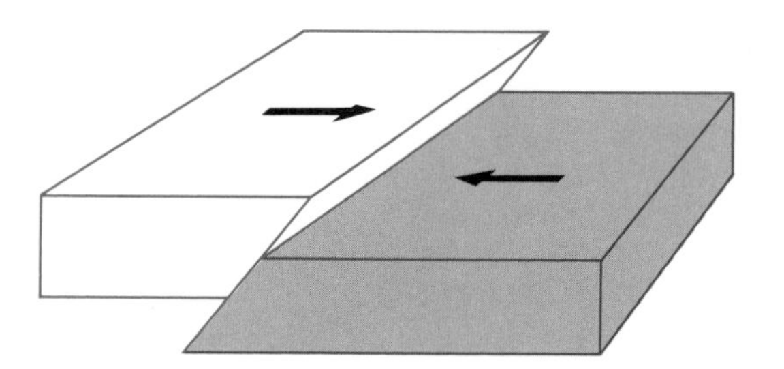

Thrust

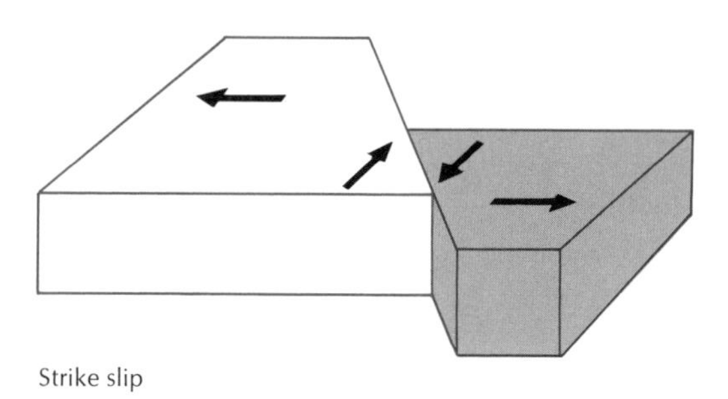

Strike slip

15-9 *The three major types of faults. The arrows indicate the horizontal motion of the two sides relative to each other.*

placement process is a very gradual one in which most of the organic detail is preserved faithfully by the silica. The end result is, perhaps, an opalized bone or log in which the original cell structure, growth rings, and the like are clearly discernible even though the organic material has disappeared completely. If on the other hand a tree cast is subsequently filled with silica, the resulting agate has the shape of the original tree, but none of its organic detail.

The reader who is interested in agates, geodes, and petrified wood and how they form will find good summaries in many of the rock and mineral books cited for Chapters 1 through 3. Now the narrative must turn toward problems with more significance for understanding the structural evolution of eastern Oregon.

FAULTS AND WHY THEY FORM

Faulting has played a major role in the evolution of southeastern Oregon. Some of the relationships between faults and stress in the earth must be understood in order to consider the possible significance of the structures found in this region and elsewhere in the Northwest.

Geologists recognize three major types of faults—thrust faults, normal faults, and strike-slip faults. Each type represents a rock fracture formed in response to a particular stress field. The fundamental relationships between faults and stresses is illustrated in Figure 15-9. Thrust faults were discussed in Chapter 7 but some additional comment may be helpful. *Thrust faults* form in response to horizontal compressive stress in the earth. The strata can yield in several ways. One is by folding, the folds forming at right angles to the compression. A second response is thrust faulting, in which one section of rock is thrust up and over the other. In most instances, the strata yield to horizontal compression by both folding and faulting, and this explains why folds and thrust faults normally occur together in deformed mountain ranges.

In thrust faulting, the strata are compressed; but in *normal faulting* they are stretched. Note in Figure 15-9 that the total length of the rock section has been increased as the result of faulting. Another way to look at normal faults is to regard them as having formed in response to vertical compression. If the diagram illustrating a normal fault is rotated 90° in a clockwise direction, the geometry becomes identical to that of the thrust fault. Since the principal vertical stress in the earth's crust is gravity, normal faults are sometimes called *gravity faults.*

The motion in a *strike-slip fault* differs from that of the other two types of faults in that the rock masses on the two sides of the fracture move horizontally, and there is no vertical displacement. Consequently, the strata are stretched in one direction and compressed in the other. This relationship must be understood in order to consider the possible causes of some of the structures found in the Cordillera.

The fact that parts of the earth's crust in the past have undergone major horizontal and vertical displacements has been apparent throughout the book. Fault movement is fundamental to much of the motion. The geologist studying an area uses the nature, orientation, and age of faults and folds to interpret past displacements and the nature of the stresses that produced them. For example, if he finds numerous normal faults that trend north-south across his area, he concludes that they formed in response to a stretching or lengthening of the strata in an east-west direction. North-trending normal faults are the dominant Late Cenozoic structures throughout the entire Basin and Range Province. (The province extends from the Blue Mountains and Snake River Plain south to essentially the Mexican border and from the Cascade–Sierra Nevada front eastward to the Rocky Mountains.) The total east-west lengthening of the region as the result of normal faulting has been estimated as 30 to 50 miles, calculated across Nevada and western Utah. Note that this calculation does not explain why the region stretched, which, of course, is essential to understanding the deformational history of the Basin and Range Province.

ONE MODEL FOR LATE CENOZOIC DEFORMATION

The development of normal faults in the Basin and Range Province may have been produced by horizontal rotation or bending of the earth beneath the region rather than by any simple stretching. In other words, the structures seen at the surface may be side effects of a quite different type of motion occurring at depth. The importance of the east-west lengthening and north-south shortening produced as a result of the strike-slip faulting in Figure 15-9 was noted. What would happen at the earth's surface if just this kind of displacement was going on at depth? A simple experiment serves to illustrate the possibilities. Two boards are placed against one another on a tabletop and are covered with moist clay. (The consistency and thickness of the clay is critical, so if you want to try this experiment, be prepared to spend some time with it.) One board is then moved parallel to the other so that the boards remain in contact. To discuss the results, let us assume that the boards are oriented northwest-southeast and that the board on the southwest side is moved to the northwest. A little displacement of the board may produce most interesting structures in the clay immediately above the contact between the two boards. They consist of tiny normal faults oriented north-south and folds and/or thrust faults oriented east-west. In other words, strike-slip displacement at depth has stretched the clay in an east-west direction and compressed it in a north-south direction, and the clay has responded by folding and faulting. If the clay is too stiff, or if the board is moved farther, the strike-slip "fault" at depth may carry to the surface as a strike-slip fault in the clay.

In order to apply this simple experiment, we must imagine that the Cordilleran region is underlain by an almost infinite number of narrow boards, each oriented northwest and each moving to the northwest a tiny bit faster than its neighbor to the east. The more numerous the boards, the more evenly the stress is distributed in the overlying crust. With enough boards, the effect on the clay is the same as if the underlying material is rotating, but the clay cannot rotate with it. This type of displacement at depth could explain the following Late Cenozoic structures in the Cordillera:

1. The normal faulting of the Basin and Range Province, the western part of the Snake River Plain, and the Blue Mountains.
2. The east-west folding and thrust faulting of the Blue Mountains Province, the Columbia Plateau, and parts of the Cascades and Coast Range Province.
3. Large northwest-trending strike-slip faults such as the San Andreas and Hayward Faults in California, the Las Vegas Shear Zone and Walker Lane in western Nevada, and the Lewis and Clark Line in northern Idaho and northwestern Montana (Appendix E, Trip 8).
4. The apparent bending of the pre-Cenozoic margin of the continent along the south side of the "Columbia Arc" (Chapter 11).
5. Deep faults responsible for many of the region's earthquakes.

Note that this section is entitled "One Model for Late Cenozoic Deformation" and not "The Model, etc." This model has some problems—for instance, it does not explain such fundamental structures as the Cascade and Coast Range Uplifts. It has been sufficiently useful, however, so that it has been suggested in one form or another by geologists working in almost every major province in the Cordillera. Furthermore, it has been gaining in stature in recent years, supported by new concepts of global tectonics. The many structural problems that are not explained by the model may reflect our inability to fully understand the rotational process and how it relates to the interaction between the continental crust, the oceanic crust, and the asthenosphere.

PLEISTOCENE LAKES AND GLACIERS

Many large lakes formed in the Pleistocene Epoch in what is now the arid Basin and Range Province. The stratigraphy of the entire Cordilleran region demonstrates that the overall climate slowly changed during the Cenozoic Era from warm and subtropical to generally cool and dry. This change was more pronounced to the east of the Sierra

15-10 *A simple model to illustrate surface effects of subsurface horizontal shear. The insets are magnified cross sections of structures formed near the surface of the clay when the left-hand board is moved away from the viewer.*

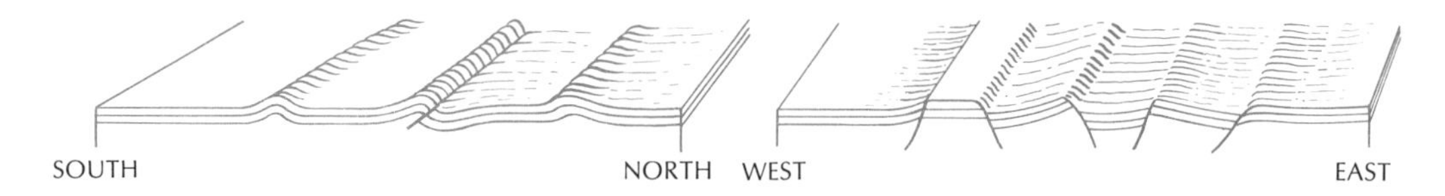

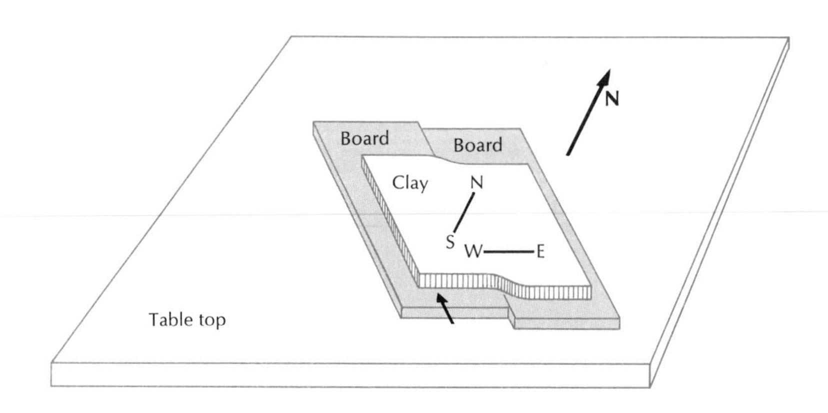

15-11 *Kiger Gorge in Steens Mountain, looking north. The U-shaped profile of the valley results in part from Pleistocene glaciation. Note the numerous small block faults that displace the layers of Steens Basalt in the far wall of the canyon. (Photograph courtesy of Oregon State Highway Division.)*

Nevada-Cascade-Coast Mountains chain, which formed late in the era and blocked the onshore flow of moist, marine air. Clearly, however, it is drier in the interior today than it was even a few tens of thousands of years ago, when the large lakes last existed.

Vestiges of these Pleistocene lakes include the Great Salt Lake of Utah, Pyramid and Walker Lakes in western Nevada, and the many small lakes in southeastern Oregon (such as Upper Klamath, Silver, Summer, Abert, Harney, Malheur, Goose, and the Warner Lakes.) The higher levels of these lakes in the past are proved along the valley sides, where prominent shoreline terraces are seen as much as 350 feet above present lake levels. Associated with the terraces are many other shoreline features, such as bars, spits, deltas, and beach deposits. The former extent of these lakes can be determined simply by "filling up" the present topography with water to the level of the highest shoreline. This method demonstrates, for instance, that Summer Lake and Lake Abert were once joined (via Valley Falls) as part of a single lake, known as Lake Chewaucan. (To avoid confusion, geologists use different names for the older, enlarged lakes than they do for their modern remnants. Thus Lake Bonneville preceded the Great Salt Lake and Lake Lahontan was the large lake that covered many of the basins of northwestern Nevada.)

The levels of these ancient lakes fluctuated with periods of Pleistocene glaciation, the highest levels corresponding to periods of maximum glaciation. Such periods resulted presumably from worldwide climate changes. Moreover, the existence of continental ice sheets in the northern Cordillera and extensive alpine glaciers in the Cascades and Sierra Nevada would have had some influence on the climate of the western interior of the continent. At any rate, several times in the Pleistocene Epoch, southeastern Oregon received considerably more precipitation than it does today, and the scene was quite different.

At least four different mountains in southeastern Oregon had small alpine glaciers in the Pleistocene Epoch. These were Steens Mountain, Newberry Volcano, Gearhart Mountain (located south of Summer Lake), and Yamsay Peak, which rises to an elevation of 8,242 feet across Klamath Marsh from Crater Lake. The most pronounced effects of glaciation are in Kiger Gorge on Steens Mountain (Figure 15-11).

Erupting volcanoes, immense lakes, alpine glaciers—it seems hard to believe that these features were an important part of this desolate corner of Oregon just a few thousand years ago, and yet the evidence is unmistakable.

16 snake river country

The Snake River starts at the continental divide in the Teton Range of western Wyoming and flows northwest to join the Columbia in south-central Washington. Its path is not direct. First it loops to the southwest, following the structural downwarp known as the Snake River Plain around the southern end of the many mountain ranges of central Idaho. Then it turns north and runs through the eastern end of the Blue Mountains along the Idaho-Oregon border. At Lewiston it turns west again, across southeastern Washington to Pasco where it empties into the Columbia. As tributaries go, this is a dirty one, for although its volume at Pasco is considerably less than that of the Columbia, its sediment load is greater. At present, this sediment comes largely from the lower reach of the river, for dams upstream have blocked much sediment transport. Before the dams were built, the sediment consisted of particles from five different geologic provinces—the Northern Rockies, Basin and Range, Snake River Plain, Blue Mountains, and Columbia Plateau Provinces.

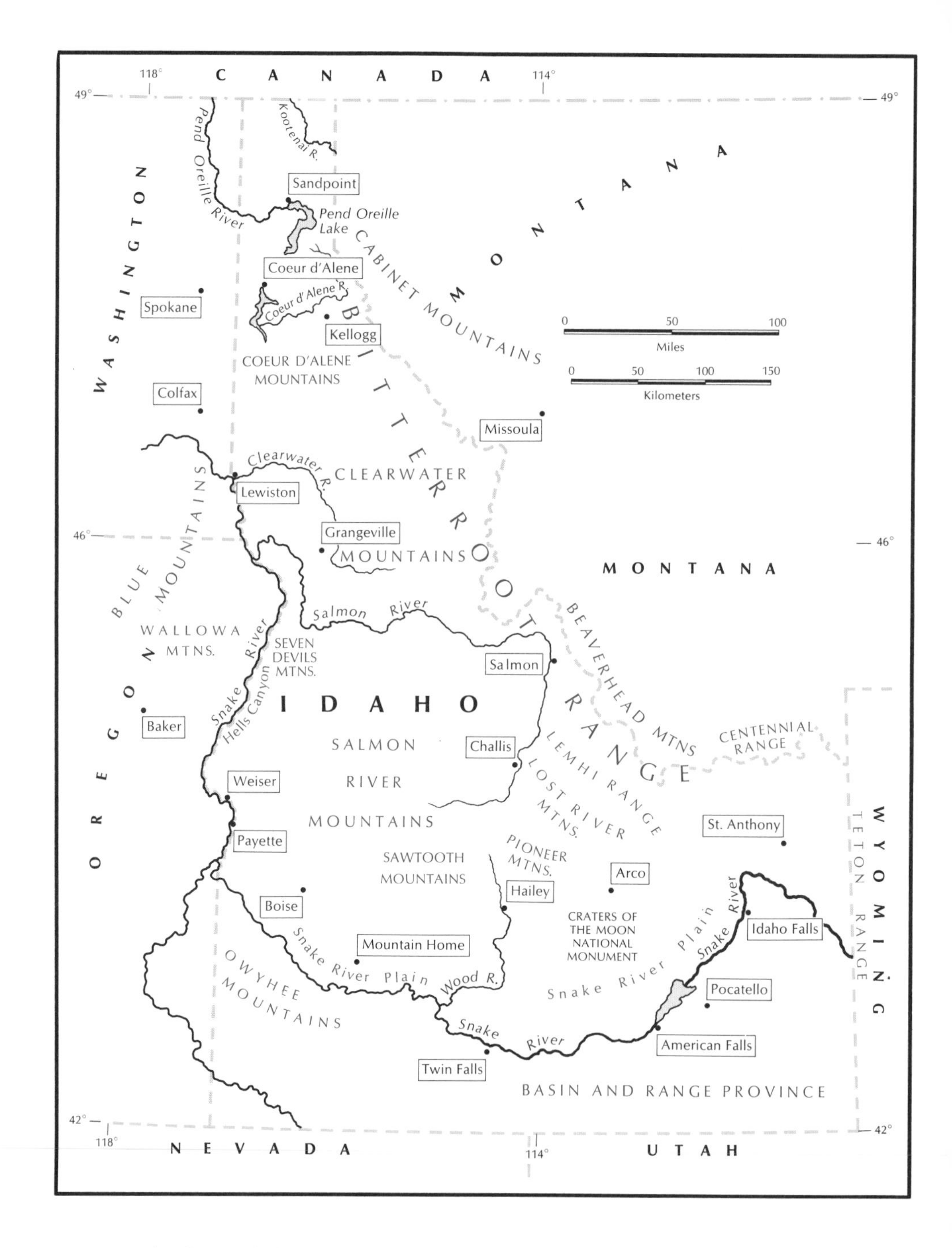

16-1 *Major landforms of Idaho.*

HARD TIMES ON THE SNAKE RIVER

The dams built by Man in the last several decades are nothing compared to the problems that the Snake River has handled in the past several million years. For one thing, it had to battle the rising Blue Mountains Uplift which grew across its path. A lesser river might have given up, but the Snake had no place else to go. It had to cut the deepest canyon in North America, Hells Canyon, and not through sedimentary strata like those found in the Grand Canyon but through hard, massive greenstone.

Another problem that the Snake River met successfully was Late Cenozoic volcanism on the Snake River Plain. At first glance, the basaltic lava, exposed in canyon walls where the river has incised into the plain, looks much like the Columbia River Basalt. There is, however, an important difference. The Columbia River Basalt flows can be traced for many tens of miles, and the succession of flows seen on one wall of a canyon matches perfectly that found on the opposite wall. (This is true in most river canyons cut in flat-lying strata.) In many places along the Snake River, the section on opposite sides of the canyon is entirely different—nothing matches at all. The Snake River Basalt was not erupted entirely from fissures, as was the Columbia River Basalt. Instead, many of the flows came from vents that built very low, broad shield volcanoes. The Snake River was trying to maintain a course westward through this maze of volcanoes, but new cones kept growing in its path, filling its channel with lava and forcing it to go around them. There were so many cones that finally they began to overlap one another. This explains some of the stratigraphic mismatch between opposing canyon walls, for in some cases the two sections are parts of different volcanoes.

During the last glacial maximum, the Snake River gained a measure of revenge on the geological obstacles placed in its path. Lake Bonneville, the immense lake that covered much of western Utah and eastern Nevada, rose until it spilled over into the Snake River Plain. For a brief time the drainage area of the Snake River was approximately double what it had been. There was no question then that anything was going to block the river's path to the sea.

IDAHO GEOGRAPHY

The Snake River and its tributaries drain much of the state of Idaho. Exceptions are the northern panhandle, whose runoff reaches the Columbia via the Spokane, Pend Oreille, and Kootenai Rivers, and the southeastern corner of the state, which drains into the Great Salt Lake.

The northern two-thirds of Idaho lies largely within the Rocky Mountains Province. The principal ranges, from north to south, are the Selkirk, Bitterroot, Clearwater, Salmon River, and Sawtooth Mountains. The Sawtooths are flanked on the east by the Pioneer, Lost River, Lemhi, and Beaverhead Mountains.

The smoothest topography in the state is found on the Snake River Plain, but parts of it are cut by impressive canyons or dotted with recent, small volcanic cones. South of the Snake River Plain, the topography is that of the Basin and Range Province (Chapter 15), although somewhat subdued compared to southeastern Oregon.

The Blue Mountains pinch off the Snake River Plain at its west end. The western part of the state from the Seven Devils Mountains north to Lake Coeur d'Alene is the high eastern edge of the Columbia Plateau Province. The basaltic lava flows of the plateau end against the Clearwater Mountains to the east.

THE GEOLOGIC FRAMEWORK

The geology of central Idaho is dominated by the Idaho Batholith. This Cretaceous intrusion is one of the largest in the Northwest and has produced much of the state's mineral wealth. Its emplacement represented another nail in the coffin of the Cordilleran Geosyncline. The country rock peripheral to the Idaho Batholith has suffered intensive metamorphic recrystallization. Consequently, for

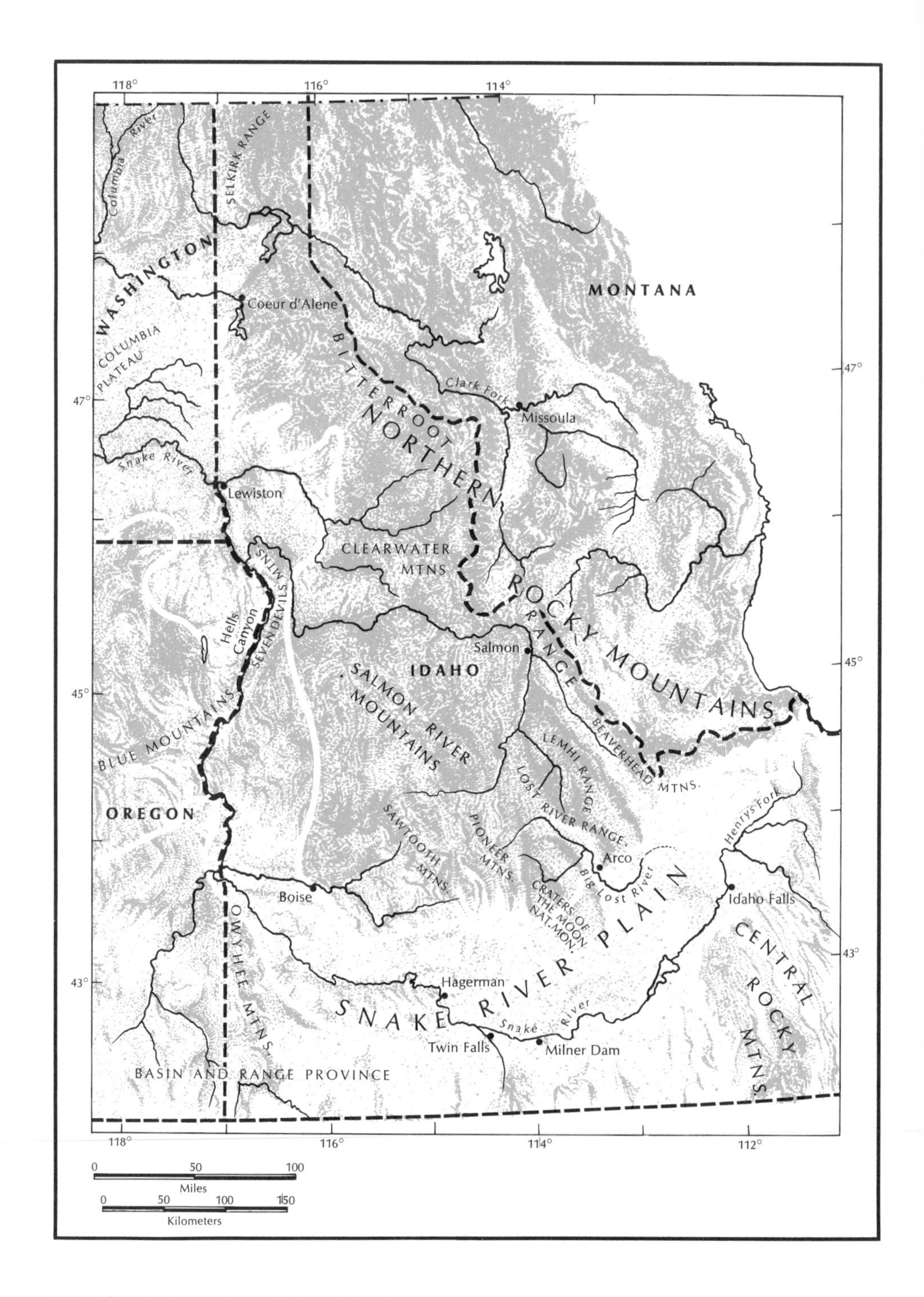

16-2 *The physiography of Idaho. (After Raisz, 1941.)*

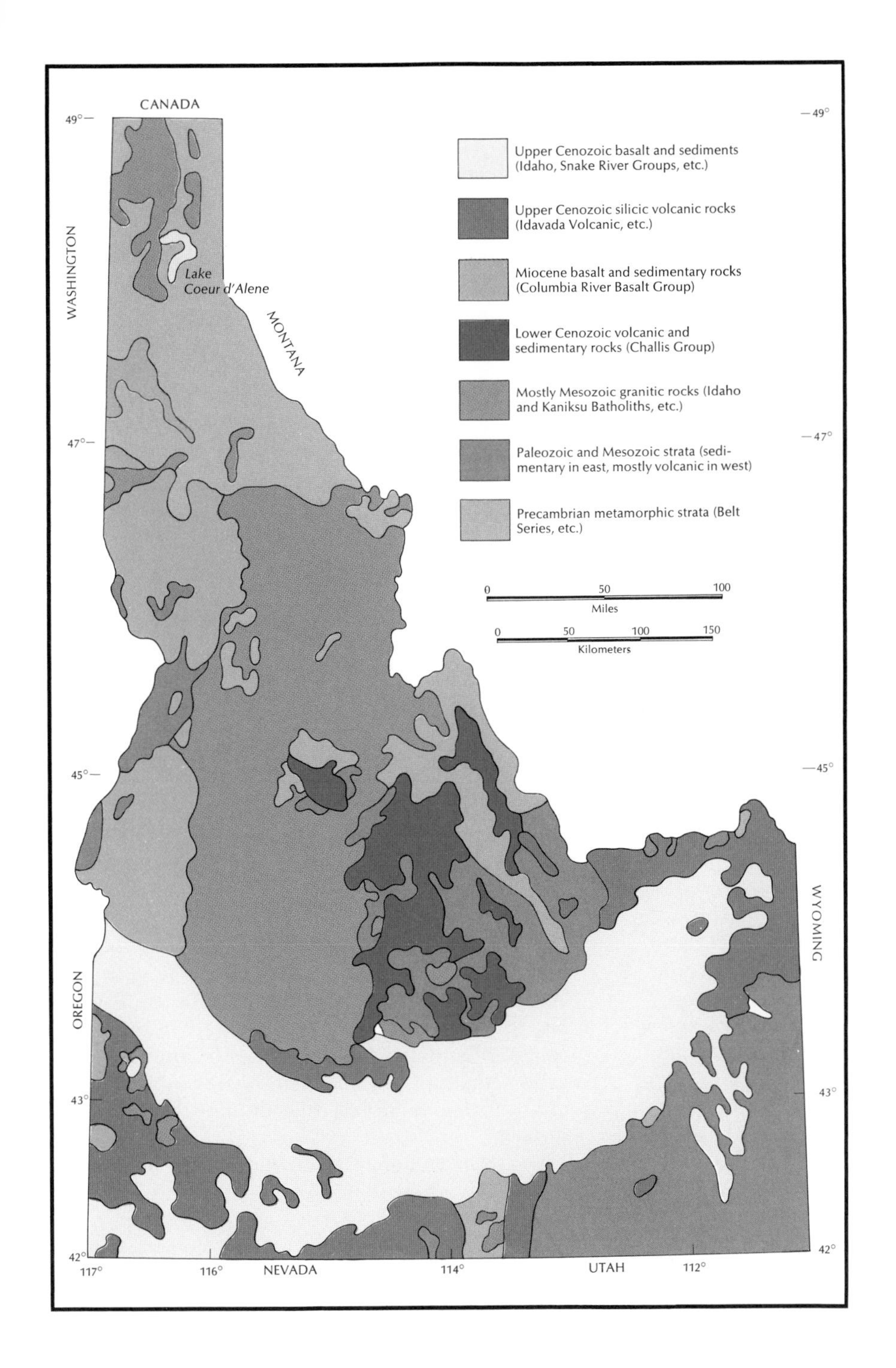

16-3 *Generalized geology of Idaho. (After Ross, Reid, and Weissenborn, 1964, Fig. 6.)*

information about the pre-Cretaceous history of the state, it is necessary to study areas well removed from the intrusion.

Precambrian strata of the Belt Series (Chapter 6) underlie much of the panhandle and adjacent parts of British Columbia and Montana. Paleozoic and Mesozoic rocks crop out primarily in the east-central and southeastern parts of Idaho, both north and south of the Snake River Plain. Upper Paleozoic and Lower Mesozoic volcanic formations crop out in the Seven Devils Mountains west of the Idaho Batholith. They bear little resemblance to contemporary units from the eastern part of the state, which are entirely of sedimentary origin. Unfortunately, the intrusion of the Idaho Batholith and the downwarp of the Snake River Plain have concealed all evidence of the original nature of the boundary between these two distinct parts of the Cordilleran Geosyncline.

The Idaho Batholith was the largest but not the only Cretaceous intrusion in the state. The Kaniksu Batholith occurs in the panhandle region, and several granitic plutons are found in some of the ranges and uplands south of the Snake River Plain. Most of the batholiths had several pulses of emplacement, some occurring early in the Tertiary Period. The batholithic intrusions and associated metamorphism were not the only manifestations of the disruption of the Cordilleran Geosyncline. The orogeny also involved major folding, faulting, and uplift of the entire region above sea level.

The Cenozoic history of Idaho, like that of Oregon, Washington, and much of British Columbia, was dominated by voluminous volcanic eruptions. The major volcanic sequences have been termed (oldest to youngest) the Challis Volcanics, the Columbia River Basalt, the Idavada Volcanics, and the Snake River Group. The Challis Volcanics were produced early in the Tertiary Period by eruptions from numerous vents in central and eastern Idaho. The Columbia River Basalt issued in the Miocene Epoch from many long fissures in northeastern Oregon and southeastern Washington. The lava spread almost like water and ponded against the subdued highlands of west-central Idaho.

The Idavada Volcanics occur along both sides of the Snake River Plain. They are of Pliocene age, and consist primarily of siliceous flows and ash layers interbedded with stream and lake sediments. The Snake River Group is of late Pliocene and Pleistocene age and covers much of the Snake River Plain. It is a complex pile of basaltic lava flows and interbedded gravels, sands, and silts. The oldest flows are probably Pliocene in age, but most of them erupted in the Pleistocene Epoch. The youthful volcanic features of the Craters of the Moon area indicate that eruptions have continued virtually to the present day.

The Snake River volcanism was accompanied by subsidence of the Snake River Plain. The subsidence was due in large measure to downwarping but it also involved, at least at the west end, significant downfaulting. As the plain subsided, the surrounding mountains rose. Idaho's present physiography results largely from Late Cenozoic deformation plus Pleistocene glaciation.

PRE-TERTIARY STRATA

Much of northern Idaho is underlain by rocks of the Belt Supergroup (Chapter 6), which is at least 20,000 feet thick in the panhandle. The Belt Supergroup is the most extensive and best studied Precambrian unit in Idaho, but it is not the only one. Gneisses in the Clearwater and Coeur d'Alene regions locally underlie (and hence predate) Belt strata. Other metamorphosed Precambrian formations occur in the Sawtooth and Pioneer Mountains of central Idaho and in the Albion Range south of the Snake River Plain. Upper Precambrian lava beds and conglomerates, correlative with the Windermere Group of southeastern British Columbia, rest unconformably on Belt strata in the Selkirk Mountains. Sills and dikes of diorite and gabbro that cut Belt rocks in this region may represent Windermere lava that did not reach the surface.

The most complete Paleozoic section in any of the northwestern states occurs in southeastern Idaho. All seven of the Paleozoic periods are represented by one or more formations, and the total section is as much as 50,000 feet thick (Table 16-1).

The region did not see continuous sedimentation throughout the entire era, for there are significant unconformities in the record. Likewise, the thickness is somewhat misleading, since individual formations are thinner in some areas than in others. Probably no single area has as much as 50,000 feet of Paleozoic rocks. The formations are primarily marine sandstones, shales, and limestones, which are locally quite fossiliferous. Many of the units can be traced for very long distances through the Rockies and the mountains of the Basin and Range Province. They represent strata deposited in the slowly subsiding, generally shallow, and nonvolcanic nearshore part of the Cordilleran Geosyncline (the miogeosyncline). The unconformities indicate periods of uplift and erosion, but on the whole these represent times of rather mild deformation compared to disturbances in the eugeosyncline to the west.

Thus we can establish that southeastern Idaho was in the miogeosyncline where Paleozoic conditions were fairly stable and that west-central Idaho was in the eugeosyncline, at least toward the end of the era. The Paleozoic record for the rest of the state is either buried under younger strata, obscured by metamorphism around the Idaho Batholith, or stripped away by post-Paleozoic uplift and erosion. Some geologists think that much of northern Idaho stood above sea level and provided sediment to adjacent regions during the Paleozoic Era. This would explain the absence there of Paleozoic strata, but so would subsequent large-scale uplift and erosion. The stratigraphic record for the Mesozoic Era in Idaho is not very complete. Triassic volcanic and sedimentary formations, plus a few hundred feet of possibly Lower Jurassic strata, crop out in the Seven Devils Mountains. In southeastern Idaho, more than 12,000 feet

TABLE 16-1 Paleozoic and Mesozoic stratigraphy of southeastern Idaho. (From U.S. Geological Survey, 1964.)

AGE	THICKNESS, FEET	LITHOLOGY	PRINCIPAL FORMATIONS
Cretaceous	13,800	Mostly nonmarine sandstone and shale	Gannett, Wayan
Jurassic	6,700	Marine sandstone, limestone	Nugget, Twin Creek, Preuss, Stump
Triassic	5,500	Marine sandstone, shale, limestone	6 named units
Permian	1,000	Marine phosphatic shale, chert	Phosphoria
Pennsylvanian	2,500	Marine shale, sandstone	Wells
Mississippian	3,500	Marine limestone	Madison, Brazer
Devonian	1,100	Marine limestone, minor shale and sandstone	Jefferson, Three Forks
Silurian	1,000	Marine dolomite (magnesian limestone)	Laketown
Ordovician	2,500	Marine limestone, shale, quartzite	Garden City, Swan Peak, Fish Haven
Cambrian	7,000	Marine limestone, quartzite	7 named units

of marine sandstones, shales, and limestones were deposited in the Triassic and Jurassic Periods. The only sedimentary rocks of Cretaceous age found in Idaho also occur in the southeastern part of the state. Some are of Early Cretaceous age and were deposited partly in fresh water lakes and along streams. This suggests that uplift of the region had begun by this time. The sedimentary Frontier Formation of Late Cretaceous age crops out east and southeast of Idaho Falls. It is of nonmarine origin and contains much of Idaho's coal reserves, which are not extensive.

THE IDAHO BATHOLITH AND IDAHO MINING

The Idaho Batholith of central Idaho is about 250 miles long and has a maximum width of about 100 miles. The most prominent rock type is light-colored quartz monzonite, but there are phases ranging in composition from true granite to quartz diorite. The contact of the batholith with the surrounding country rock is complex. It typically consists of a migmatite zone many miles across in which igneous and metamorphic rocks are intimately mixed. Much of the batholith itself has faint gneissic layering. The Idaho Batholith probably crystallized at a considerable depth, where high temperature and pressure conditions allowed considerable reaction between the magma and the country rock.

Radiometric age dating of the batholith indicates that the main stage of emplacement occurred around 100 million years ago, in the middle of the Cretaceous Period. This coincided approximately with the crystallization of the Wallowa Batholith in the Blue Mountains, and the Kaniksu Batholith along the Idaho-British Columbia border. Smaller stocks and batholiths were intruded after the main phase. Some of these clearly were associated with the extrusion of the Challis Volcanics during the first half of the Tertiary Period.

The rocks marginal to the Idaho Batholith were deformed as well as metamorphosed by the emplacement of the magma. The principal style of deformation was folding and thrust faulting. Perhaps this was due to the lateral compression of the intruding magma, but the structures are too complex to be explained so simply. Certainly, in

16-4 *The Lost River Range from the northwest. The mountains consist primarily of Paleozoic sedimentary strata. Borah Peak (elevation 12,665 feet) at left is the highest peak in Idaho. (Photograph by Bates McKee.)*

16-5 *City of Rocks in the Albion Range of Idaho, 55 miles southeast of Twin Falls. The numerous rounded rock spires consist of granitic rock that was emplaced in the Cretaceous Period at about the same time as the Idaho Batholith. (Photograph by Bates McKee.)*

southeastern Idaho, the thrusting was part of the regional eastward thrusting of the Rocky Mountain Province (Chapter 6). It is wisest to regard batholiths as one manifestation of orogeny and not as its cause.

The story of mining in Idaho closely parallels that in Washington and Oregon, except that mining activity remains higher in Idaho. The Coeur d'Alene area is the state's most important mining district. Since 1884, it has produced metals worth 2 billion dollars, which is more than three-quarters of Idaho's total. The ore deposits are mostly in Belt strata near the contact with the Idaho Batholith. Lead and silver have been the primary metals mined, but important tonnages of gold, zinc, copper, and antimony have also been produced, in part as by-products of the mining and smelting of the lead-silver ore.

The second significant mining district in the state is along the southeast margin of the Idaho Batholith, primarily in Custer and Blaine Counties. The geology of the various ore deposits is much the same as in the Coeur d'Alene district. Clearly, however, not all the mineralization can be related to the Idaho Batholith. Many of the deposits are in Tertiary strata, especially the Challis Volcanics, which clearly postdate the intrusion by tens of millions of years. Some of the ore occurs around the margins of Tertiary stocks and batholiths, with which it is related genetically. Placer deposits of gold and silver have been mined extensively from Upper Cenozoic sands and gravels in many parts of the state.

The list of nonmetallic resources mined in Idaho is most impressive, and various deposits are gaining importance each year. The reader interested in pursuing this subject further is referred to the reference list at the end of the book.

CENOZOIC ROCK SEQUENCES

The Idaho landscape in the early and middle part of the Cenozoic Era closely resembled that of its neighboring states to the west. Volcanic eruptions

16-6 *Massive beds of tuff, capped by river and lake sediments, in the Bitterroot Range north of Salmon, Idaho. These strata are part of the Challis Volcanics of Early Tertiary age. (Photograph by Bates McKee.)*

were commonplace, and the volcanoes spewed forth a great variety of flows and pyroclastic debris, including much volcanic ash. The climate was warm and moist, as proved by the fossil flora and fauna.

Lower Tertiary volcanic strata, which range in composition from basalt to rhyolite, are interbedded with mudflow deposits, and with well-bedded sediments. The latter consist mostly of volcanic debris reworked by running water and deposited along stream channels and in lakes. The volcanics and associated sedimentary rocks are known as the Challis Volcanics; they have an estimated maximum thickness of approximately 5,000 feet. The first Challis eruptions probably occurred late in the Eocene Epoch, and the last early in the Miocene. The age of the formation is thus comparable to that of the Clarno and John Day Formations in the Blue Mountains, and the Little Butte and Keechelus Volcanic Series in the Cascades. The lithologies of these various formations are also quite similar, as were, presumably, the environments of deposition.

The large, fluid lava flows of the Columbia River Basalt Group ponded against the subdued ranges that surrounded the Columbia Plateau late in the Miocene Epoch. This relationship is apparent in many places in western Idaho. The great dark layers of basalt rise from the lowlands of eastern Washington high into the Clearwater and neighboring mountains and end finally against the granitic rocks of the Idaho Batholith. Excellent exposures of the irregular base of the Columbia River Basalt Group are found in the walls of the Snake River Canyon, where the pronounced even layering of the lavas contrasts markedly with the rough, massive aspect of the underlying Seven Devils Volcanics. The eastward rise of the Miocene section reflects the Late Cenozoic uplift of the Rocky Mountains.

The Columbia River Basalt Group contains numerous sedimentary interbeds—primarily sands, silts, clays, and layers of diatomite deposited in lakes that existed between periods of volcanic eruption. The sediments have been given various formational names, but many are not well defined. For instance, the term "Latah Formation" is used in northeastern Washington and northern Idaho for any sedimentary unit associated with the basalt, even though the Columbia River Basalt Group is at least 3,000 feet thick in this region, and sediments occur at all horizons. The term "Payette Forma-

TABLE 16-2 Cenozoic stratigraphy of south-central Idaho. (After Malde and Powers, 1962.)

AGE	GROUP	DESCRIPTION	MAXIMUM THICKNESS, FEET
Late Pleistocene to Holocene	Snake River	Basalt flows and interbedded stream gravels	>900
Early Pliocene to middle Pleistocene	Idaho	Siliceous volcanic ash beds; lake and stream beds of clay, silt, sand, gravel; basalt flows.	>3,000
Early to middle Pliocene	Idavada Volcanics	Silicic volcanic rocks, mostly welded ash flows. Minor clastics	3,000
Middle to late Miocene	Columbia River Basalt	Basalt flows and interbedded clastic rocks (mostly lake beds) of "Latah" and "Payette" Formations	>3,000
Eocene (?) to early Miocene	Challis Volcanics	Flows of andesite, minor basalt and rhyolite. Prominent tuffaceous beds locally. Minor sedimentary rocks	5,000

tion" is applied to a sequence of tuffaceous lake beds that is several hundred feet thick. These beds are exposed in bluffs in the western part of the Snake River Plain. In part, they underlie the basaltic flows, but they are considered a part of the Columbia River Basalt Group.

Silica-rich volcanic rocks found north and south of the Snake River Plain constitute the Idavada Volcanics. Welded tuff is the most common rock type, but bedded airfall tuffs and lava flows are important constituents. The Idavada Volcanics have been dated as early to middle Pliocene, based mostly on fossils from sedimentary interbeds. The fossils include snails, diatoms (tiny siliceous plant remains), plants, seeds, pollen, and fragments of teeth. Other formations in the Northwest that are similar in age and type to the Idavada Volcanics include the Rattlesnake, Dalles, and Troutdale Formations of Oregon and the Ellensburg Formation of south-central Washington.

The Snake River Plain is structurally a very large trough. It has been sinking and filling with lava and sediment for the past several millions of years. Some of the subsidence results from withdrawal of support and loading of the surface by the large quantities of lava that have been extruded.

Some geologists have referred to the entire basaltic section of the Snake River Plain as the Snake River Group. In recent years, however, this name has been applied to only the Late Quaternary part of the section, and the older lavas and sediments have been called the Idaho Group. The thickness of the Idaho Group in some areas exceeds 3,000 feet. The strata consist of thin basaltic lava flows (not unlike those of the Columbia River Basalt) and interbedded sands, silts, clays, and beds of ash. The sediments erode to form "badland" topography. Fossils suggest that the group ranges in

16-7 *Lake and stream sediments of the Idaho Group, 30 miles northwest of Twin Falls. (Photograph by Bates McKee.)*

age from Pliocene to middle Pleistocene. Presumably volcanism and sedimentation kept up with subsidence, and the Snake River Plain retained a low, even surface (broken locally by faulting) as the surrounding ranges rose.

Downcutting by the Snake River and its various tributaries had begun by the time of extrusion of the flows of the younger Snake River Group. They built the shield volcanoes that were mentioned at the beginning of the chapter. An awareness of the stratigraphic complexity of the Snake River Group has been provided by geologic investigations for groundwater supplies. The remarkable storage capacity of the basalts and interbedded river and lake deposits is displayed dramatically in the Thousand Springs area along the Snake River west of Twin Falls.

THE SNAKE PLAINS AQUIFER

The Snake River Plain north of the river is underlain by one of the world's great aquifers. (An *aquifer* is a formation from which an appreciable quantity of groundwater can be pumped.) Known as the Snake Plains Aquifer, it is not technically a single geologic formation. Instead, it includes the lava flows and interbedded alluvial sediments of the Snake River Group, part of the underlying Idaho Group, and some recent stream sediments that locally overlie the Snake River Group.

The eastern half of the Snake River Plain has a rather remarkable drainage pattern. The river lies close to the south edge of the plain, generally 40 to 50 miles southeast of the ranges of central Idaho. These ranges receive fairly heavy amounts of precipitation, especially during the winter at high elevations. The mountain valleys contain some moderately large rivers, such as the Big Lost

16-8 *The canyon of the Snake River, 8 miles east of Twin Falls, exposing basaltic flows of the Snake River Group. Note the proximity of the youngest flow to the surface of the plain. (Photograph by Bates McKee.)*

16-9 *The surface of the Snake River Plain south of Arco. Young basalt flows are exposed at the surface. The hills in the background, known as Twin Buttes, are ancient volcanic cones that were partly buried by Snake River basalt flows. Surface water in this region disappears quickly into the fissured surface of the basalt. (Photograph by Bates McKee.)*

River, and the total amount of runoff toward the Snake River Plain is quite considerable. Yet none of these rivers is a tributary to the Snake River. They all disappear into the surface of the Snake River Plain close to the mountain front. Along a distance of 268 miles, downstream from the junction of the Henrys Fork River and the Snake River, not one perennial stream joins the Snake River from the north. For most of this distance the Snake is losing water, both by evaporation and by diversion for irrigation. During periods of maximum irrigation, the Snake River below Milner Dam (25 miles east of Twin Falls) does not flow at all. The river recovers, however, mostly by addition of water from gigantic springs along the north bank. In the next 94 miles of river, downstream from Milner Dam, an estimated total volume of approximately 200 billion cubic feet (1.4 cubic miles) of water enters the channel from springs each year. More spring water enters in just the Thousand Springs area near Hagerman than reaches there via the Snake River; sometimes the ratio is as high as 10:1.

The groundwater flows in the Snake Plains Aquifer generally to the southwest, which is down the regional dip of the strata (Figure 16-10). The Snake River has cut its channel through the aquifer, and consequently it is gravity and the weight of the water in the strata updip that drives the gushing springs. Since irrigation of the Snake River Plain began, some of the springs have more than doubled their rate of flow. This is due to the infiltration into the aquifer of irrigation water, but of course this does not mean that the entire region has more water now than it did before the first settlers arrived. In areas in which irrigation depends largely on water diverted from the Snake River, the water table may be higher now than it was 50 years ago. In areas in which wells provide most of the water, however, the water table may have been drawn down hundreds of feet due to heavy pumping.

The Snake Plains Aquifer is one of Idaho's most valuable resources. As such, it deserves careful management, and various state and federal agencies are concerned with this problem. Unlike many natural resources, water is a replenishible commodity; and the Snake Plains Aquifer is a reservoir that can hold hundreds of times more water than can be contained in surface reservoirs in the region. The major problem with any aquifer is recharge—how to replenish the water that is withdrawn by pumping or lost via springs. Nature accomplishes recharge naturally through downward percolation of surface water. To date, this has been adequate for the Snake River Plain, but planners can foresee a need arising soon to give nature a helping hand. One possible method would be the diversion of water from the Snake or some of its tributaries during periods of heavy flow into special percolation ponds, much as is done in parts of California. Along with artificial recharge will probably come stricter control measures on the drilling and operation of wells, thereby attempting to maintain an equilibrium groundwater condition throughout the region.

This is not just Idaho's problem; it is becoming everyone's problem. It is probably no exaggeration to say that water is our most important natural resource; and, as our demands for water increase, the efficient utilization of our limited water supply becomes an increasingly vital challenge.

THE CRATERS OF THE MOON

Volcanism has continued on the Snake River Plain almost to the present day, although no eruptions have occurred during the past several hundred years. The youthfulness of some of the basaltic flows on the surface of the plain is apparent from the absence of much soil or other sediment on flow surfaces and the nearly perfect preservation of some flow-top features. It is particularly obvious from the air, where the dark, barren surfaces of recent flows stand in marked contrast to the fertile, cultivated land marginal to the lava.

The Craters of the Moon National Monument lies close against the north edge of the Snake River Plain southwest of Arco. It is one of the best places in the Northwest to observe a variety of young volcanic features in a small, readily accessible area. Most of the eruptions have occurred

16-10 *Generalized cross section of the northern part of the Snake River Plain, showing the movement of groundwater through the Snake Plain Aquifer.*

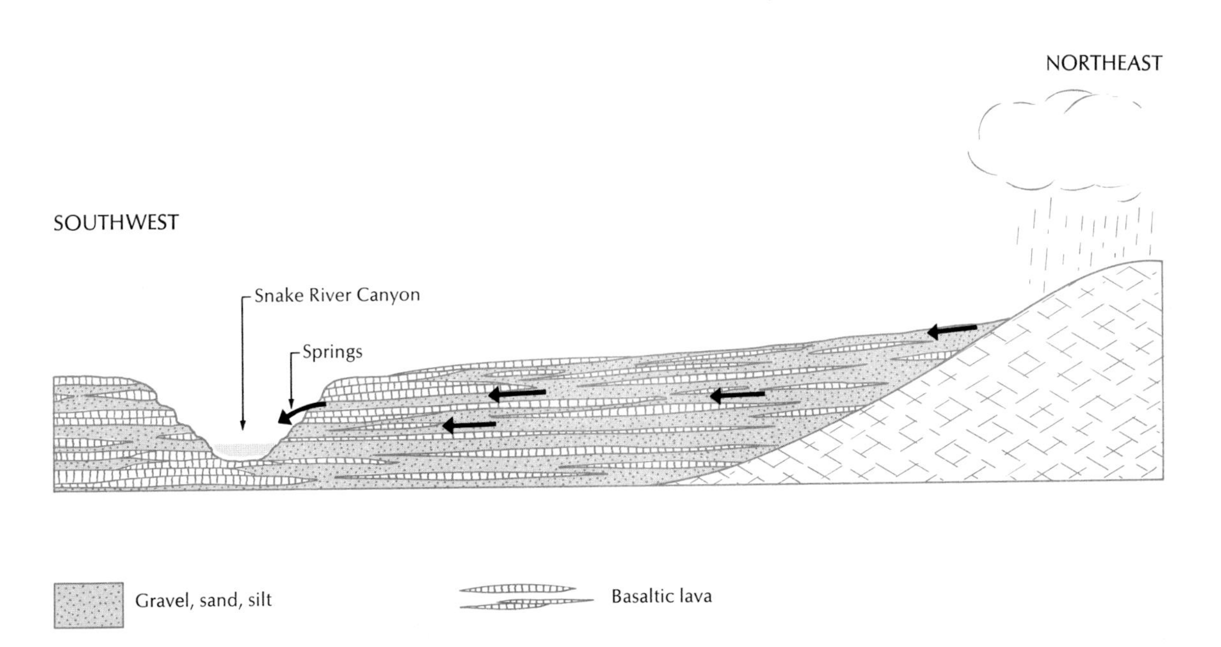

16-11 *The smooth margin of a recent basaltic lava flow, Craters of the Moon National Monument.* (Photograph by Bates McKee.)

along the so-called Great Rift, a series of parallel fractures that run for a dozen miles or more southeast from the foot of the Pioneer Mountains. Nineteen cones, rising from 120 to 800 feet above the plain surface, have been built along the Great Rift and a few more away from it. These are cinder cones, formed by the buildup of loose, fragmental ejecta thrown out from central vents, and their slight elongation toward the northeast suggests that southwest winds prevailed during the eruptions. Lava flows issued from the bases of some of these cones. Some of the more fluid lava flows solidified with a smooth or evenly wrinkled surface. Others were little more than moving rubble piles and have a very rough, jagged surface. Small spatter cones formed around some vents, while quiet eruptions from others built broad "mini-shield" volcanoes only a few tens of feet high. Lava caves, tree molds, volcanic blocks, and bombs of all descriptions are among the many features available for inspection in the Craters of the Moon area. Several excellent geologic pamphlets and some informative displays are available in the Monument Visitor center.

Geologists viewing the Craters of the Moon area have been struck by the apparent freshness of the volcanic terrain. Until recent years, it was assumed that the youngest flows are no more than several hundred years old. Some of the flows that appear so fresh have small, stunted pines growing out of crevices in their surface, and counts of their annual growth rings indicate a surprising antiquity—461 years in one case and 1,500 years in another. The youthful appearance of the flows must be due to the relatively arid climate and slow rate of weathering in this part of Idaho.

Like the volcanoes of the Cascades, the volcanic activity of the Snake River Plain should be considered dormant rather than extinct, for future eruptions are quite probable.

17 the columbia plateau

■ The Columbia Plateau is most exciting geologically, for the country has been shaped by geologic processes that have acted on a gigantic scale. If we had to sum it all up in one word, it would have to be "floods." First, great outpourings of basalt lava built one of the world's most impressive volcanic piles. Later, gigantic floods of water raced across the plateau and produced one of the world's most intricately channeled surfaces. To conservative geologists, both events were offensive, because in the truest sense of the word they were catastrophes. Nowhere today can we see flooding of either water or lava on this scale. It is no wonder that many geologists were skeptical of the explanation of these events when it was first proposed.

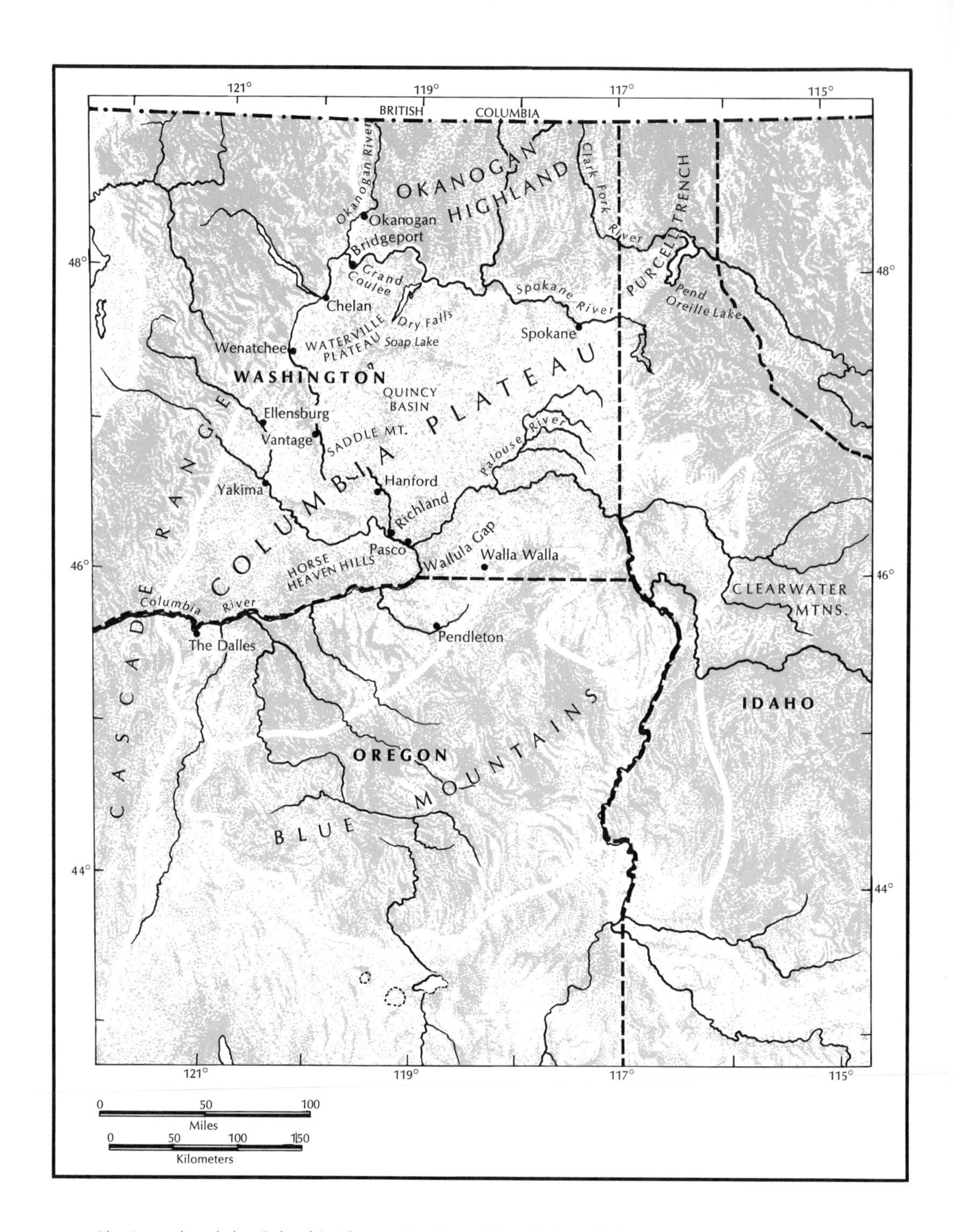

17-1 *Physiography of the Columbia Plateau Province. (After Raisz, 1941.)*

SUMMARY OF PLATEAU GEOGRAPHY AND GEOLOGY

The Columbia River enters the region known as the Columbia Plateau from the Okanogan Highland Province northwest of Spokane and exits through the Cascades via the Columbia River Gorge. The Columbia Plateau is surrounded on all sides by mountains—the Okanogan Highland to the north, the Clearwater Mountains to the east, the Blue Mountains to the south, and the Cascade Range to the west. Many rivers carry water into this interior lowland from surrounding ranges, and all, even the mighty Snake River, join the Columbia before entering the gorge. There just is no other way out.

The Columbia River flows around the north and northwest part of the plateau, then edges away from the Cascades at Wenatchee before turning west again into the gorge. The low point of the natural basin is near Pasco. From there, the plateau surface rises gently toward the surrounding ranges.

The pre-Quaternary history of the Columbia Plateau can be traced back only about 15 million years. There are no significant windows eroded through the Columbia River Basalt, which erupted in the Miocene Epoch and buried everything. Obviously there is a much older story beneath the basalt, but geologists can only guess what it might be by studying the older rocks that disappear under the flows around the edge of the plateau. The Columbia River Basalt is such an effective screen that even various geophysical methods of "seeing" underground are relatively useless.

The bedrock is not entirely volcanic, for the Late Tertiary landscape resembled in some ways that which we see today. Rivers flowed out onto the lava plain from surrounding highlands. They carried sediment which was deposited along their beds

17-2 *Flows of Columbia River Basalt exposed in cliffs in Moses Coulee, 20 miles east of Wenatchee, Washington. The scene results largely from floods of lava in the Miocene Epoch and water in the Pleistocene Epoch. (Photograph by Bates McKee.)*

17-3 *Many separate lava flows of the Columbia River Basalt exposed in the canyon walls of the Grande Ronde River in the southeastern corner of Washington. The river has incised into the plateau surface as the region has been uplifted in Late Cenozoic time, and its former meandering course has been preserved. (Photograph courtesy of Washington State Department of Commerce and Economic Development.)*

and in shallow lakes that existed between periods of eruption. Consequently, there are layers of sedimentary rock interbedded with the lava flows. The Pliocene strata consist primarily of stream and lake sediments with a few basaltic flows interbedded. Deformation on a large scale has occurred in the region during the past 10 million years. The Columbia River Basalt dips away from the surrounding mountains, as a result of both the uplift of the ranges and the sinking of the plateau. In addition, many folds and minor faults have warped and broken the flows.

THE LAVA FLOWS OF THE COLUMBIA RIVER BASALT

A "typical" flow of the Columbia River Basalt was approximately 100 feet thick, although some flows were more than 200 feet thick, and others had a thickness of just a few feet. The eruptions were not from a single vent but from very long cracks or fissures, each of which was many miles long. An individual eruption was probably fed by many fissures erupting simultaneously.

The flows spread almost like water for great distances. This fluidity is suggested by the very even tops of the flows and the fact that they are traceable for tens of miles or more without significant changes in thickness. One of the most studied units, the Roza Basalt Member, has been traced from the Grand Coulee area east to Spokane, south to Pendleton, Oregon, and southwest into the Columbia River Gorge over an area of approximately 20,000 square miles. Its outer edge appears to have been eroded away so that its original extent may have been even greater. A recent theoretical calculation suggests that the average spreading velocity of a Columbia River Basalt flow may have been 25 to 30 miles per hour. Thus a single eruption could have inundated much of the plateau area in a few hours.

Spreading of the lava may have been rapid, but cooling was not. This conclusion is based on studies of recent basaltic lakes in the craters of Kilauea Volcano on Hawaii. The lava pool in these craters crusts over very quickly. Thereafter solidification is quite slow—perhaps on the average no more than ten feet downward per year for pools several hundred feet deep. This suggests that the typical flow of the Columbia River Basalt may have taken several decades to solidify completely, even allowing for the fact that the lava would have been crystallizing from the bottom up as well as from the top down.

The crystallization of a flow inward from both the bottom and top surfaces produces two distinct layers, which in a cliff exposure look like two separate flows. The bottom part is known as the *colonnade* and the top as the *entablature*. As the lava cools and solidifies, shrinkage cracks develop. They impart to the rock a columnar structure, which is referred to as *columnar jointing* (Figure 17-4). The individual columns normally are larger and more uniform in the colonnade than they are higher in the flow. In some flows, the columns in the entablature radiate downward in peculiar fanlike structures (Figure 17-5).

The *vesicle* is another structure found in most congealed basaltic lava flows. This is a cavity produced by the entrapment of gas bubbles in the cooling lava, and it is most common near the top of the flow. Vesicles range in size from tiny holes the size of a pinhead to cavities a foot or more in diameter. Many agates and thunder eggs (Chapter 15) have been formed by groundwater precipitation of silica in vesicles.

The first flows of Columbia River Basalt were erupted onto a landscape of rolling hills. Gradually, the lava from successive eruptions filled in the lowland of the Columbia Plateau Province and lapped ever higher against the surrounding highlands. The Miocene topography beneath the center of the province lies completely concealed by basalt, but it is exposed around the plateau margin. The best exposures of the prebasalt topography are in and

17-4 *Columnar jointing in the Columbia River Basalt near Vantage, Washington. Each column has a diameter of approximately 2 feet. Similar jointing occurs in many parts of the world, including The Devil's Postpile in the Sierra Nevada, Devil's Tower in Wyoming, and the Giant's Causeway in Ireland. (Photograph courtesy of A. S. Cary.)*

17-5 *Fanlike arrangement of joints on the Oregon side of the Columbia River Gorge. The origin of this structure is unknown. (Photograph courtesy of A. S. Cary.)*

around the upper part of Grand Coulee near the dam, at the north edge of the Columbia Plateau. The earliest flows to reach this area came from the south. They ponded against the base of hills of granite as much as 1,000 feet high. Succeeding flows filled the valleys and rose higher against the flanks until finally some of the hills were buried completely and others nearly so. Some of the younger flows spread a considerable distance north of the modern Columbia River.

THE LANDSCAPE BETWEEN ERUPTIONS

The landscape that developed on the top of one lava flow was buried beneath the next flow. When successive eruptions were close in time, the buried surface was relatively smooth and "naked," that is, it consisted of the barren treeless top of the older flow. In many instances, hundreds or even thousands of years elapsed between the cooling of one flow and the spreading of the next. Then the surface was markedly different. The Miocene climate was considerably moister than that of the region today. Consequently, the processes of weathering, erosion, and deposition could quickly transform the initially barren flow top to a more hospitable environment with a rich soil, abundant vegetation, lakes, and streams. Physical evidence of these features was subsequently buried beneath the next flow. Although geologists cannot accurately date the time of eruption of each flow, they can gain a qualitative idea of the interval of time between eruptions by a careful study of features found beneath each flow. Every eruption reset the clock; it buried the topographic irregularities and all the various sediments and organisms of the landscape. The top of each new flow was virtually flat and utterly barren.

Wood is preserved beneath the bottom of many of the flows of the Columbia River Basalt, especially in buried soil or sediment. The richest horizon is near the top of the extensive Vantage Sand-

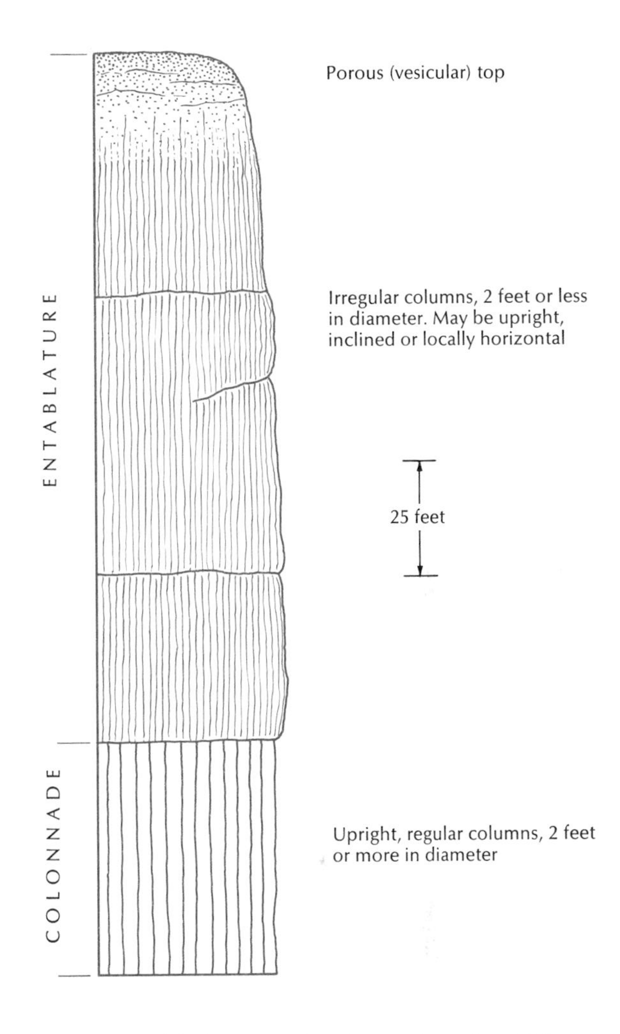

17-6 *A simplified cross-sectional view through a thick basalt lava flow.*

stone Member (Table 17-1). Several hundred species of wood and leaf material have been found around the town of Vantage in central Washington. Part of this area has been set aside as the Ginkgo Petrified Forest State Park (named for one of the more unusual types of deciduous trees preserved). The museum at Vantage has a magnificent display of the fossilized flora, as well as exhibits that explain the geology of the region. The large size of many of the petrified logs and the great variety of the flora indicate that there was a long period without volcanism prior to the eruption of the Frenchman Springs Member that followed.

The most bizarre fossil from the Columbia Plateau is the outline of a small rhinoceros cast in basalt. Found near Dry Falls in the lower part of Grand Coulee, the "Blue Lake Rhino" was trapped beneath the advancing front of one of the basalt flows. Nothing remains of the bones, flesh, or skin of this creature. Its outline is faithfully preserved, however, in the congealed lava at the bottom of the flow. Vertebrate fossils are not common in strata of Miocene age in Washington, but the remains of rhinoceroses, camels, and other animals not found in North America today are preserved in Miocene beds in Oregon and elsewhere in the West. The paucity of such fossils in Washington is probably due to the relative scarcity of Miocene sedimentary rocks favorable to their preservation.

PLIOCENE AND EARLY PLEISTOCENE EVENTS

The Pliocene Epoch witnessed the start of the uplift that produced the Cascade Range. The rising mountains shed volcanic sediment, some of which was deposited by rivers along the western edge of the Columbia Plateau as the Ellensburg Formation (Chapter 12 and Table 17-1). By this time, flood eruptions of basaltic lava occurred much less frequently than in the Miocene Epoch, but at least three flows are interbedded with the Ellensburg sediments.

The White Bluffs along the Columbia River north of Hanford are composed of light-colored sediments known as the Ringold Formation. About 600 feet of loosely cemented sandstone and siltstone are exposed in the bluffs. Subsurface investigations have shown, however, that exposed strata constitute only the upper half of the unit. Beneath them are conglomerate beds that have been traced from Sentinel Gap in the Saddle Mountains south to the Wallula Gap in the Horse Heaven Hills. The lowermost part of the Ringold Formation consists of beds of bluish siltstone and local gravel deposits. The Ringold strata were deposited in lakes and along stream beds. The conglomerate beds in the middle of the formation have been interpreted as possible deposits of an ancestral Columbia River.

The exposed portion of the Ringold Formation was deposited probably in the Pleistocene Epoch. This conclusion is based on a few vertebrate fossil remains. The basal siltstone may be correlative with the Pliocene Ellensburg Formation.

The eastern part of the Columbia Plateau is often referred to as the Palouse Country, a name derived from one of its few prominent rivers. Without a doubt, its most important mineral resource is the Palouse soil, which is responsible for the large wheat production of the region.

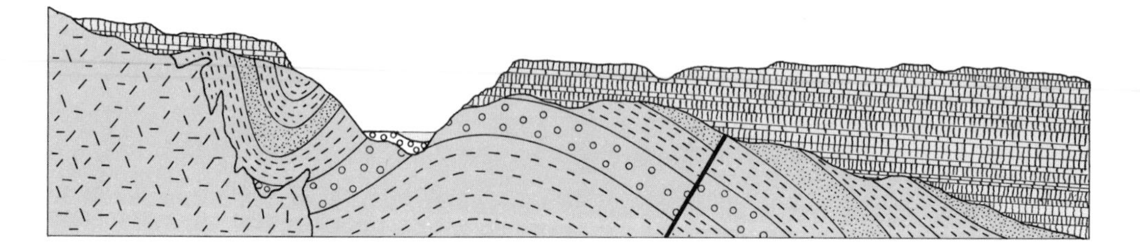

17-7 *A generalized cross section through the margin of the Columbia Plateau, illustrating the relationship of the Columbia River Basalt to the older rock in the surrounding highlands. The basalt has been tilted away from and partly stripped off the adjacent uplands.*

Throughout most of the Palouse Country, this rich soil rests directly on the Columbia River Basalt.

The Palouse soil is not a residual soil. This means it was not derived by the weathering of the underlying bedrock. Instead, it is *loess,* or wind-deposited silt. The Palouse soil is more than 100 feet thick in places. Much of the material was derived from the Pasco Basin and was carried to the east and northeast by prevailing winds. Wind erosion of the Ringold formation supplied some of the silt. Other grains were eroded from sediment lying on the floodplains of the Columbia and Snake Rivers.

The Palouse soil did not form all at once. Much of it was deposited before the last glaciation, but some of the younger beds postdate that event.

The Late Cenozoic deformation of the Columbia Plateau and surrounding regions has been referred to in several previous chapters. Within the plateau proper, the most obvious manifestation of this event are the folds, most of which trend approximately east-west. Erosion has not had sufficient time to wear away the anticlinal ridges. Consequently, the structure is mirrored in the topography. Most of the folds are asymmetrical, with one limb dipping more steeply than the other. Frequently the steeper flank has been broken by one or more faults, but faulting appears to be of relatively minor significance.

The folds have formed as the result of north-south regional compression, possibly caused by deep-seated rotational movement beneath much of the Cordillera (Chapter 15). The deformation appears to be continuing today, judging from the involvement in the folding and faulting of Quaternary formations and some very limited data on re-

TABLE 17-1 Late Cenozoic stratigraphy of south-central Washington. (After various authors.)

AGE	FORMATION	MEMBER
Holocene	Alluvium	
Pleistocene	Glacial outwash and flood deposits	
	Palouse Formation	
	Ringold Formation	
Pliocene	Ellensburg Formation	Upper Ellensburg WARD GAP* ELEPHANT MOUNTAIN Rattlesnake Ridge POMONA Selah
Late Miocene	Yakima Basalt Formation†	UMATILLA Mabton PRIEST RAPIDS Quincy Diatomite ROZA Squaw Creek Diatomite FRENCHMAN SPRINGS Vantage Sandstone YAKIMA BASALT (undifferentiated)

*Members in capital letters are basaltic lava flows.
†Name for the Columbia River Basalt in eastern and central Washington.

17-8 *The Palouse country from Steptoe Butte north of Colfax, Washington. The rolling topography is typical for this part of the plateau, which was not swept by floods. The hills in the distance and Steptoe Butte are composed of Precambrian rocks that project like islands above the general level of the Columbia River Basalt. (Photograph courtesy of Washington State Department of Commerce and Economic Development.)*

cent changes of level of the ground surface.

Some of the major rivers of the region cut through these young folds in spectacular gorges. Most notable is the Yakima River between Ellensburg and Yakima, where four major anticlines (the Manastash, Umtanum, Selah Butte, and Yakima Ridges) have been breached by the meandering river. The Yakima River is considered to be a classic example of an antecedent river, which was able to maintain its course across growing folds. (The antecedence of the Columbia River relative to the Cascades and of several rivers relative to the Coast Mountains of British Columbia has been discussed in earlier sections.)

GLACIATION OF THE COLUMBIA PLATEAU

The Columbia Plateau is a magnificent region in which to study the many ways in which glaciation can modify the landscape. Glaciation means not only the effects produced directly by the passage of glacial ice over the ground. The term also includes all the modifications brought about by glacial meltwater and glacially diverted rivers far beyond the terminus of the ice.

The preglacial topography along the northern edge of the plateau was probably not too different from that of today. Glaciers moving south down the valleys of the Okanogan and Columbia Rivers encountered the high basaltic rim of the plateau along the east-west segment of the Columbia River–Spokane River axis. This north-facing escarpment was an effective barrier to the ice. Only at the west end was a glacial lobe able to top the plateau rim. There the Okanogan Lobe spread al-

17-9 *The Withrow Moraine, which marks the southern limit of glacial advance of the Okanogan Lobe on the Waterville Plateau. View is to the northwest, toward the Cascade Range. Note the large basaltic blocks (haystack rocks) concentrated along the moraine front and the marked difference in the topography on the two sides of the moraine. (Photograph courtesy of John Whetten.)*

most 30 miles south across the Waterville Plateau, a subdivision of the Columbia Plateau.

The surface of the Waterville Plateau preserves many classic glacial features. The veneer of glacial deposits is generally thin, and the basalt bedrock is exposed in many places. Glacial grooves and striations are evident on some rock surfaces (Figure 2-10). These features, plus larger lineations in the glacial deposits, indicate that the Okanogan Lobe fanned out from the plateau rim near Bridgeport. It flowed generally southeast toward Coulee City, south toward Mansfield, and southwest toward the foot of the Cascades at Chelan.

The glacier carried some sediment derived from the Okanogan Valley, including granitic boulders. These light-colored glacial *erratics,* scattered on the plateau surface, contrast markedly with the dark basaltic bedrock. The most spectacular glacial boulders on the plateau, however, consist of great blocks of the basaltic bedrock. Many of these were peeled from the north rim of the plateau as it was overrun by the ice. Some of the blocks are as big as a house. The residents of the Waterville Plateau call them "haystack rocks," and their resemblance in size and shape to some of the haystacks of the region is quite remarkable, although they are not nearly so welcome in the fields.

The southern limit of the ice advance on the Waterville Plateau is well defined by the Withrow Moraine. It consists of a series of irregular hills up to several hundred feet high and several miles across. The moraine extends across the plateau from the lip of the Columbia River opposite Chelan to the rim of Grand Coulee near Coulee City. South of the Withrow Moraine, the topography is

17-10 *Dry Falls in Grand Coulee, looking east toward Coulee City. Only a small part of this immense former cataract is seen. The total length of the falls is approximately 4 miles, and its average height greater than 300 feet. (Photograph courtesy of Washington State Department of Commerce and Economic Development.)*

smooth and thickly mantled with sediment. On the north side, the plateau surface is quite irregular, with numerous haystack rocks and many small ridges and hills of glacial sediment.

THE ORIGIN OF GRAND COULEE

The southwest trending channel known as the Grand Coulee is the Columbia Plateau's most famous topographic feature. It is also one of the most important, as its upper end is used for the storage of water diverted from the Columbia River at Grand Coulee Dam. This water irrigates a large part of the Columbia Plateau.

The coulee itself is really two coulees at different levels, connected end to end. The upper part heads above Grand Coulee Dam and trends southwest for approximately 20 miles. The walls are nearly vertical and rise about 400 feet above the flat coulee floor. The upper coulee is about 4 miles wide at its widest point. The lower Grand Coulee heads at Dry Falls, a dry waterfall almost 4 miles long southwest of Coulee City. The lower coulee extends southwest for 7 miles to Soap Lake, where it opens into the Quincy Basin. The lower coulee is no more than a mile wide. The east wall is less than 200 feet high, but the west wall rises steeply more than 500 feet above the coulee floor. This difference is due to the Coulee Monocline, a fold that parallels the lower part of Grand Coulee. It has raised all the rocks on the west side of the coulee relative to the east side.

The origin of this great channel has been the subject of some debate. Clearly, whatever waters were responsible for the cutting of the coulee ultimately abandoned this channel. Most geologists have favored a glacially diverted Columbia River. The Okanogan Lobe, when it extended onto the Waterville Plateau, must have dammed the Columbia River. Geologic mapping has shown that during the time of farthest ice advance, the blockage was essentially where Grand Coulee Dam stands today. (North of the Columbia Plateau region there was no Columbia River as such, since the Okanogan Highland was buried under thousands of feet of ice. The segment of the Columbia Trough marginal to the plateau, however, must have carried much meltwater from the glaciers, plus ice marginal drainage from the east.)

The immediate result of the ice dam was the formation of a large lake in the valley of the Columbia River east of Grand Coulee. The evidence for this is an impressive sequence of sand and silt beds that were deposited in the lake. The sediments are plastered today along the valley sides high above the level of the modern lake (Figure 1-1). The water rose higher and higher in the lake until finally it spilled over the rim of the plateau at the head of Grand Coulee.

Once onto the plateau, this water flowed southwest down the natural slope of the surface. Near Coulee City, this diverted Columbia River joined a small river that was located along the axis of the Coulee Monocline. Presumably, the Columbia River followed this course for as long as the Okanogan Lobe blocked its regular channel. Once its regular channel was free of ice, the river returned to it and abandoned Grand Coulee.

No geologist has seriously questioned the glacial diversion of the Columbia River down Grand Coulee. One expert, J. H. Bretz, did challenge the presumption that the flow of the Columbia River alone could explain the coulee and its features. Initially, his ideas were not taken too seriously, but now they are accepted as essentially correct.

THE BRETZ FLOODS

The mechanism Bretz proposed for the cutting of Grand Coulee (and the many dry scabland channels to the east) was erosion by gigantic floods of water. The theory was inspired by evidence from many parts of the plateau. This consisted mostly of features that are found along rivers today, but of such large dimensions as to demand the passage of unimaginable volumes of water. The evidence included stream boulders weighing many tons scattered widely across the plateau, especially

17-11 *The west wall of lower Grand Coulee and Lake Lenore. The islands in the lake are composed of basalt layers that are dipping eastward along the center of the Coulee Monocline, a young fold. (Photograph courtesy of S. C. Porter.)*

at channel mouths; giant ripple marks ten feet or more high and several hundred feet from crest to crest; and huge boulder gravel bars as much as 2 miles long and a hundred feet or more high. Bretz believed that the intricate and complex network of coulees in the Channeled Scabland indicated simultaneous flooding of the entire region.

The theory was not well received when Bretz first outlined it almost fifty years ago. The region was not well known at the time. Soon, however, other geologists came to look for themselves. For the most part they discounted the "Bretz Floods." They proposed more complicated but less catastrophic theories that relied heavily on normal runoff augmented by meltwater from the icecap to the north. None of these geologists spent as much time studying the evidence as did Bretz. Bretz was unable to explain the origin of his floods, however, and the geologic profession was for the most part unwilling to accept his theory.

General acceptance of Bretz's ideas has come for two principal reasons. First, intensive study of the plateau during the past thirty years (largely in connection with the construction of Grand Coulee Dam and the vast network of irrigation canals) has found much additional evidence of flooding. Second, the origin of the water has been explained. The floods came from Glacial Lake Missoula in northern Idaho and northwestern Montana.

Glacial Lake Missoula developed when a lobe of Cordilleran ice moved south down the Purcell Trench and blocked the Clark Fork Valley, near Pend Orielle Lake in the Idaho panhandle. This ice dam reached a maximum height of several thousand feet above the valley floor. When the ice dam broke, as much as 50 cubic miles of water rushed westward toward the Columbia Plateau. The lake drained completely in less than two weeks, and the immense volume of water was more than the existing river valleys could handle. The water rushed out onto the surface of the Columbia Plateau near Spokane. From there, the floods flowed west and southwest, ripping off the cover of loess along the main channels and cutting into the basalt bedrock. Some water also moved west down the Spokane River and into the Colum-

17-12 *The Drumheller Channels south of Moses Lake, Washington. This was one of the principal Pleistocene flood channels. (Photograph courtesy of Washington State Department of Commerce and Economic Development.)*

bia. This water was also diverted—via Grand Coulee—across the Columbia Plateau.

The existence of Glacial Lake Missoula has been proved by substantial evidence gathered during the past thirty years, especially the many "fossil" shorelines and beach deposits perched high on the valley sides of the Clark Fork and the Bitterroot Rivers. Rapid draining of the lake is shown by the evidence downstream of many of the flood features first noted by Bretz on the Columbia Plateau. Proof of as many as seven separate cycles of lake formation, ice dam failure, and flooding has been found, both in Montana and in Washington.

The debate has shifted from the reality of past floods to their significance in the development of the physiography of the Columbia Plateau. To cite one controversy, Bretz maintains that Grand Coulee was essentially cut by these floods and that any diversion of the Columbia River was incidental to the coulee's origin. Clearly, this calls for spectacular amounts of erosion during relatively short-lived events, including cataract retreat of Dry Falls a distance of 7 miles (from the coulee mouth to the upper end of the lower coulee). Yet the coulee contains unmistakable evidence for the passage of flood water. Are these flood features essentially the "icing on the cake," added after the Columbia

17-13 *The relation of Glacial Lake Missoula to the Channeled Scabland. (After Thornbury, 1969, p. 455.)*

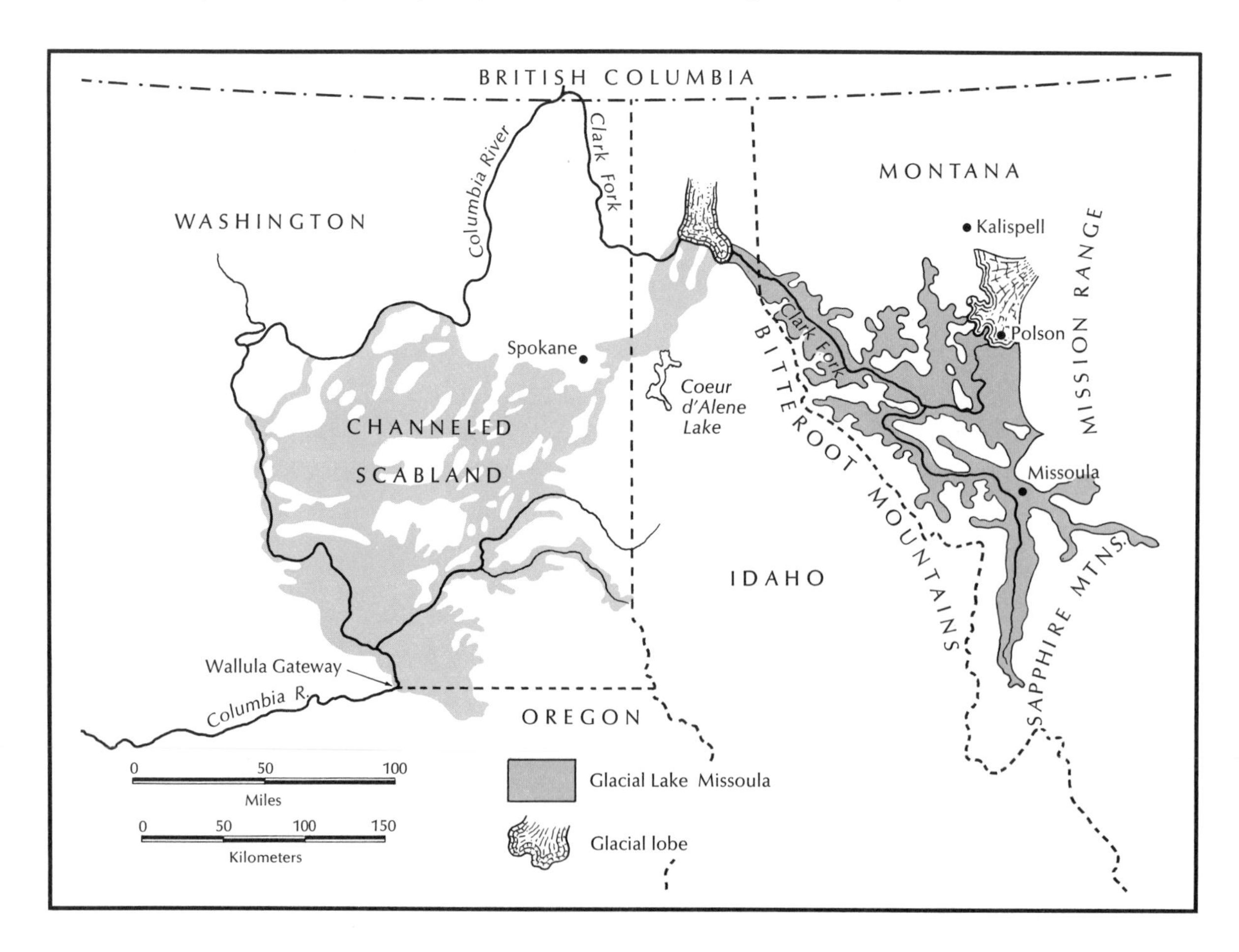

River had done most of the hard work, or is Bretz correct? Much detailed work remains to be done before this question can be answered.

DOWNSTREAM EFFECTS OF THE FLOODING

The establishment of great floods pouring across the Columbia Plateau and flowing on to the Pacific Ocean has explained some features in the Pasco Basin, the Columbia River Gorge, and the lower reaches of the Columbia River that have puzzled geologists for a long time. First there are the erratics, found in the Pasco Basin and in the Willamette Valley, more than 100 miles south of the glacial margins. East of the Cascades these erratics, which consist largely of granitic rock, are found at elevations up to 1,100 feet above sea level. The only explanation would seem to be rafting by icebergs in a large lake. Before the floods were recognized, the origin of such a lake was a problem. Now, considering the amount of water released by Glacial Lake Missoula, such a lake is easily ex-

17-14 *The Palouse River (right) and tributary channels along the Cheney-Palouse floodway in southeastern Washington. The floodwaters sweeping across the surface eroded the basaltic bedrock along linear joints in the strata. The Palouse Soil forms the distant hills, beyond the edge of the floodway. (Photograph courtesy of John Whetten.)*

17-15 *The Columbia River looking downstream toward Mount Hood and the Cascade Range. Maryhill, Washington, and Biggs, Oregon, are in the middle distance. Note the scabland surface on basaltic terraces in the foreground, produced by flood erosion. The Columbia Hills along the Washington side of the river have been uplifted by Late Cenozoic folding and faulting. (Copyrighted photograph courtesy of Delano Photographics.)*

plained. The immense volume of flood water could not have passed through the constriction of the Columbia River Gorge all at once, and a lake would have formed. This was an example of *hydraulic damming,* which is the partial blockage of stream flow by channel-wall constriction. Once through the Cascades, a second hydraulic dam developed along the Columbia River, near Castle Rock north of Portland. This produced a temporary lake in the Willamette Valley, which reached a maximum elevation of approximately 400 feet above sea level. The flood waters carried icebergs, which in turn contained the erratics. The icebergs were stranded as the lake drained, and the erratics were released when the ice melted.

The floods also explain thick deposits of cross-bedded sands and gravels in the Pasco Basin along the banks of the Columbia River. These beds are called the Spokane Flood Gravels. (The term "Spokane Flood" has replaced "Bretz Flood" in the geologic literature. This is one indication of the acceptance of the floods as real events and not merely the product of Bretz's imagination.) Fine-grained silt and clay beds lie along many of the tributary streams away from the main channel of the Columbia. They are thought to represent lake beds deposited in relatively quiet water away from the main stream of the floods. The existence of erratic boulders in some of the clay beds ties this formation to the floods.

Finally, the floods explain the scabland erosion on some basaltic benches along the walls of the Columbia River Gorge. The scarred surface opposite The Dalles is particularly striking, but there are many other good examples of the erosive ability of the floodwater.

RADIOACTIVE WASTE DISPOSAL—A POSSIBLE ANSWER

Man has taken advantage of some of the geologic characteristics of the Columbia Plateau Province to transform the region into a prime asset. Many of the projects, such as the harnessing of the energy of the Columbia and Snake Rivers to provide electricity and the diversion of water for irrigation, are too well known to need comment here. There is, however, another possible use for the region which is just now being investigated. This is the disposal of radioactive waste material.

The problem of safely disposing of radioactive wastes is becoming more critical each day. The problem is aggravated by the relatively long half-lives of many of the by-products of radioactive processes, for some of the waste must decay for several hundred years before it "cools" down and becomes no hazard. In the past, scientists have disposed of some waste material in deep, abandoned salt mines or caverns. (Salt provides good natural shielding against radioactive radiation.) Clearly, however, there is a limit as to how much material can be stored in this way. Another disposal method has involved sinking the waste in thick, lead cannisters in the deep ocean, but there is always some possibility of leakage, and this could have very serious consequences.

In recent years, some geologists and engineers have been studying the feasibility of placing radioactive waste substance far beneath the surface of the Columbia Plateau. The region has some natural characteristics that make this possibility very attractive. For one thing, basalt, like salt, is a good shield against radiation. Furthermore, the plateau is underlain by a tremendous thickness of lava. (One well near Richland, Washington, was drilled through 12,000 feet of basalt, although some of this was older than the Columbia River Basalt.) The basin structure beneath the plateau is also favorable for underground disposal, for contaminated groundwater could not easily reach the surface. Finally, there are numerous sites in the region, including parts of the large Hanford Atomic Reservation near Richland, where additional nuclear energy plants could be located.

The possibility of waste disposal in the Columbia River Basalt looks so promising that millions of dollars are currently being spent to study various plans. Some of this investigation involves regional study of the unit's chemistry and stratigraphy, so the "pure" science benefits will be considerable as well. Geologists will continue to learn new things about this fascinating region as they help to meet particular engineering challenges.

18 the puget-willamette lowland

■ A "hard rock" geologist in the middle of the Puget Lowland of western Washington is like a fish out of water. The lowland south from the San Juan Islands to Olympia contains an exceptionally thick fill of glacial sediments that conceals all but a very few pre-Quaternary rock exposures. Outcrops of bedrock are somewhat more numerous from Olympia south to the floodplain of the Columbia River. The hills around Portland offer some rock exposure, but the main part of the Willamette Valley, from Oregon City south to Eugene, is covered mostly by late Pleistocene and Holocene alluvium. South of Eugene, the Coast Ranges and Klamath Mountains begin to merge with the Cascades, and exposures are much more numerous.

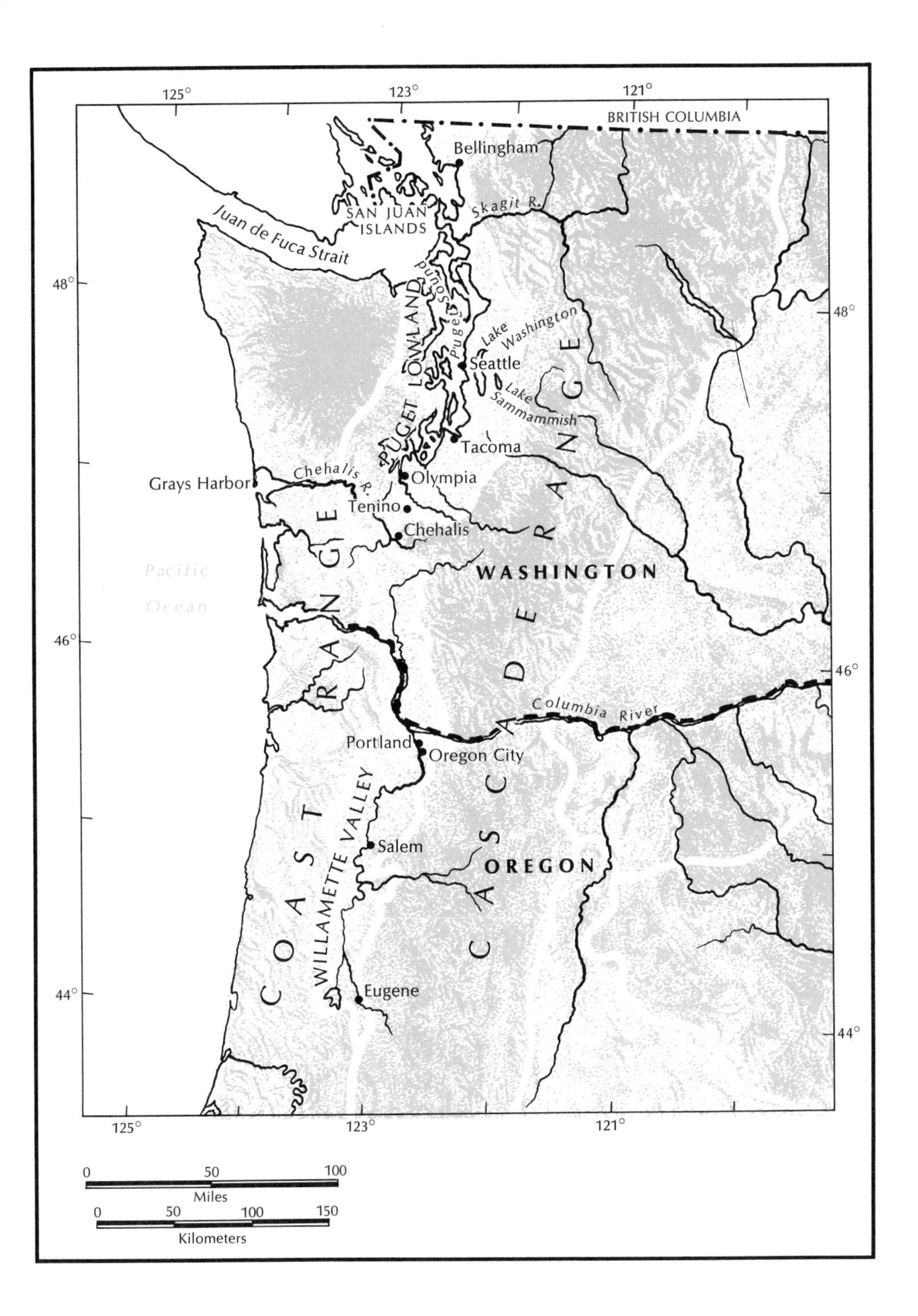

18-1 *Physiography of the Puget-Willamette Lowland. (After Raisz, 1941.)*

TERTIARY ROCK EXPOSURES

Rock outcrops in the Lowland Province relate closely to the Tertiary bedrock exposures in the adjoining Cascade and Coast Range Provinces. The overall structure of the lowland is a broad, young downwarp between the Cascade and Coast Range uplifts. In detail, the structure is much more complex, consisting of numerous mid-Cenozoic folds and faults that trend transverse to the regional downwarp. Hills within the lowland consist most commonly of relatively resistant volcanic rock. In Washington, these volcanics are correlated with the Eocene basalts of the Olympics and the Willapa Hills. Along the eastern side of the lowland, they are correlated with nonmarine andesitic and basaltic flows in the Cascades. Sedimentary formations interbedded with the flows locally contain fossils that permit both dating and interpretation of the depositional environment. The scarcity of hard rock in the lowlands has made the few outcrops available valuable sources of construction rock. For this reason most of the hills contain rock quarries.

THE PUGET LOBE

Several periods of glaciation of the Puget Lowland produced a complex stratigraphic record. These Quaternary deposits have been studied quite intensively by geologists in the past several decades, for they offer a significant challenge to development of the region as an industrial-urban center. Whether a particular hill is underlain by sand or by clay can mean tremendous differences in the cost of such projects as freeway construction, the installation of sewer trunk lines, or the excavation of foundations for large buildings. The potential hazard posed by possible ground conditions unfavorable to engineering projects is balanced, in a sense, by the economic importance of commercial deposits of gravel, sand, and clay.

The northern part of the North American continent was covered at least four times in the Pleistocene Epoch by great ice sheets. The great tongue of ice that spread south from British Columbia across the Puget Lowland is known as the Puget Lobe. Ultimately, this great glacier reached all the way to Olympia. It covered all the hills of the Puget Lowland and the San Juan Islands and lay high against the flanks of the Olympics and the Cascades. This happened at least four times. The record of the older glacial events has been obscured greatly by the most recent glaciation, and consequently our knowledge of the early glacial history of the lowland is poor.

The effects of glaciation on the Puget Lowland were so great that a preglacial reconstruction of the topography is quite difficult. We can assume that the basic distribution of major landforms was as we see it today with the Cascades and Olympic Mountains separated by a broad valley not unlike, perhaps, the present Willamette Valley. Drainage was to the north and then west to the ocean via the Strait of Juan de Fuca. As the ice wall moved south across the Gulf Islands and San Juan Islands the drainage of the Puget Lowland was probably affected very little. An entirely different situation prevailed, however, as soon as the advancing glacier reached the northeast front of the Olympics. At that time, the wall of ice effectively dammed the entire lowland, for little drainage could escape past the ice to reach the ocean. The result, inevitably, was the formation of a lake. Eventually, this lake filled the entire lowland between the ice wall, the mountains, and the divide between the Puget Sound and Chehalis River drainage basins. This lake received runoff from the mountains and meltwater from the Puget Lobe. When the basin filled, it drained to the ocean via the Chehalis River. This explains why the modern Chehalis River, which is not very large, occupies a valley suggestive of a much larger river. At its largest, the river probably had a discharge that was several times greater than that of the modern Columbia River. This glacial discharge also explains the presence of very coarse gravel in the Grays Harbor area and on the adjacent continental shelf.

The ice front continued to advance, ultimately covering almost all the lake basin and reaching very close to the Chehalis River itself. Evidence suggests that the glacier did not occupy this maximum position for long. When it started to melt back, it created a new lake, which persisted until

18-2 *The Willamette River, looking east toward the Cascade Range and Mount Hood. (Photograph courtesy of Oregon State Highway Division.)*

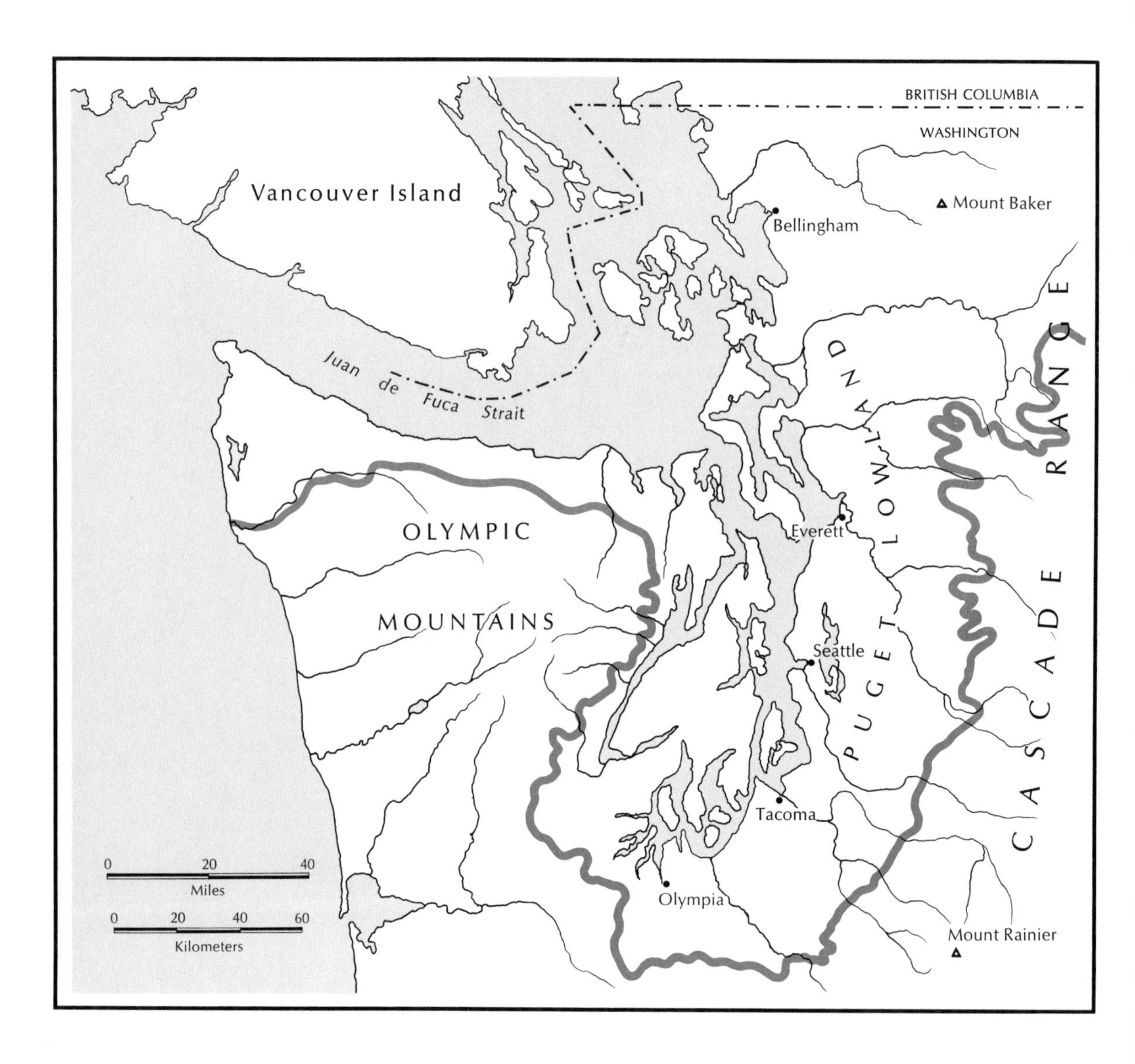

18-3 *The maximum extent of the Puget Lobe during the last glacial period. (After Easterbrook, 1969, p. 2274.)*

retreat freed the natural drainage of the lowland via the Strait of Juan de Fuca. As normal drainage patterns were reestablished, the streams began eroding the sediments left in the lowland from the lake stage and from direct glacial deposition. These sediments must be considered in more detail in order to understand the present stratigraphy and physiography.

GLACIAL SEDIMENTS

Sedimentation patterns during the lake stages are easy to imagine. The streams flowing into the lake from the Olympics and the Cascades carried considerable amounts of sediment. This ranged in size from *rock flour* (that is, fresh, fine-grained rock and mineral particles produced by glacial abrasion—the cause of the milky appearance of streams fed by alpine glaciers) to coarse cobbles and locally even boulders. Deltas grew where these rivers entered the lake. As the lake level fluctuated so did the position of the lake shoreline and hence the position of deltaic sedimentation. The deltas are easy to recognize in the stratigraphic record, for they consist primarily of sand and gravel layers, many with well-developed inclined layering or *cross-bedding*. The finer-grained fraction of the sediment was not deposited on the deltas but was transported farther into the lake. There it settled out of suspension as finely layered beds of silt and clay.

This relatively simple pattern of sediment distribution would be much more apparent today if the Puget Lobe, once having dammed the outlet of the lowland, had advanced no farther. It continued south into the lake, however, and ultimately overrode the lake bottom entirely. What effect did this have? Once again, a rather predictable one. The ice front was essentially a migrating shoreline, for it too contributed vigorous meltwater streams that flowed off the surface of the glacier. The meltwater transported sediment that ranged in size from clay to boulders. The same size grading of sediment that was effected along the east and west sides of the lake was produced along this northern ice margin. Thus the history of sedimentation at a point situated near the middle of the lake (near the present site of Seattle) would have

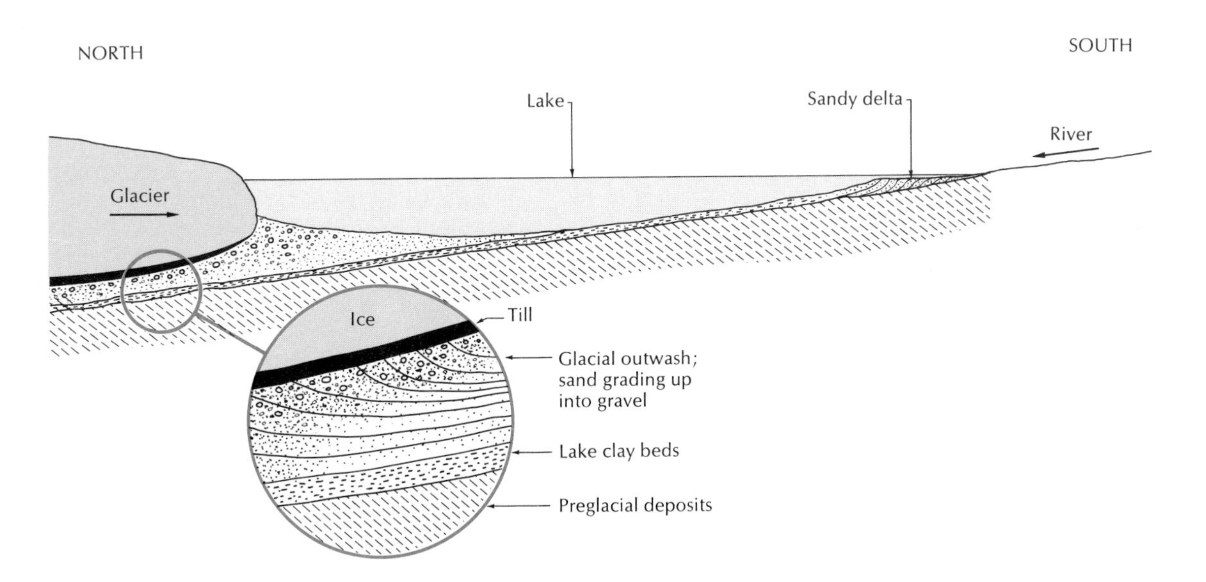

18-4 *Simplified cross section of sedimentation patterns in the ice-dammed lake that covered the Puget Lowland. Enlargement suggests the stratigraphy beneath the advancing glacier.*

18-5 *Glacial till along Interstate Highway 5, 1 mile north of the Stillaguamish River. The boulder is about 3 feet in diameter. Note the numerous pebbles that have weathered out of the clay and accumulated at the base of the cut. (Photograph by Bates McKee.)*

been something like this. When the lake first formed, the area was submerged; but, as the lake rose, the site was soon far from any shoreline. Sedimentation at that time consisted primarily of silt and clay deposition. As the front of the Puget Lobe advanced and approached the Seattle area, the sediment being deposited gradually coarsened, first to sand and finally, just prior to actual ice override, perhaps to gravel.

The effect of the passage of ice across a particular site is difficult to predict. The Puget Lobe did much erosional excavation in some areas while elsewhere it did very little. Thus the lake sediment might have been removed completely or it might have been preserved. If preserved, it would have been compacted under the weight of the overriding ice, which in the Seattle area attained a maximum thickness of about 4,000 feet and exerted a pressure at the base of approximately 8,000 pounds per square inch. The lake beds, if preserved, also would have been buired under a layer of glacial *till.* This is essentially a bouldery clay material deposited directly by the glacier, either at the bottom of the advancing ice (in which case the till would also be compacted) or let down on the surface as the ice melted away and its load of sediment dropped.

The final stratigraphic section, then, for this lake phase of the site, would consist of lake clays at the bottom, grading upward into silts, then sands, then perhaps gravel, and finally a capping of glacial till. Glacial erosion might have stripped off any amount of this, so that the till might cap any of the sediment types—the gravel, the sand, the silt, the clay, or even the underlying, prelake units.

If only it had been this simple. We do find bluffs that reveal just this stratigraphy (Figure 18-6), but the overall patterns are complicated by such factors as:

1. The formation of a lake or lakes in the lowland as the ice front retreated. This reversed the above stratigraphic sequence, for an emerging site on the lake bottom would have been progressively farther from the northward migrating ice margin, until the Strait of Juan de Fuca reopened and the lake drained.

2. Glacial advance and retreat not once but at least four times.

3. Differential glacial erosion that produced a markedly irregular ground surface. The surface was carved out of lake sediments, plus the varied sediments derived from older glacial and interglacial periods, plus preglacial bedrock. Glacial erosion deepened preexisting river valleys. This produced the Lake Washington and Lake Sammamish basins and deepened Puget Sound off Seattle to as great a depth as 1,000 feet below sea level. Where the ground was not markedly excavated by the overriding ice, it was shaped. In both the Seattle area and elsewhere in the Puget Lowland, many of the elongate hills are parallel to the general direction of ice movement (which was for the most part approximately north-south). This elongation resulted from glacial shaping.

4. Postglacial erosion and sedimentation, both by streams and by waves and currents in Puget Sound.

Multiple glaciation of the Puget Lowland has produced, therefore, a Quaternary stratigraphic record characterized by marked lateral and vertical changes in sediment type. A complex, although

18-6 *Bluff along Puget Sound at Fort Lawton, Seattle. The beds at the base of the cliff are glacial lake clays. Higher in the section, they grade into outwash sand and gravel. A thin veneer of till caps the bluff and the slope to the left. Tipped and downed trees on the face of the bluff reflect active landslides. (Photograph courtesy of A. S. Cary.)*

18-7 *The Lake Washington Ship Canal and north-central Seattle looking west toward Puget Sound and the Olympics. The complex topography of the lowland results from a combination of glacial erosion and deposition, plus postglacial processes. (Copyrighted photograph courtesy of Delano Photographics.)*

somewhat lineated, topography was another result of lowland glaciations.

GLACIATION AND CONSTRUCTION COSTS

The glacial history of the Puget Lowland is of very great interest to scientists and scholars. It is of more direct concern, however, to engineers and local homeowners. A few examples of the Seattle area alone can illustrate the point. The homeowner is affected by such factors as the stability of his building site, which in the Seattle area is likely to be sloping; the availability and cost of construction materials such as sand, gravel, and crushed rock; and perhaps the depth of the water table. The nature of the glacial sediment at or near the surface of a particular lot also influences excavation costs for sewers and drainfields and the yearly expenditure for mulch and fertilizer. The ground condition can differ markedly from one lot to the next. This glacial inheritance is seldom considered by the average landowner.

The engineer charged with designing a freeway, a large building, or a major sewer system such as the multimillion-dollar Metro Project of Seattle cannot ignore the local stratigraphy. The structures must be designed to fit the ground conditions. For this reason, much of what is known about the geology of the Puget Lowland has come from engineering geology studies financed by commercial firms or by various state or federal government agencies.

Any city or state, faced with the problem of locating a large freeway in a densely populated region, must make difficult decisions. They involve such factors as the areas to be served, the value of the land and structures to be condemned, the engineering problems expected along possible routes and their effects on the cost of construction, and (increasingly these days) the attitude of the people. Assuming a freeway is feasible and desirable, how is the best route chosen?

In the case of the Seattle Freeway (Interstate Highway 5), the glacial history played an important role, especially in the central and southern part of the Seattle area. Certain hillsides had not been developed commercially, because of the high costs of building and because of notorious past histories of slope instability. As one might predict, the least stable slopes are those that are underlain by beds of glacial lake clay. The north end of Beacon Hill near the Spokane Street and Dearborn Street exits of the freeway is a good example. Interstate 5 is located on some of the least stable slopes in the Seattle area. This is not because the engineers enjoyed the design challenge but because they calculated that it was more economical to pay more per mile for construction costs than to condemn valuable land. The construction costs of a freeway across the heavily industrialized valley floor of the Dunwamish River would have been less than what it cost to build the existing freeway, perched as it is on the steep flank of Beacon Hill. The amount of money required for land acquisition, however, would have been disproportionately greater.

An alternative solution for highways is to place them underground. This approach is gaining favor as underground excavation techniques are perfected. Although this eliminates certain sociological objections it fixes an even greater importance on the local stratigraphy.

THE FLANKS OF THE RANGES

The estimate that the Puget Lobe had a maximum thickness of 4,000 feet at Seattle (measured from the bottom of Puget Sound) is based on data obtained at the front of the Cascade and Olympic Mountains. Clearly, the ice was thick enough to bury all of the hills in the lowland. The maximum elevation attained by the Puget Lobe against the adjacent mountains is determined in several ways. First, the topography below that elevation shows glacial smoothing, which contrasts with the rougher, less regular topography above. Second, the highest ice stand is recorded by the highest position of glacial deposits on the mountain flank. Glacial erratics are helpful, especially in the Olympics where the light-colored granitic boulders derived from western British Columbia contrast markedly with the darker volcanic and sedimentary

18-8 *Beds of glacial lake clay, exposed in Beacon Hill in Seattle. (Photograph courtesy of A. S. Cary.)*

rocks that are native to the range. Finally, the contact between the Puget Lobe and the mountain front was the location of large ice-contact rivers, confined on one side by rock and on the other by ice. This condition is due to the fact that virtually all glaciers are higher in the middle than at their margins. As the Puget Lobe melted away, these ice-marginal rivers were lowered. Their successive positions were recorded by both river deposits and channels notched into the adjacent rock wall. The highest position of the ice against the mountain fronts is then used to estimate the probable cross-sectional configuration of the Puget Lobe and to determine maximum ice thickness in the adjacent lowland.

Even a glance at the valleys in the Cascades and the Olympics offers convincing proof that they were glaciated in the recent past. They served as passageways for large alpine glaciers that originated near the range crests and flowed out into the lowlands. A close look along the Cascade mountain front reveals an interesting relationship, however. South of the Skagit River, the front of the alpine glaciers had receded well back into the range by the time the Puget Lobe made its last major advance. At that time, the lower reaches of the valleys were ice free. The appearance of the Puget ice across the valley mouth produced a very efficient dam. This created a lake in each of the valleys. Each lake rose until it spilled around the intervening mountain spur and into the lake in the next valley to the south. Eventually a large river, carrying all of the meltwater and runoff from the Cascades and the eastern part of the Puget Lobe, flowed south along the mountain front. It swung west around the ice terminus and flowed into the lake, which filled the southern (ice-free) part of the lowland. At the time of maximum extent of the Puget Lobe, the river flowed directly into the Chehalis River.

The existence of ice-dammed lakes in the Cascade valleys is proved by evidence found within the valleys. Tremendous deltas as much as 1,000 feet high were built at the mouth of each valley. They consist of very coarse beds of gravel and sand dipping upvalley. Farther back in the range, the lake deposits get progressively finer grained,

18-9 *Downtown Seattle looking southeast toward Lake Washington (middle distance), Lake Sammamish (to the left, beyond Lake Washington), and the Cascade Range. Elliott Bay, a part of Puget Sound, is in the right foreground. This photograph, taken in 1964, shows the Seattle Freeway under construction. South of the central district it now follows the undeveloped, vegetated strip visible beneath Beacon Hill at the right edge. (Copyrighted photograph courtesy of Delano Photographics.)*

until thick lake clays floor the valley. Fortunately, postglacial erosion has not completely removed the lake beds, for they emphasize the age discrepancy between the times of maximum lowland and maximum alpine glaciations.

GLACIAL DATING

The sequence of glacial events, recognized and named in the Puget Lowland, is shown in Table 18-1. Good dating control is provided by the radiocarbon method for the most recent major glacial advance, the Vashon Stade. (A *stade* is an interval of glacier advance within an overall period of glaciation.) Radiocarbon dates from organic material immediately under and just above the Vashon Till prove that the Seattle area was covered by ice between approximately 15,000 and 13,500 years ago. The absolute age of the older Salmon Springs Till is less certain, although it probably is somewhat older than 35,000 years. The ages of the Stuck and Orting Glaciations are unknown. The age of the maximum recent major advance of glaciers in the Cascades, the Evans Creek Stade, is probably about 17,000 years, or slightly older than the Vashon Stade.

The dates of glaciations in western Washington correlate reasonably with data from other parts of the continent and from Europe and Asia. The last major glacial stage in the midcontinent region of North America is termed the Wisconsin Stage. It began earlier than 40,000 years ago and continued, with several glacial maxima, to about 10,000 years ago. The Vashon Stade and the Evans Creek Stade, which are both included in what has been termed the Fraser Glaciation, would correspond to late Wisconsin time. The Salmon Springs Glaciation is considered to be early Wisconsin in age.

THE ENIGMATIC MOUNDS

Lest we conclude that geologists now understand in a general way the recent history of the Northwest, a closing discussion of the many mounds that dot the prairies near Tenino south of Olympia is appropriate. They are particularly numerous and

TABLE 18-1 Pleistocene sequence in the Puget Lowland. (After Easterbrook, 1969.)

GEOLOGIC CLIMATE UNITS		APPROXIMATE CARBON-14 AGE, YEARS
Holocene		
		10,000
FRASER GLACIATION	SUMAS STADE	
		11,000
	Everson Interstade	
		13,000
	VASHON STADE	
		18,000
Olympia Interglaciation		
		29,000
SALMON SPRINGS GLACIATION		
		>40,000
Puyallup Interglaciation		
		?
STUCK GLACIATION		
		?
Alderton Interglaciation		
		?
ORTING GLACIATION		

18-10 *The mounds on Mima Prairie, south of Olympia. View is to the west. Are these the former homes of gophers, or can they be explained in some other way? (Photograph courtesy of V. B. Scheffer.)*

well developed on Mima Prairie, where they are known as the Mima Mounds. These features have interested observers for more than a hundred years, and numerous theories, both geologic and nongeologic, have been proposed to explain them.

The mounds are as much as 7 feet tall and 70 feet in diameter, but most of them are only about 40 feet wide. The area between the mounds is approximately equal to that of the mounds. The mounds consist of pebbly silt and sand, although the surface of the prairie beneath the mounds is composed of gravel. The sediment was deposited by glacial meltwater during the recessional stage of Vashon glaciation. A few mounds have been excavated or incised. In these, the mound silt projects down into the prairie gravel near the center of the mound.

One theory relates the mounds to the formation in the ground of polygonal ice wedges, such as those that produce so-called "pattern ground" in arctic regions today. The close association of the Mima Mounds with Vashon glacial gravels supports this hypothesis, but the mechanism whereby mounds this large could form is obscure. The absence of other characteristic arctic-type landforms does not support this hypothesis. Furthermore, the Mima Mounds are not unlike mounds found locally in the Great Valley of California, where the frozen ground theory seems less plausible.

A more colorful theory suggests that zoologists rather than geologists should be working on the problem. It blames the mounds on the past activity of energetic pocket gophers. Gophers live in nest burrows and forage for food outward around the burrow opening. They are colonial with definite territorial rights, and a mound may have grown as the animal dragged food (and dirt) toward home. The silt "root" projecting down into the gravel is supposed to be the remains of a burrow. Gophers live today in some of the mounds, but this does not prove that their ancestors built them. The chief difficulty in this theory is imagining a mound 7 feet tall and 70 feet across built by small gophers, but perhaps we underestimate the energy of gophers.

Other theories call upon flood deposition (around trees or bushes, or ice wedges) or differential erosion or solution action to explain the mounds. Until scientists do detailed research on the mounds, they will remain the objects of much speculation.

19 rocks as crystal balls

■ The book began by emphasizing that geology seeks to interpret the past history of the earth by careful examination of the rock record and the landscape as we see it today. Processes at work today can be projected back into time to understand events in the past. Geologists must operate in a time framework that encompasses millions and even billions of years. We have developed the Northwest scene in a somewhat modified sequential fashion, starting with those regions that contain the most ancient rocks and finishing where there is the most complete record for the past few million years. Geologic history, like other kinds, seems most meaningful when we start in the past and finish with the present. This sequential approach identifies certain significant trends and carries them up to the present. Such trends provide opportunity for predicting the future. Statistically, we are on reasonably firm ground when we presume that a certain circumstance which has prevailed for the past million years or more will probably continue for a long time to come (at least a long "man" time if not a long geologic time).

A review of the geologic history of the Northwest indicates that some rather spectacular events have occurred in relatively recent time. It is natural to question if they can or will happen again. If our conclusion is that the volcanoes, for instance, will erupt again or that glaciers will return to cover much of the region, we appropriately speculate about our ability to challenge and perhpas to change geologic events—to control our environment on a gigantic scale. This kind of conjecture seems a bit esoteric, but we are approaching the time when our technology may make the impossible possible. Present studies on the feasibility of triggering minor earthquakes with the hope of preventing major tremors represents a move in this direction. You might think that glacial control is out of the question, but consider the crash programs that would be launched if major ice caps began once again to form in Canada and northern Europe. However, before we speculate about our potential ability to control such events, let us look backward at past trends in deformation, volcanism, and glaciation in the Northwest, and see if we can obtain some basis for prediction.

DEFORMATIONAL TRENDS

The geologic history of the Northwest is one of crustal instability and deformation for as far back as we can see. Stratigraphic sections contain numerous unconformities, great thicknesses of sedimentary and volcanic strata, and evidence of significant orogeny. The record seems to get more confused and more deformed to the west, toward the continental margin. Because of limited exposure to pre-Cenozoic rocks, however, one cannot obtain a very coherent picture of older events. The stratigraphy indicates overall subsidence and marine sedimentation in the Northwest throughout much of the Paleozoic and Mesozoic Eras, with perhaps major volcanism and deformation in the Devonian, Permian, Triassic, and Cretaceous Periods and general unrest in this part of the Cordilleran Geosyncline at all times. The climax of orogeny came late in the Mesozoic Era. Major folding, faulting, metamorphism, igneous intrusion, and uplift occurred; and the Cordilleran structural and stratigraphic patterns were altered dramatically. All but the present Coast Range region was raised permanently above sea level by the beginning of the Cenozoic Era, after which the deformational style of the entire region was changed. The new deformational patterns are the ones of interest here, for they are the ones that continue today.

The Cenozoic record is not easily interpreted. The most obvious major structural elements in the western Cordillera are the broad uplifts and intervening downwarps, which trend generally north-south. These are the Vancouver Island-Olympic Mountains-Coast Range Arch, the Fraser-Puget-Willamette Trough, and the Coast Mountains (British Columbia)-Cascade Arch. These structures are of Late Cenozoic age and are probably active today. Smaller structures trend across these major warps. Examples include folds and faults in the Columbia Plateau, Blue Mountains, Cascades, and Coast Range Province, and the large Snake River Plain downwarp. Many of these structures may have been active throughout much of the Cenozoic Era and continue to be active today. As yet our knowledge of the tectonic history of the Northwest is insufficient to explain all these various features in some simple way. Models have been proposed that relate many of them to a gigantic subcrustal rotation that could produce both horizontal and vertical displacements over much of the Cordilleran region.

A few years ago, such master schemes seemed quite wild. New evidence for continental drift, seafloor spreading, and underthrusting of oceanic crust at continental margins has not only opened up new worlds of speculation, but it has emphasized the necessity for analysis on a very large scale. A critical key is knowing what is going on at the present time. New techniques in geophysics now permit analysis of deep crustal structure, the sense of movement on buried faults, and both vertical and horizontal strain of the earth's surface. While these types of measurements are now possible, very few have been made in the Northwest (especially in comparison to California).

EARTHQUAKE PREDICTION

Man will not be able to prevent earth deformation, but he will be able to predict with some accuracy its most immediate and hazardous manifestation—the earthquake. Scientists can measure the accumulation of strain that produces earthquakes. Such measurements, plus other techniques now being perfected, will allow more precise prediction of earthquakes. We certainly can minimize damage in this way. The day may not be too distant when earth strain can be released by inducing frequent small fault displacements, thereby preventing major movement and destructive earthquakes.

PREDICTING VOLCANIC ACTIVITY

It is impossible to identify a single geologic period when the Northwest was not the scene of some volcanic activity. The very great frequency of eruptions during the Cenozoic Era certainly suggests that volcanism will occur in the future. Potential sites of future eruptions can be identified by locating areas of recent volcanism. The dormant volcanoes of the Cascades and the Coast Mountains of British Columbia clearly are suspect—some more so than others. The probability of future eruption of a particular cone is based ideally on a detailed knowledge of its past history, including the age of the most recent eruptions and the total time span over which the cone was built. On this basis, volcanoes like Mount Rainier and Mount Saint Helens, with eruptive activity during the past several hundred years, are more dangerous than are Mount Baker or Glacier Peak, which probably have not been active for thousands of years. Even the "more dormant" volcanoes of the Cascades should not be considered extinct, however, especially since relatively few eruptions have been dated accurately. Volcanoes considered extinct in other parts of the world have erupted quite unexpectedly. Some new vents could open up also and build new cones in the Cascades.

Two regions outside the Cascade-Coast Mountain chain are likely to see future volcanic activity. One is southeastern Oregon with its abundant evidence of late Pleistocene and Holocene eruptions, some dated as no older than a few thousand years. These recent eruptions have issued, for the most part, from well-defined centers, especially vents along young faults. Extensive Pleistocene fissure flows underlie the High Lava Plains region. None of these appear to be very young, but the thought of future eruption of fluid basaltic lava capable of covering hundreds or even thousands of square miles in a few hours is a sobering one.

Another province that is a likely site for future volcanism is the Snake River Plain. Flows could be quite localized, as they have been at Craters of the Moon; or they could be quite extensive, as they have been south of there.

Although three areas that have had recent eruptions have been identified as likely to see additional volcanic activity, other regions should not be ruled out. Provinces like the Coast Range, the Columbia Mountains, the Rocky Mountains, the Puget-Willamette Lowland, the Klamath Mountains, and Vancouver Island have not witnessed eruptions, however, for tens of millions of years. They must be considered less probable sites of future volcanism.

What can be done to control future eruptions? Almost certainly, nothing. We might imagine artificially cooling regions within the earth where magma is being generated. The total amount of energy involved in volcanic activity, however, is so immense that this seems unlikely. Our technical skills could be applied more profitably to devising better techniques for predicting eruptions and issuing warnings to those in danger. Volcanic eruptions are the earth's safety valves; and, should we invent some means of preventing volcanism, who could predict what might result?

MAGNETIC REVERSALS

The fact that the earth's magnetic field has been reversed in the past has been well substantiated by measurement of rock magnetism (Chapter 4). Reversals have been fairly frequent in the Cenozoic Era—some scientists believing that they may

19-1 *The wrinkled top of a young lava flow in the Three Sisters region of the Oregon Cascades. Broken Top, a deeply eroded volcano, is in the distance. A flow as fresh as this makes a prediction of future volcanic activity in the Cascades seem safe. (Photograph courtesy of Oregon State Highway Division.)*

occur every 200,000 to 300,000 years on the average. The process of reversal is not well understood, although most geophysicists believe it is probably caused by changes in the flow pattern of currents in the earth's core. Another issue centers on the effect the reversal process produces at the earth's surface.

A reversal probably involves a gradual diminution to zero in the strength of an existing field of earth magnetism, followed by a slow increase in strength in the new direction. The entire process might take 1,000 years or so. Geophysicists believe that the earth's magnetic field shields the earth from some extraterrestrial radiation, so that presumably this protection diminishes and for a while disappears during the reversal process. The results are difficult to gauge. Some scientists believe that the effects would be relatively minor. Some geologists, however, have suggested that certain magnetic reversals in the past may have brought about the extinction of some organisms. This theory has been supported recently by an apparent coincidence between Late Cenozoic reversals and the disappearance of certain marine microorganisms. Perhaps certain reversals have been slower than others and have brought about significant changes in the earth's organic population. Scientists simply cannot be sure, for they have never witnessed a magnetic reversal and can only speculate on its possible consequences.

Two facts are certain. First, it has been 700,000 years since the last reversal, and the earth may be due for another one. Second, if another reversal does occur, disastrous as it might be, scientists will not be able to do anything to prevent it.

ARE WE OUT OF THE ICE AGES?

This is one of the more interesting speculations. When geologists first accepted, in the nineteenth century, the conclusion that large areas in the higher latitudes of the northern hemisphere had been covered by glaciers in the past, it was generally assumed that the Ice Ages were over. The time scale reflected this assumption, for the Quaternary Period was divided into two epochs, the Pleistocene (glacial) and the Recent or Holocene (postglacial). The Pleistocene record in Europe and North America showed that continental glaciation had occurred at least four times, and that the glacial deposits were separated by sediments and soils formed between glacial episodes. Close examination of these interglacial deposits suggested that they represented long periods of time and a climate at least as warm as today. Various relative and absolute dating techniques were applied to the Pleistocene record, as were chemical techniques that permitted good estimates of the past climatic conditions. The more that geologists learned about the Pleistocene Epochs, the less certain seemed the assumption that we were out of the Ice Ages. It is now estimated that the first major Pleistocene glaciation may have occurred more than 2 million years ago and that interglacial stages may have been several hundred thousand years long. The last major glacial advance was only 10,000 to 15,000 years ago, so we have not been free from ice for very long. Perhaps the Recent or Holocene Epoch is merely an interglacial episode and glaciers will return in the next few thousand years. (Geologists now believe that the climate was somewhat warmer 6,000 years ago than it is today.) Suddenly the measurement of glaciers has taken on new meaning, for the "health" of glaciers may indicate a significant trend.

The history of past glaciations certainly provides no support to an assumption that we are out of the Ice Ages. On the other hand, the glacial record alone is not an adequate basis for predicting the future. What is required is a better understanding of the causes of glaciation or, to be more precise, of the causes of worldwide climatic variations that in turn produce glaciation. Many theories have been proposed. Each has some merit; each could be a factor; but no single theory is as yet acceptable. Some of the variables that might influence climate are changes in the earth's atmosphere, variations in solar energy, the position of continents

19-2 *The Challenger Glacier on Mount Challenger in the North Cascades, as it looked 20 years ago. Note the melted sinkhole in the stagnant ice on the left and the cleaner, active lobe on the right. Since 1950, the active lobe has covered the stagnant ice and now covers much of the wallrock to the left. Are such short term variations in glacial activity significant for predicting possible future glaciations? (Photograph courtesy of Peter Misch; previously published in* The Mountaineer, *vol. 45, 1952.)*

relative to the earth's axis of rotation, volcanism, patterns of oceanic circulation, and orogenic uplift of land regions. The very diversity of this list indicates how far we are from understanding the causes of glaciation.

The onset of a new glacial interval would be gradual. Scientists would have ample time to ponder the problem and consider possible remedies. For example, even though the Vashon advance is considered rapid (inasmuch as Seattle was covered by up to 4,000 feet of ice and then uncovered in no more than 1,500 years), the probable rate of advance of the ice front was perhaps no more than one-tenth of a mile per year. When we know more about the causes of long-term climatic variation, perhaps we can influence or even control them and not surrender the higher latitudes to ice. However, the thought that we might exert this type of control over our environment is a bit frightening. Mankind is at last realizing that some of the changes it has wrought unthinkingly on the environment over the past several centuries have had serious and, in some cases, irreversible results. With this awareness has come the obvious need to understand the present better in order to optimize the future. No longer can we presume that controlling nature is necessarily a desirable goal, for seemingly beneficial short-range solutions can produce unforeseen harmful long-range results. Environmental awareness can only benefit the science of geology, however, because the present is truly the key to the past *and* the future.

APPENDIXES

SEDIMENTARY ROCKS

Definitions: a *sedimentary rock* forms by the solidification of sediment, the chemical precipitation of solids from solution, or the secretion of organisms.

Sediment is solid material both mineral and organic that is in suspension, is being transported, or has been moved from its site of origin by air, water, or ice, and has come to rest on the earth's surface.

Sedimentary rocks are grouped loosely into three major categories: *clastic rocks* (that is, fragmental or detrital rocks, formed by the accumulation and solidification of solid particles, *chemical rocks* formed by the precipitation of inorganic mineral matter, and *organic rocks* produced by the accumulation and solidification of organic fragments or by direct organic precipitation. A fourth type, intermediate between sedimentary and igneous rocks, is represented by the *pyroclastic rocks,* which form by the accumulation and solidification of solid fragments ejected from volcanoes. They are included here because their classification parallels that for clastic sedimentary rocks.

IGNEOUS ROCKS

Definition: an *igneous rock* forms by the solidification of a rock melt (magma).

Many different classification schemes for igneous rocks are in use. Most are based on two parameters: mineral (or chemical) composition and texture (the size, shape, and arrangement of a rock's component particles). The scheme presented here is relatively simple but encompasses most of the igneous rock types found in the Northwest.

The composition is judged principally by *color.* The lighter shades generally represent rocks that are rich in silica and contain major amounts of the minerals quartz and feldspar. Darker varieties contain less of these minerals and more of the minerals that contain iron and magnesium. (An exception is *obsidian.* Although it is very siliceous, it is dark-colored, because its glassy and translucent nature permits rich coloring by very small amounts of iron.)

A rock classification and geologic time scale

TYPE	ROCK NAME	MAJOR CONSTITUENT
Clastic	Conglomerate	Boulders, gravel, pebbles
	Breccia	Angular blocks
	Sandstone	Sand
	Siltstone	Silt
	Claystone	Clay-sized particles
	Shale	(Siltstone or claystone with distinct cleavage)
Chemical	Limestone	Calcite (calcium carbonate)
	Dolomite (dolostone)	Dolomite (calcium-magnesium carbonate)
	Chert	Quartz, chalcedony, opal (silica)
	Rock salt	Halite (sodium chloride)
	Gypsum	Gypsum (calcium sulfate)
Organic	Diatomite, radiolarian chert, etc.	Siliceous shells of microorganisms
	Coal, peat	Plant remains
	Fossiliferous limestone	Calcareous shells, coral, etc.
Pyroclastic	Agglomerate	Volcanic blocks, bombs
	Tuff	Volcanic lapilli, ash, dust

GRAIN SIZE \ COLOR	LIGHT	INTERMEDIATE	DARK
Glassy (nongranular)	Pumice (porous)		Obsidian Basalt glass
Fine grained	Rhyolite* Rhyodacite* Dacite*	Andesite	Basalt
Coarse grained	Granite* Granodiorite* Quartz diorite*	Diorite	Gabbro

*The light-colored rocks, both coarse- and fine-grained varieties are subdivided according to the type of feldspar mineral contained. Each group is arranged in order of a decreasing potassium feldspar (increasing calcium-sodium feldspar) content. Feldspar determination is difficult or impossible to make in the field. Consequently, geologists sometimes use the term "felsite" to denote all light-colored, fine-grained igneous rocks. Similarly the term "granite" is commonly used to include granite proper, plus granodiorite, quartz diorite, and diorite.

The aspect of texture emphasized here is *grain size.* This characteristic reflects primarily the rate of solidification of the rock. Slow cooling produces relatively large crystals (coarse grain size), rapid cooling results in very small crystals (fine grain size) or even glass (no crystals at all). The rate of cooling is determined by the size of the molten mass and the temperature of the surrounding medium. In general, fine-grained and glassy rocks are volcanic (formed at the earth's surface), while coarse-grained rocks are intrusive (crystallized at depth).

Classification schemes depend on relatively rigid and arbitrary divisions, but the rocks themselves represent a full gradational spectrum of particular parameters. Geologists hedge their bets by using terms like "basaltic andesite."

METAMORPHIC ROCKS

Definition: a *metamorphic rock* forms by the recrystallization in the solid state of preexisting rock in response to pronounced changes in temperature, pressure, or chemical environment below the earth's surface.

The classification of metamorphic rocks is based on several types of criteria and in detail is quite complex. Fortunately most metamorphic rocks can be described adequately by a simple textural classification, modified somewhat by compositional variants. A major distinction is drawn between *foliate* and *nonfoliate* types, foliation referring to lamination or layering along which the rock tends to cleave. The foliate types are most common and are subdivided according to grain size. Nonfoliate types are distinguished for the most part by mineral composition.

Modifying adjectives are used extensively in describing metamorphic rocks. Thus the mineralogy may be shown by adding the minerals before the textural name (for example, a garnet-muscovite schist), or the texture may be designated in front of a compositional term (for example, a gneissose marble). The prefix "meta-" is used sometimes to denote parentage (for example, a metabasalt), but by itself such a term indicates neither the texture nor the composition of the rock.

TYPE	ROCK NAME	TEXTURE AND/OR COMPOSITION
Foliate	Slate	Very fine grained. Dull luster
	Phyllite	Fine grained (most minerals invisible). Shiny
	Schist	Medium to coarse grained (minerals visible). Shiny
Slightly foliate	Gneiss	Coarse grained. Normally light colored
	Amphibolite	Coarse grained, dark. Mostly the amphibole mineral hornblende
Nonfoliate (massive)	Quartzite	Coarse grained; mostly quartz. Recrystallized quartz-rich sandstone or chert
	Marble	Coarse grained; mostly calcite or dolomite. Recrystallized limestone or dolostone
	Greenstone	Field term for meta-andesite or metabasalt. Green color, fine grained

GEOLOGIC TIME SCALE

ERA	SUBDIVISIONS: PERIOD	SUBDIVISIONS: EPOCH	APPROXIMATE AGE BEFORE PRESENT, MILLIONS OF YEARS
Cenozoic	Quaternary	Holocene (Recent)	0.015
		Pleistocene	2–3
	Tertiary	Pliocene	10–13
		Miocene	25
		Oligocene	36
		Eocene	58
		Paleocene	63
Mesozoic	Cretaceous	Many epochs recognized	135
	Jurassic		180
	Triassic		230
Paleozoic	Permian		280
	Pennsylvanian*		310
	Mississippian*		340
	Devonian		400
	Silurian		430
	Ordovician		500
	Cambrian		570
Precambrian	No subdivisions recognized worldwide		3,500+

*Outside the United States, the Mississippian and Pennsylvanian Periods are not recognized as such, the two together comprising the Carboniferous Period.

B
british columbia trip guides

The following guides presume some familiarity with Chapters 5, 6, 7, 8, and 18.

TRIP 1. TRANS-CANADA HIGHWAY

vancouver–hope–kamloops–revelstoke–banff (alta.)

Vancouver to Hope (*93 miles*). The first 75 miles is across the flat surface of the immense delta of the Fraser River, gradually narrowing as the precipitous glacier-scoured walls of the Coast Mountains to the north and Cascade Mountains to the south close in on the river. Sumas Mountain, nearly 3,000 feet high, rises from the center of the valley east of Abbotsford. The last 50 miles of highway leading into Hope lies close along the south bank of the Fraser River. The valley wall consists mostly of Upper Paleozoic marine rocks of the Chilliwack Group (Chapter 7) and then massive granitic rocks of middle Tertiary age (20 million years old). The latter intrude the Chilliwack Group and probably represent one of the last pulses of intrusion of the Chilliwack Batholith. The rock exposures are good. The evidence for massive glacial erosion and scour of the valley walls is most impressive.

Hope to Kamloops (*173 miles*). The 67 miles from Hope to Lytton is along the bottom of the canyon of the Fraser River—one of the most spectacular gorges on the continent. This is the only escape route to the sea for most of the drainage of the southern half of the Interior Plateau Province and one of the few places along the western margin of the continent where a river cuts through the Coast Mountains-Cascades-Sierra Nevada mountain system. It is also the dividing line between the Cascades and Coast Mountains Provinces. The outcrops are a mixture of crystalline igneous and metamorphic rocks of the two systems, plus some less metamorphosed slates, phyllites, and marbles of Late Paleozoic age.

The Fraser River Canyon is one of the most accessible regions for the study of the types of complexly deformed rocks that compose the core of many of the world's great mountain chains. However, this is not a good highway for "rubbernecking" while you drive along. The geologically minded motorist is strongly advised to locate safe pullouts in which to park his car before commencing his investigations. Those who prefer to keep moving should take the Canadian National or Canadian Pacific Railroad train and sit in the dome car! The rocks are all around you (but sampling is difficult).

The highway leaves the Fraser River at Lytton and turns up the Thompson River, climbing up onto the Thompson Plateau (a part of the Interior Plateau Province). The route provides a good cross section of the pre-Tertiary stratigraphy of south-central British Columbia. Major rock sequences (Chapter 8) that are crossed between Lytton and Kamloops are, in order:

1. Kingsvale-Taseko: Upper Cretaceous marine sedimentary and volcanic strata;
2. Cache Creek: Upper Paleozoic marine sedimentary and volcanic rocks, somewhat metamorphosed;
3. Fraser: Upper Jurassic and Lower Cretaceous marine strata;
4. Nicola-Rossland: Upper Triassic to Middle Jurassic marine volcanic rocks with minor sedimentary interbeds.

Kamloops to Revelstoke (131 miles). The first 25 miles is along the valley of the South Thompson River; it offers intermittent exposures of the Cache Creek Assemblage. A narrow extension of the crystalline schists and gneisses of the Shuswap Metamorphic Complex (Chapter 6) is crossed near the town of Shuswap, where the highway enters the subprovince referred to as the Shuswap Highlands. Then more Cache Creek strata crop out along the route until at Sicamous the main body of the Shuswap Complex is entered. The remaining 44 miles to Revelstoke and the Columbia River lies within the Shuswap terrane. The highway crosses the Monashee Mountains, a segment of the Columbia Mountains Province.

Revelstoke to Golden (92 miles). The highway offers many exposures and panoramic views as it climbs across the high northern end of the Selkirk Mountains. The strata are Late Precambrian (Windermere Series) to Devonian in age. Originally sedimentary, they have all been more or less metamorphosed and deformed by a complex series of northwest-trending folds and faults. Golden lies within the impressive, linear Rocky Mountain Trench, east of the Columbia Mountains and west of the Rocky Mountains.

B-1 *Index map to trips in British Columbia.*

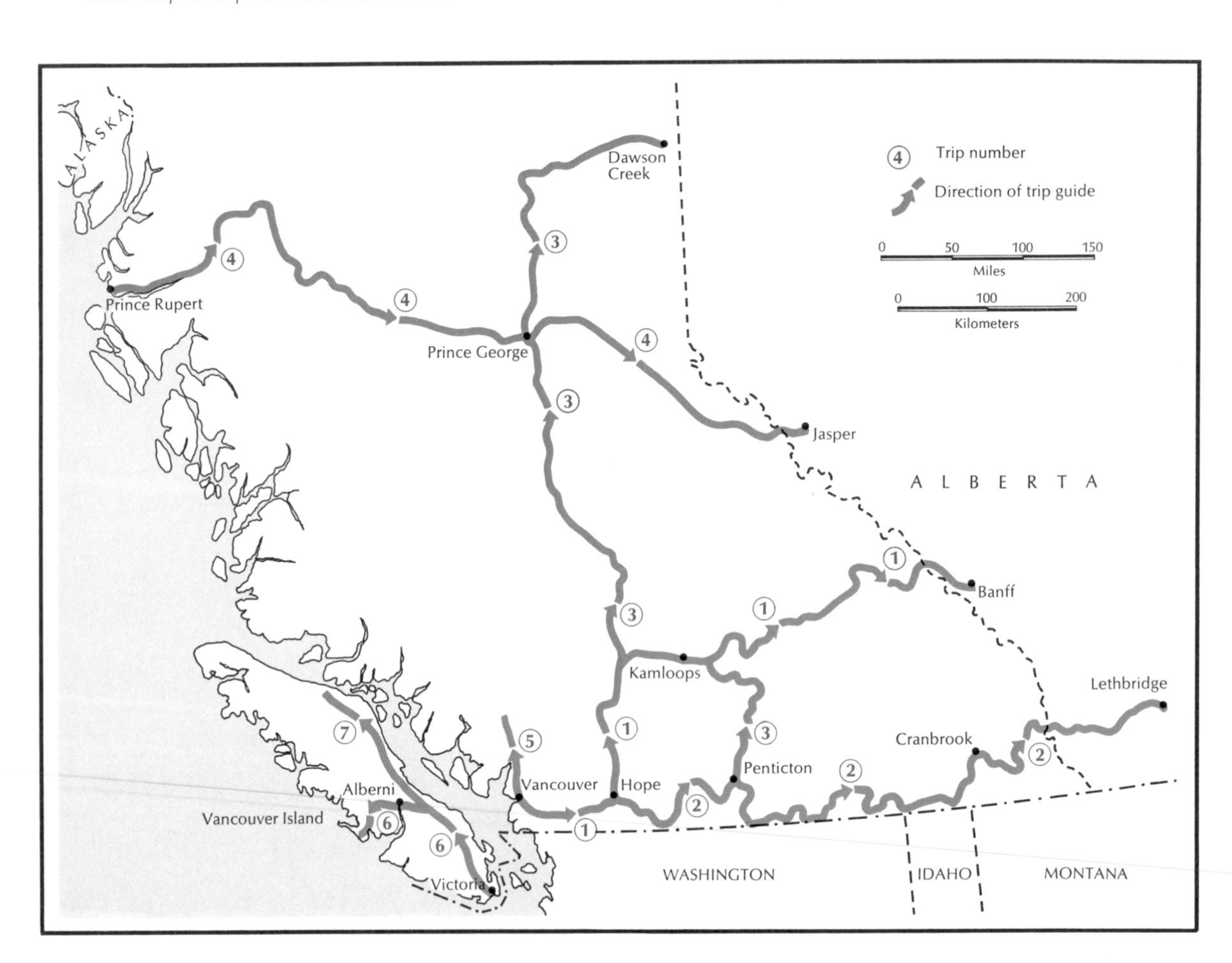

Golden to Banff (*86 miles*). This segment of highway provides some of the most celebrated scenic vistas in Canada. The Rocky Mountain crest (continental divide) is crossed at Kicking Horse Pass (elevation 5,400 feet) and from there the road descends past Lake Louise to Banff. The strata are marine sedimentary rocks. They are primarily of Cambrian, Ordovician, and Silurian age, although Windermere (Upper Precambrian) beds crop out extensively around Banff. The strata are well bedded, unmetamorphosed, and only moderately folded. The section is sliced by numerous west-dipping thrust faults, as this area lies within the Rocky Mountain thrust system.

TRIP 2. PROVINCIAL HIGHWAY 3

hope–princeton–osoyoos–trail–cranbrook–fernie–lethbridge (alta.)

Hope to Princeton (*83 miles*). Most outcrops for the first 15 miles are in the Hozameen Group. These strata consist of somewhat metamorphosed sedimentary and volcanic rocks of Late Paleozoic age, equivalent to the Chilliwack Group to the west and Cache Creek Assemblage to the north. They are moderately folded and fractured. Near the entrance to Manning Provincial Park, the highway crosses the steeply dipping Hozameen Fault. This defines the western side of the Methow-Pasayten fault trough (graben). For the next 20 miles, all exposures are of conglomerate, sandstone, and shale of the Jurassic and Cretaceous Ladner, Dewdney Creek, and Jackass Mountain (Pasayten) Groups, plus minor amounts of lava and small Tertiary granitic intrusions.

The eastern edge of the graben (the Pasayten Fault) is crossed near the eastern park boundary. The remaining distance to Princeton offers outcrops of Triassic and Jurassic submarine volcanic rocks (Nicola and Rossland Groups), and Cretaceous and younger intrusive granitic rocks. Here the highway is leaving the Cascades and entering into the southern part of the Thompson Plateau.

Princeton to Osoyoos (*71 miles*). The highway travels upstream along the valley bottom of the Similkameen Valley for most of this stretch. The hillsides and road cuts expose more granitic rock and Nicola-Rossland strata. Terraces and other valley glacier features are well developed. The large Okanagan Valley is entered at Osoyoos. It was a major avenue along which glaciers flowed southward into Washington (Chapter 17).

Osoyoos to Trail (*141 miles*). The route is through the Okanagan Highland and western part of the Columbia Mountains. Glaciation has helped to smooth out the landscape, although some of the higher mountains still rise to elevations well over 2 miles above sea level. Roadside exposures are not so numerous as in the more rugged country to the north and west. Most outcrops are of somewhat metamorphosed Paleozoic strata (Cache Creek equivalent) and younger granitic intrusive rock, but gneisses correlated with the Shuswap Metamorphic Complex are found in a fault block 25 air miles west of Trail.

Trail to Cranbrook (*146 miles*). This segment crosses the south end of the Selkirk and Purcell Mountains. The first half of the trip crosses the Kootenay Arc—a complex structural and lithologic mix of crystalline rocks not conducive to casual observation. From Creston east, the geology is somewhat more simple (and monotonous), consisting of good exposures of the Precambrian Purcell Group, not too severely deformed. The Purcell, equivalent to the Belt Supergroup of Montana and Idaho, consists of thick sequences of pale sandstone (now quartzite), shale (now slate or phyllite), and limestone (now fine-grained marble). Occasional outcrops reveal well-preserved ripple marks or other sedimentary structures, even though these strata are as old as 1 billion years.

Cranbrook to Lethbridge (*195 miles*). A few miles east of Cranbrook, the highway crosses the Rocky Mountain Trench, here occupied by the south-flowing Kootenay River. The Purcell strata are left behind as you enter the Rockies, and the road passes through nearly vertical Paleozoic and then Lower Mesozoic beds on the gradual climb to Crows Nest Pass (4,453 feet elevation). This is one of the lowest parts of the Canadian Rockies and certainly is not so spectacular as the Banff or Jasper areas to the north. As you descend the eastern flank to the plains west of Lethbridge, the strata are Upper Cretaceous sedimentary rocks. The characteristic Rocky Mountain folds and thrust faults diminish in magnitude toward the range front.

TRIP 3. PROVINCIAL HIGHWAY 97

osoyoos–penticton–kamloops–prince george–dawson creek

Osoyoos to Kamloops (183 miles). This stretch skirts the eastern and northeastern edge of the Thompson Plateau, running first along the glaciated Okanagan Valley. It then turns northwest around the upper end of Okanagan Lake and crosses the plateau to the valley of the South Thompson River east of Kamloops. The topography is subdued. The lithologies are a mixture of intrusive granitic rocks, sedimentary and volcanic strata of the Paleozoic Cache Creek Assemblage, and, along the east side of the Okanagan Valley, crystalline gneisses of the Shuswap Complex.

Kamloops to Prince George (319 miles). A long drive across the relatively low-relief surface of the Interior Plateau Province. The second half lies along the Fraser River. Much of the plateau surface is underlain by flat-lying basaltic lava flows of Miocene age. Where rivers have cut through the basalt, older strata are seen. They consist of sedimentary and volcanic rocks of the Cache Creek (Upper Paleozoic) and Nicola-Rossland (Lower Mesozoic) Assemblages. The route crosses only the eastern edge of this great lava plateau. To the west, the flows gradually rise until they disappear (due to erosional stripping) well up on the east flank of the Coast Mountains.

Prince George to Dawson Creek (256 miles). 90 miles east of Prince George, the British Columbia-Alberta border leaves the crest of the Rocky Mountains and runs north along the 120° meridian line to the Northwest Territories. Thus one crosses the Rockies on the drive to Dawson Creek but remains in British Columbia. The highway runs north to the edge of the Rocky Mountain Trench, which it crosses near McLeod Lake, and then turns east into the Rockies (here referred to as the Hart Ranges). The pass is low (only 2,850 feet above sea level) and the mountain valleys quite broad. The peaks are more or less smooth and rounded—in marked contrast to the sharp relief of the higher, more heavily glaciated ranges to the south. The bedrock, which is not well exposed along the highway, gets younger in age as you cross the range, with Precambrian and Paleozoic strata dominant on the west side and Mesozoic (largely Cretaceous) sediments to the east. The structure consists of gentle, broad folds and a few west-dipping thrust faults.

Dawson Creek is regarded as the "jumping-off point" for the Alaska Highway.

TRIP 4. PROVINCIAL HIGHWAY 16

prince rupert–prince george–jasper (alta.)

Prince Rupert to Prince George (456 miles). This route provides a very good cross-sectional view of the geologic framework of western and central British Columbia. Prince Rupert is surrounded by typical Coast Mountains topography and rock types. The town is situated at the base of steep ranges that rise up out of the sea, slashed here and there by glacially scoured U-shaped valleys, many of which extend below sea level as great fiords. The mountains are formed of crystalline rock which, from a distance, appears to be quite uniform and massive. When they are viewed more closely, this impression is dispelled. The rock is not uniform but consists of a complex mixture of granite, gneiss, and schist. Some of the units had an igneous intrusive origin, but much of the rock was derived by the metamorphic recrystallization of Mesozoic and Paleozoic sedimentary and volcanic strata. This then is the reality of what was once referred to simply as the Coast Mountains Batholith (Chapter 8).

The road eastward ascends the broad, glaciated valley of the Skeena River—a valley floored with so much gravel left over from its recent glacial occupation that the river hardly knows which path to follow. The mountains here are referred to as the Kitimat Ranges, and their lower slopes are mantled by a thick vegetative cover.

Near Terrace, the highway passes out of the crystalline core of the Coast Mountains. From here to Hazelton, it runs northeastward across the south end of a very large covering of Upper Jurassic and Lower Cretaceous volcanic sedimentary rocks known as the Bowser Group. (This is correlative with the Fraser Group to the south, the two together referred to as the Bowser-Fraser Assemblage.) Some granitic bodies are traversed, but they are of Tertiary age and lack the gneissose structure that is typical of many of the older granitic rocks of the Coast Mountains. Here the surrounding mountains are known as the Hazelton Mountains—the Bulkley Ranges to the southeast and the Nass Ranges to the northwest.

At Hazelton, the highway arcs around to the southeast and climbs up the valley of the Bulkley River onto that part of the Interior Plateau Province known as the Nechako Plateau. Here the landscape is not unlike parts of central and eastern Canada—a glaciated surface of generally low relief with some residual rounded mountains rising above the plain surface and some rather large and quite linear glacial lakes. The bedrock consists mostly of sedimentary and volcanic strata of the Triassic and Jurassic Takla-Hazelton Assemblage, cut here and there by Tertiary granite. Near Houston, the highway crosses the northern edge of the great basalt sheets of Miocene age that cover so much of the Interior Plateau. The basalt edge is approximately followed almost to Prince George, but Takla-Hazelton and Cache Creek (Upper Paleozoic) strata are traversed also, and old gneissose granitic rocks are prominent around Fraser Lake.

Prince George to Jasper (234 miles). From Prince George, the highway runs east to the Fraser River and the Rocky Mountain Trench around the northern tip of the Cariboo Mountains. It turns southeast along the trench for about 100 miles, offering many excellent views of this enigmatic and geologically controversial linear feature. The Rockies to the east and Cariboo Mountains to the west, each with peaks almost 2 miles high, seem to be closing in on the valley, and yet it persists as a remarkably straight trough. Little bedrock is seen along the valley floor, but the distinct layering of the strata in the mountain walls on either side reveals impressive folds and faults. Most geologists agree that the Rocky Mountain Trench must be underlain by a fault or series of faults, but the structure is concealed by the thick alluvial fill of the valley floor. Some geologists feel it is a great rift where the crust has split apart. Others maintain that it may be a zone like the San Andreas Fault in California along which major translational (horizontal) movement has occurred. Still others feel it is a relatively insignificant structural feature. At any rate, the trench is lovely to drive along and offers wonderful geologic vistas, whatever its origin may be.

At Tete Jaune Cache the highway turns east (as does the Fraser River) and climbs up to the crest of the Rockies at Yellowhead Pass. The climb is not steep, but the route passes right under the nearly vertical south face of Mount Robson, at 12,972 feet the highest peak in the Canadian Rockies. Mount Robson and the nearby peaks provide magnificant cross-sectional views of the internal architecture of the range—layers of light-colored sedimentary strata arranged generally in broad, open folds but locally steeply tilted. The westward-dipping faults that characterize Rocky Mountain structure are present but are not immediately obvious. The Fraser River valley here is a classic example of a glaciated, alpine valley. From Yellowhead Pass, the highway runs 16 miles east to Jasper and then continues down the east flank of the Rockies toward the Alberta Plains and Edmonton. The geologic panoramas are of textbook caliber.

TRIP 5. PROVINCIAL HIGHWAY 99

vancouver–squamish–garibaldi

Vancouver to Garibaldi (*61 miles*). This trip up the east side of Howe Sound to Garibaldi Provincial Park (and beyond if you are adventuresome) provides many excellent exposures representative of the Coast Mountains Province. Outcrops are primarily of schist and gneiss, unmetamorphosed granitic rock around Squamish at the head of the Sound, and occasional patches of Mesozoic sedimentary and volcanic strata only slightly recrystallized. Study of roadside cuts is more valuable for gaining a feel for Coast Mountain lithologies than it is for understanding the history of the region. Mount Garibaldi itself and some of its neighboring peaks are the southernmost representatives of British Columbia's Quaternary volcanic cones. The eruptive history is complex, and the region has been heavily modified by glaciation.

TRIP 6. TRANS-CANADA HIGHWAY

victoria–nanaimo–port alberni–tofino

Victoria to Nanaimo (*69 miles*). This trip begins in the oldest rocks of Vancouver Island—perhaps as old as any in western British Columbia. These are the rather dark gneisses that crop out so extensively in downtown Victoria (especially along the shore) and northward along the west side of Haro Strait. The gneisses correlate apparently with the pre-Middle Devonian Turtleback Complex of the nearby San Juan Islands (Chapter 9). They have not been studied thoroughly. The road to Nanaimo also crosses outcrops of Late Paleozoic sedimentary and volcanic strata (Sicker Group), submarine volcanic greenstone of the Triassic Karmutsen Formation, and marine sandstones and shales of the Upper Cretaceous Nanaimo Group. Thus many of the most important units of Vancouver Island are represented, although the relief along the route is low and exposures are not impressive.

Nanaimo to Tofino (*132 miles*). For the first 20 miles, the highway follows along or near the Strait of Georgia, and exposures are of Nanaimo strata. At Parksville, it turns west, into the hills and into the Karmutsen core of the island. These volcanic rocks are particularly well exposed in the park at Little Qualicum Falls—certainly worth a stop and a short walk. Port Alberni lies at the head of a long fiord, the upper part of which is underlain by Upper Mesozoic granite. From here to Tofino, the outcrops (such as they are—here one gains an appreciation of the province's timber resource) are again mostly of Karmutsen Group. The drive is more scenic than geologically instructive.

TRIP 7. PROVINCIAL HIGHWAY 19

nanaimo–campbell river–kelsey bay

Nanaimo to Kelsey Bay (*146 miles*). By continuing north from Parksville along the Strait of Georgia, the traveler remains in the Nanaimo Group until a few miles past Campbell River. From here to the end of the road, the route is a little inland, close under the steep slopes that flank this part of the island. Outcrops are somewhat more abundant than farther south, in part because the road is back again in the relatively resistant greenstones of the Karmutsen Formation. The view east to the mainland is magnificent.

BRITISH COLUMBIA BY TRAIN

Many of the routes discussed herein are followed by trains as well as highways—the natural result of rugged topography of the region, the limited number of canyons and mountains that are suitable for transit, and the relatively few towns and cities to be served. (This, of course, is a "chicken or egg" relationship, inasmuch as the location of towns is controlled in part by the availability of adequate transportation facilities.)

The major railroads are the Canadian Pacific (CPR), the Canadian National (CNR), and the Pacific and Great Eastern (PGE). The major CPR route runs from Vancouver to Calgary, essentially following the Trans-Canada Highway (Trip 1). The CNR line from Vancouver eastward also follows this route as far as Kamloops; from there it runs north up the North Thompson River, across the northern part of the Shuswap Highlands, and between the Cariboo and Monashee Mountains, essentially following Provincial Highway 5. It joins Highway 16 (Trip 4) in the Rocky Mountain Trench at Tete Jaune Cache, and together they cross Yellowhead Pass into Jasper. Another main line of the CNR follows Highway 16 westward all the way to salt water at Prince George (Trip 4).

The PGE main line runs from Vancouver north to Dawson Creek, with side lines out to Fort Saint James and Fort Saint John. The route is via Squamish and Garibaldi (Trip 6), through the Coast Mountains to Clinton, and essentially along Highway 97 (Trip 3) through Prince George to Dawson Creek.

The most obvious advantage of geologic sightseeing by train (or bus) is having someone else do the driving; and, if vista-dome viewing is available, some magnificent panoramic views are seen—especially upward in the canyons. The disadvantages are obvious, most notably the predetermined pace, the impossibility of stopping for closer investigation of geologic features, and the inability of sampling any of the rocks exposed.

SALTWATER TOURING

Little need be said, really. Geology aside, the British Columbia coastline provides some of the world's finest cruising, although the weather is always something of a risk. Marine transportation is the only way to reach the major part of the coastline (including Vancouver Island and the Queen Charlotte Islands). Consequently, the geology is better known along the shore than inland. The combination of glacial scour and wave erosion has guaranteed good rock exposure almost anywhere along the thousands of miles of British Columbia's coastline, and close approach by boat is often possible. The geologist, our geologically minded tourist, with a small boat has much to work with. If the boat can be moved around with a larger ship, his scope is obviously greater. Passage on a coastal freighter or steamer provides a look at a lot of spectacular country in a short time, but of course very little feel for the rocks themselves.

Particular trips to recommend for geologically oriented yachtsmen are so numerous that I hardly know where to start. The Gulf Islands southeast of Nanaimo provide the best exposures of the marine sedimentary rocks of the Nanaimo Formation and the opportunity to collect fossils by dinghy. Barkley Sound on the west side of Vancouver Island offers excellent exposures of crystalline granites and gneisses in protected waters, and sounds farther north display well the Mesozoic volcanic strata that compose most of the island. Geologic variety is more concentrated in the Queen Charlotte Islands, but getting there is more of a problem. The many sounds and inlets on the mainland offer outstanding cross-sectional views of the Coast Mountains Crystalline Complex. In short, there is something geologic to see wherever you go by boat in British Columbia, and a few minutes spent looking at the rocks after you have collected the oysters can be quite rewarding.

C washington trip guides

The following guides presume some familiarity with Chapters 5, 6, 7, 11, 12, 13, 17, and 18.

TRIP 1. U.S. HIGHWAY 101

astoria (ore.)–aberdeen–port angeles–olympia

This highway, a continuation of the Oregon Coast Highway, travels most of the distance around the Olympic Peninsula of western Washington. Salt water lies outside the loop, and the inaccessible core of the Olympic Mountains lies in the center. The highway runs, generally, inland from the shoreline and somewhat out from the base of the range, but various secondary roads (many of them dead end) lead to the water or part way into the Olympics. Bedrock outcrops are relatively few and far between on the main highway.

Astoria to Aberdeen (*82 miles*). The Columbia River is crossed via the Megler-Astoria toll bridge, which provides an excellent view of the lower reaches of the Columbia. Outcrops of submarine basalt of the Eocene Crescent Formation are quite numerous around Megler on the north bank. Northeast of Ilwaco the highway turns north and runs along the eastern margin of Willapa Bay to Raymond. Some outcrops of marine sandstones and shales of Oligocene and Miocene age are passed in the first 15 miles—north of there the highway runs entirely on Quaternary sands and gravels.

The 25-mile stretch from Raymond to Aberdeen runs across the northwestern corner of the Willapa Hills. Marine sedimentary rocks of Oligocene and Miocene age are exposed in some road cuts.

Aberdeen to Port Angeles (*145 miles*). Almost no bedrock is seen along the route until the west end of Lake Crescent is reached. The hills immediately north of Aberdeen consist of massive sandstones of late Miocene and Pliocene age, deposited in a coastal embayment as the Pacific Ocean shoreline moved westward (Chapter 11). These are well exposed in bluffs at the east edge of Aberdeen, north of the mouth of the Chehalis River and also north of the highway to Hoquiam. From Hoquiam, the road runs inland north to Quinault Lake underlain by Quaternary coastal lowland gravels. North of the Humptulips River, some of these sediments were deposited by alpine glaciers that flowed west and southwest down the flank of the Olympics and onto the lowland (Chapter 18). These glacial deposits are deeply weathered and their glacial origin is not immediately obvious. Quinault Lake was formed behind a low glacial moraine produced during a temporary stand of the ice margin as it retreated.

From here the road turns west across more glacial sediments to Queets. From here to Ruby Beach, south of the mouth of the Hoh River, the highway lies along the ocean—the only stretch of the route that does. Some exposure of Miocene marine sediment is seen along the road north of Kalaloch, but better outcrops are found along the sea cliff beneath the road. From here northward, the coastline is spectacular with countless rock stacks carved out of the Tertiary bedrock by the pounding waves and numerous bold headlands (Figures 11-11, 11-12). Unfortunately, the highway turns inland to Forks, and from there curves around to the east along the Soleduck River valley, and no bedrock is seen along the road. The most scenic stretch of Washington's coast remains the domain of the hiker, not the driver. Good cuts through the dominantly Oligocene sandstones, siltstones, and shales of the Twin River Formation (Chapter 11) are found north of Sappho on the side road to Clallam Bay.

The best exposures of the entire trip occur near Lake Crescent. They include Crescent Basalt at the west end and along the east shore, "Soleduck" basalt and marine sedimentary strata along the south shore, and Aldwell shale near Lake Sutherland. All beds dip very steeply north here, this area being on the north flank of the Olympic "dome" and the south side of the Clallam Syncline. From Lake Aldwell to Port Angeles, the highway crosses sediments of the Fraser Glaciation (Chapter 18).

Hurricane Ridge Side Trip. A worthwhile side trip is to the Hurricane Ridge Lodge south from Port

C-1 *Index map to trips in Washington.*

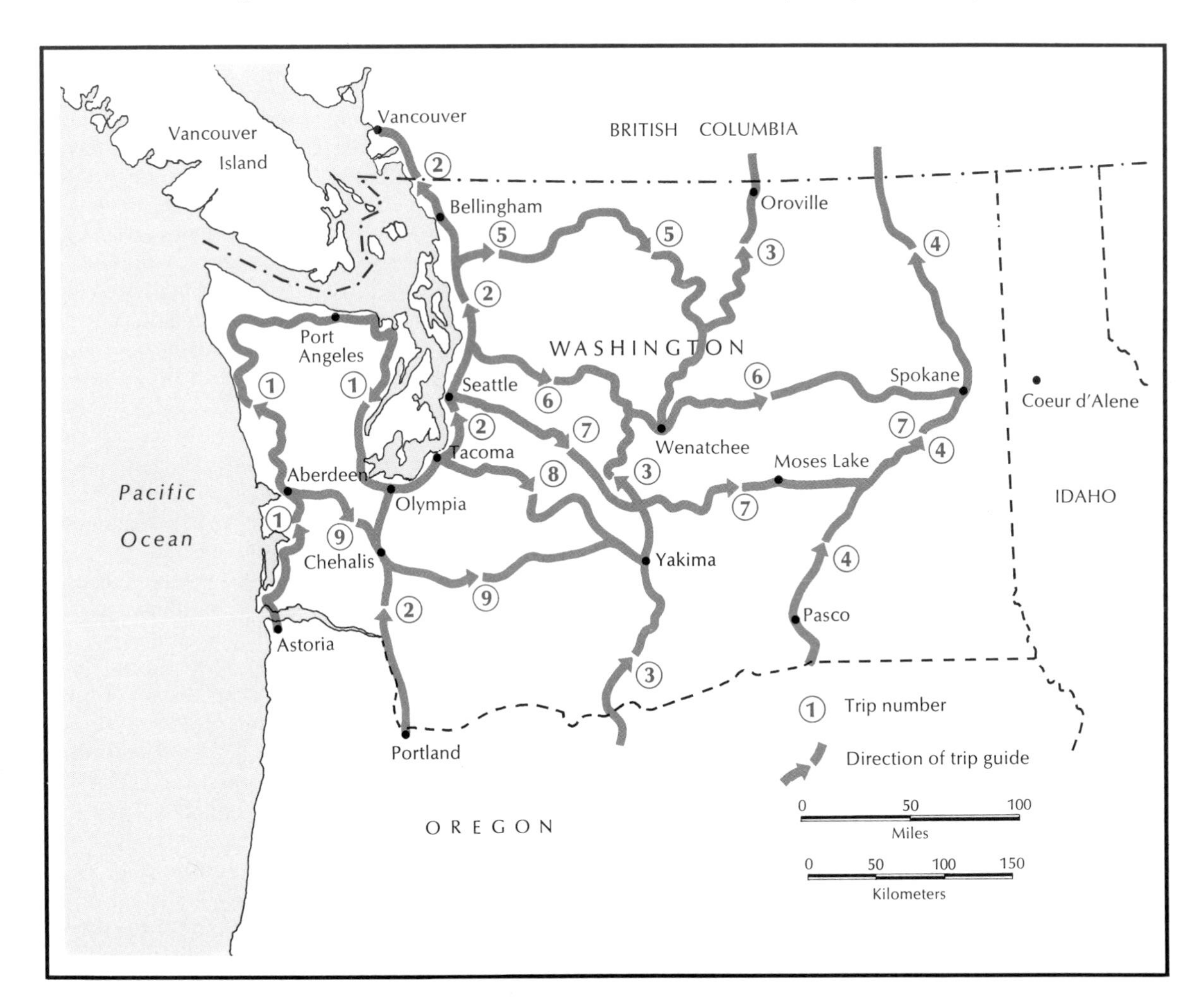

Angeles. The road climbs up the steep wall of Crescent lava that forms the north flank of the Olympics, and road cuts are numerous. Beyond the tunnel (5 miles from the road end) is an excellent place to study the pillow structure that is characteristic of much of this great volcanic pile (Figure 11-6). Shortly beyond this, the road enters the slightly metamorphosed and highly deformed sedimentary and volcanic strata (the "Soleduck Formation") that compose the core of the Olympics. The view south into the heart of the range is magnificent from the lodge.

Port Angeles to Olympia (120 miles). Just east of Port Angeles, the highway crosses a small valley where Pleistocene sand and gravel beds dip up to 60° in the road cuts. Not explicable by landsliding, these dips suggest a considerable amount of late Pleistocene deformation in this area. From here to Sequim, the road crosses a glaciated lowland surface, mantled locally with glacial erratics. Some of these are granite—a rock type not found in place on the Olympic Peninsula—and must have come from British Columbia or north-central Washington.

Crescent Basalt crops out along the road at the south end of both Sequim and Discovery Bays. The rest of the stretch to Quilcene is across upper Eocene and Oligocene marine sediments, but they are buried beneath the glacial veneer. From Quilcene to Hoodsport the route is along Hood Canal at the foot of the steep east wall of the Olympic Mountains. Outcrops are fairly numerous and consist wholly of Crescent Basalt. Hood Canal is a "blind" channel, carved out by glacial erosion but not connected at its south end to Puget Sound. Lake Cushman (Figure 11-5), west of Hoodsport, occupies a valley scoured by alpine glaciers that flowed eastward and merged with the Puget Lobe. The road up the valley provides many excellent exposures of the Crescent Formation. From Hoodsport to Olympia, the highway lies entirely on glacial sediment except for a few outcrops of Crescent Basalt 8 miles west of Olympia.

TRIP 2. INTERSTATE HIGHWAY 5

portland–olympia–seattle–bellingham–vancouver (b.c.)

This highway runs within a few hundred feet of sea level, following fairly closely the axis of the Fraser-Puget-Willamette Trough. Pre-Quaternary strata are exposed only from Kalama to Centralia—a stretch in which the Willapa Hills and foothills of the Cascades essentially join—and south of Bellingham where the Cascade foothills extend to salt water.

Portland to Olympia (113 miles). The first 30 miles crosses Quaternary sediments, much of them deposited by the great Spokane Floods that swept repeatedly across the Columbia Plateau, through the Columbia River Gorge, and on down to the sea (Chapter 17). During each flood, the Cascade barrier produced a large, temporary lake in the Columbia Basin. As noted previously, the lower reaches of the Columbia River are hemmed in by hills near Kalama, and these formed a hydraulic dam and a temporary lake in the Portland area. Coarse gravels between Vancouver (Washington) and Woodland were deposited in this lake.

Excellent road cuts north and south of Kalama show Eocene strata typical of the Cascades. Some beds are lava flows of basalt and andesite, but more striking are the massive mudflow breccias containing large angular boulders and blocks of lava. Interbedded sandstones are rich in fragmental volcanic debris and consequently are dark in color. A link with the near-present is made around Castle Rock, where more volcanic mudflow breccia crops out. These strata are no more than a few thousand years old, having been derived on Mount Saint Helens during eruptive phases. They traveled, still hot, 50 miles westward down the Toutle River valley (Chapter 13). Other exposures in the Centralia-Chehalis area consist of sandstones and shales of Eocene and Oligocene age. They were deposited on the coastal plain or in shallow water offshore, since this was the position of the fluctuating coastline in the Early Cenozoic. Coal formed in coastal swamps is a significant resource in this area, although none is exposed along the highway.

The Tenino Prairie, home of the enigmatic Mima Mounds (Chapter 18, Figure 18-10) is crossed between Centralia and Olympia. The mounds are not well developed near the roadway but can be seen along various side roads. The prairie was the terminal position of the Puget Lobe during Fraser Glaciation, but this is not obvious.

An outcrop of Eocene Crescent Basalt is seen west of the road at Tumwater just south of Olympia. The outcrop is not very impressive, but it is the last exposure of rock this side of Seattle.

Olympia to Seattle (*60 miles*). Some large, deep cuts occur along the road, but rapid weathering and vegetation obscure most of the detail. Most exposures are of interglacial and proglacial (that is, deposited in front of an advancing glacier) sediment, capped by a thin veneer of Vashon till (Chapter 18). Some alignment of topography parallel to the direction of glacier movement (north to south) is seen but this is more apparent from the air than on the ground.

Lower Tertiary marine stiltstones and sandstones are exposed in the freeway cuts immediately east of Boeing Field at the south edge of Seattle. Glacial streamlining of the topography is quite obvious in and around Seattle, and the freeway provides some good vistas of the city.

Seattle to Bellingham (*89 miles*). On a clear day, this stretch provides good views of the Cascades and the Olympics. No bedrock is seen south of Burlington. There are numerous road cuts of glacial sediments, including two excellent exposures of Vashon till immediately north of the Stillaguamish River.

Massive sandstones and interbedded shales of the nonmarine Lower Tertiary Chuckanut (Swauk) Formation (Chapter 7) are seen along the last 10 miles south of Bellingham. An alternate route from Burlington to Bellingham—State Highway 11, the Chuckanut Drive—is worth the extra time. This features not only numerous rock exposures but also beautiful views of the San Juan Islands across Samish and Bellingham Bays. The Chuckanut beds have been tilted steeply by folding processes and in places contain plant fossils, including large palm leaves.

Bellingham to Vancouver (*55 miles*). Occasional outcrops of glacial drift can be seen from Bellingham to a few miles north of the border. The remainder of the distance to the south city limits is across tidal flats of the Fraser River Delta.

TRIP 3. U.S. HIGHWAY 97

maryhill-yakima-ellensburg-wenatchee-oroville

Maryhill to Yakima (*78 miles*). The roadway climbs immediately up the steep north wall of the Columbia River Gorge, providing an excellent view up and down the river and out across the Deschutes Plateau to the south. The Columbia Hills Anticline, a faulted east-west arch with a steep south flank, lies along the hill crest and is the reason for the higher plateau edge on the Washington side (Figure 17-15). From the hill crest to Goldendale, there are good views of Mount Adams to the northwest and Mount Hood to the southwest. From Goldendale, the road runs northeast across the east end of the Simcoe Mountains via Satus Pass. Quaternary basaltic lava, which caps the range, is well exposed along the highway, as are some Pliocene sands and gravels, equivalent to the Ellensburg and Dalles Formations.

Satus Pass is underlain by Miocene lava of the Columbia River Basalt Group, here raised up by another great east-west anticline that includes the Horse Heaven Hills to the east and extends west through the Simcoe Mountains. The road follows Satus Creek down the north flank of the arch and then rises again to cross the east end of Toppenish Ridge. This is another anticline. It has a steeply dipping north flank, exposed along the base of the ridge. This fold runs west underneath Mount Adams. From Satus Pass to the Yakima Valley, all outcrops are of Columbia River (Yakima) Basalt.

No rocks are exposed along the route from Toppenish Ridge to Union Gap, just south of Yakima. There the Yakima River has cut through the anticline of Ahtanum Ridge, again providing good exposures of the Yakima Basalt. This structure is asymmetrical, like the Toppenish fold, with a steep north flank and gently inclined south flank.

Yakima to Wenatchee (115 miles). This is a very interesting stretch of highway that follows the Yakima River north through the heart of some major folds in the Yakima Basalt. It then turns northwest across the Wenatchee Mountains (an outlier of the Cascades) where pre-Miocene rocks are well exposed. (The segment from the Teanaway Junction to the Wenatchee River is detailed in Information Circular 38 of the Washington Division of Mines and Geology.)

Just north of Yakima, the highway cuts through a notch in the Yakima Ridge Anticline. Just south of Selah Butte, the next dissected mountain, the road crosses a scarp along the river that exposes the Pomona Basalt, the lowest basaltic member of the Ellensburg Formation (Chapter 17). Across the river to the northwest are seen some of the white tuffaceous and diatomaceous sediments at the base of this formation. Selah Butte is a broad, symmetrical anticline, and more than 1,000 feet of about twelve flows of Yakima Basalt are exposed in the precipitous canyon walls. Immediately north is Umtanum Ridge, perhaps the most impressive part of the Yakima Canyon. Another anticlinal fold, this one has a gently sloping south flank, but on the north side the beds are nearly vertical. This is best viewed from the highway a mile north of the gorge, looking back at the face of Old Baldy on the east side of the river. From here to the Kittitas Valley (in which lies the town of Ellensburg), the broad but somewhat complex anticlinal structure of Manastash Ridge is traversed, still in the deep river canyon.

The Yakima Canyon is the type area for the Yakima Basalt, and this is appropriate, for nowhere else are so many flows so continuously exposed. The good exposure is due, of course, to the ceaseless erosion of the Yakima River. The anomaly of a meandering river cutting through such young, steep folds was mentioned in Chapter 17. Apparently, the river maintained its course as the folds grew across its path—erosion keeping up with uplift—and this is considered a classic example of a so-called antecedent river. A new highway is being built to the east to avoid the many bends of the river, but the geologically minded motorist will want to take the existing route at least once.

Highway 97 runs north through Ellensburg and across the Kittitas Valley east of the Yakima River. (Interstate Highway 90 runs on the west side.) Near Thorp, it crosses the southwestern flank of the Wenatchee Mountains uplift, with good exposures along the highway of brown gravels of Pleistocene age, then white pumiceous Ellensburg sandstone (the type area for this formation) and then more Yakima Basalt. At the west end of the Kittitas Valley, the gentle 8° dip of the basalt eastward off the Cascade Uplift is quite apparent. Near Teanaway Junction the highway crosses into Eocene sediments of the Roslyn Formation (Chapter 7), which underlie the valley around Cle Elum.

The road turns east across Camas Prairie, north of the basaltic rim of Lookout Mountain. It climbs across low hills of glacial till, the terminal moraines of Alpine glaciers that flowed eastward off the Cascade crest down the Yakima Valley (Chapter 18). It then drops into Swauk Creek. Twenty miles to the north lies Mount Stuart, the remains of a Cretaceous granitic batholith; and the road turns north to cross the east shoulder of the Stuart Range at Swauk Pass. The rocks exposed along the road are progressively older to the north. South of Liberty, bold outcrops of basaltic to andesitic flows and breccias of the Eocene Teanaway Formation line the road. Sandstones and shales of the Swauk Formation underlie this and are seen in cuts from Liberty to near the old mining town of Blewett, north of the pass. Many outcrops are cut by dark dikes of Teanaway Basalt, and fossil plants and leaves are common. The Leavenworth Fault, which forms the western edge of the Chiwaukum Graben, is crossed near Blewett; and for a few miles the highway is in pre-Jurassic phyllite and serpentine of the Stuart Range. Then the fault is recrossed; and massive Swauk sandstones, steeply tilted by folding and faulting, form impressive ridges along the route to Wenatchee. The Wenatchee Mountains east of Swauk Pass and south of the Wenatchee River are capped by a few flows of Yakima Basalt. These dip gently eastward and form a prominent rim near the range crest. (The flows underlie the ski slopes at Mission Ridge.)

Wenatchee to Oroville (134 miles). The first half of this stretch lies generally along the west bank of the Columbia River. From Brewster to the Canadian border the highway runs along the bottom of the Okanogan Valley.

At Wenatchee, the bedrock is the Swauk Formation. Proceeding north, one crosses immediately the contact with the Swakane Gneiss, a unit of uncertain (but pre-Jurassic) age that underlies much of the Entiat Mountains west of the Columbia River. The Yakima Basalt caps the edge of the Waterville Plateau above the east bank. North toward Chelan, the nature of the crystalline rocks changes from the evenly layered gneissose texture characteristic of the Swakane to a more complex mixture of massive intrusive rocks and heterogeneous schists and gneisses that have been referred to collectively as the Chelan Batholith. At Earthquake Point, three miles north of Entiat, dark igneous dikes that have intruded this older crystalline complex are displayed dramatically in the Ribbon Cliffs above the highway. A few miles north, the highway leaves the river and turns north through a dry coulee. It follows this to the south shore of Lake Chelan.

The lower end of Lake Chelan is an exceedingly interesting area. The bedrock is part of the Chelan Crystalline Complex, and some spectacular migmatites are exposed in cuts along the short road that runs from the east edge of town down to Chelan Falls. Lake Chelan is a classic example of a "finger lake," that is, a long lake occupying a glacial valley, and the alpine glacier that occupied this trough was as much as a mile thick farther up the lake. (The deepest part of the lake floor lies a little below sea level.) To the east of Chelan, the Waterville Plateau was also covered by ice of the Okanogan Lobe, and the huge basaltic erratics left by this event are very prominent on the east valley wall of the Columbia. The two glaciations were somewhat out of phase, for by the time the Okanogan Lobe reached a point opposite Chelan, the terminus of the Chelan glacier had retreated some miles back into the Cascades. The Okanogan ice dammed the Chelan Valley. This produced a lake that rose above the present level until finally it spilled across the south valley wall. The spillover cut the impressive coulee followed by the highway southwest of Chelan. A second escape coulee was cut to the northwest, and it is followed by the road to 25 Mile Creek Campground. The impressive sand and gravel terraces along the sides of the lower Chelan Valley were built by drainage from the edge of the Okanogan Lobe into Glacial Lake Chelan. When the Okanogan Lobe retreated, the lake drained, via Chelan Falls, down to the present bedrock lip. The side coulees were abandoned.

A trip up Lake Chelan to Stekekin on the regularly scheduled excursion boat provides spectacular scenic views comparable to the fiords of Norway or British Columbia. At Stekekin, rental cars are available, and one can drive up the Stehekin Valley almost to the crest of the North Cascades.

From Chelan north, the highway runs for several miles in an abandoned glacial valley that hangs far above the Columbia. Just north of the narrowest part, the route is past very large undrained glacial depressions or *kettles*. The route then descends down the face of the Great Terrace of the Columbia. This is a huge deposit of boulders and sand that is traceable in patches from Wenatchee far up the Columbia and Okanogan River valleys. A probable late glacial feature, its exact origin is unclear. Several large "haystacks" (Chapter 17) of basalt rest on crystalline bedrock next to the highway just south of the mouth of the Methow Valley.

Five miles east of Brewster, the road crosses the Okanogan River and turns north away from the Columbia Valley. From here to the Canadian border, outcrops along the highway are not abundant. The Cascades lie to the west, the Okanogan Highland to the east. Yakima Basalt forms a thin cap on the latter almost to Okanogan. The underlying bedrock on both sides consists mostly of old crystalline gneisses; less metamorphosed schists, phyllites, and greenstones derived from Upper Paleozoic and Lower Mesozoic strata; and Upper Mesozoic intrusive granitic rock. The hills have been smoothed and rounded by the passage of glacial ice, and glacial kame terraces are well developed in places along the valley sides.

TRIP 4. U.S. HIGHWAY 395

pasco–spokane–colville

Oregon Border to Spokane (*159 miles*). This part of the route starts at Wallula Gap, the gateway to the Columbia River Gorge, and runs across the Channeled Scablands to the northeastern corner of the Columbia Plateau. The trip provides a good feeling for the devastating power of the Spokane Floods that swept along this path 12,000 to 15,000 years ago. The bedrock is the upper part of the Yakima Basalt, but only a few flows are exposed in any one place. The surface relief is nowhere more than a few hundred feet except where the Columbia cuts through the east end of the Horse Heaven Hills at Wallula Gap.

From Wallula to Eltopia the highway crosses Spokane Flood beds and some postglacial sediments that floor the Pasco Basin. At Eltopia and at Connell, the Yakima Basalt is exposed in coulee bottoms, but otherwise the route is across the wind-deposited silty soil (loess) of the Palouse Formation. Glimpses of the volcanic bedrock beneath the loess are seen in coulees all the way to Spokane, the dark, scoured basalt contrasting markedly with the smooth, light-colored soil of the adjacent hills. The road follows one of the major channel networks of the floodway, and the large dimensions of the scoured valleys are most impressive. Near Spokane the highway descends from the plateau surface to the valley of the Spokane River, providing good exposures of the basalt bedrock.

Spokane to British Columbia (*112 miles*). Highway 395 runs north-northwest across the Okanogan Highland, generally down the Colville River to the Columbia at Kettle Falls, then up the Kettle River to the border. The first 20 miles has no exposure, but north of Clayton the Precambrian and Lower Paleozoic strata characteristic of the eastern part of the Okanogan Highland crop out in occasional road cuts. The Loon Lake Batholith, a Cretaceous granitic intrusion containing some small uranium deposits, is crossed near Loon Lake. The Cambrian Addy Quartzite crops out around Addy, providing good exposures of a rock type that is relatively uncommon around the Pacific margin. Dark phyllitic shales and greenstones of Pennsylvanian and Permian age are seen occasionally from Kettle Falls north to British Columbia.

All this area except for a few of the highest mountains was covered by glacial ice, and this explains the somewhat rounded appearance of the ranges and the locally thick sections of glacial drift in the valley bottoms.

TRIP 5. STATE HIGHWAYS 20 AND 153

burlington–marblemount–twisp–pateros

This road, not yet completed across the Cascade crest, will be a geologist's paradise. It crosses through the center of the new North Cascades National Park, providing spectacular views of the very rugged scenery and numerous outcrops of the typical lithologies. It then runs down the Methow Graben through the tremendously thick section of Upper Jurassic and Cretaceous strata and into older crystalline rocks typical of the lower Methow Valley and adjoining highlands.

Burlington to Marblemount (*47 miles*). The route travels up the bottom of the broad, glaciated valley of the Skagit River, with only occasional outcrops along the north valley side. These include Paleozoic greenstone derived from submarine pillow basalt of the Chilliwack Group, seen in several hillside quarries east of Burlington; a few exposures of massive Chuckanut sandstone east of Sedro Woolley; and dark Chilliwack sedimentary rocks rich in volcanic fragments between Concrete and Rockport. Chilliwack limestone crops out locally above the valley floor and was the source rock for the now defunct lime plant at Concrete. A 1-mile side trip up the hill toward the Lower Baker Dam (turn north just east of the bridge at Concrete) offers an excellent view of Mount Baker and of the large landslide that demolished the Shannon Lake powerhouse a few years ago.

The Shuksan Thrust crosses the valley between Rockport and Marblemount but is exposed only high on the valley walls. It separates Chilliwack strata from phyllites and greenschists (locally blueschists) of the Shuksan Suite. The latter can be seen south of the river at Marblemount and locally along the Rockport-to-Darrington road.

Marblemount of Mazama (75 miles). This stretch offers many exposures of the crystalline core of the North Cascades (Figure 7-3). The Straight Creek Fault (not exposed) is crossed just northeast of Marblemount. Outcrops between here and Newhalem include (in order):

1. Gneiss of the Yellow Aster Complex (pre-Devonian).
2. Mica schist and talc schist of the Cascade River Schist Formation.
3. Granodiorite, diorite, and quartz diorite of the Chilliwack Batholith (Early Tertiary), which contains dark inclusions of the surrounding country rock.
4. Skagit Gneiss at Newhalem.

A unique opportunity to study the complexity of the crystalline core of a mountain range is available along the next 11 miles of road. The route lies entirely within the Skagit Gneiss, and outcrops are nearly continuous. The variety seen in close inspection is frightening (Figure 7-5). Much of the unit is a migmatite with many igneous-appearing veins and dikes of various ages and compositions cutting different kinds of gneisses. Faults are numerous. Pink garnets are found in the gneiss in numerous outcrops, including some immediately south of Diablo Dam. Meanwhile, the mountains are close in around the Skagit River (and its reservoirs) and seem to rise straight up for many thousands of feet. (This means, of course, that the road must be carved out of the mountainsides, and hence the numerous road cuts.)

At Ross Dam, the highway turns southeast, up the glaciated valley of Ruby and Granite Creeks, toward Rainy Pass, 29 miles distant. Much of the way, the route is across the northwest end of the Golden Horn Batholith, a true granite intrusion of Early Tertiary age. Rainy Pass and Washington Pass (5 miles farther) are in the older Black Peak Batholith. The descent down Early Winters Creek crosses the Golden Horn intrusion again, with many good exposures of the granite.

Cretaceous sedimentary rocks of the Methow Graben lie against the east side of the Golden Horn Batholith. The intrusive contact is crossed about 8 miles beyond Washington Pass. From there to Mazama, the steeply dipping Cretaceous strata can be seen in the glaciated walls of Early Winters Creek, but exposures along the road are poor.

Harts Pass Side Trip. If the weather is right and the road is open (usually July and August), a side trip from Mazama northwest to Harts Pass and Slate Peak is always worth while. The gravel road runs to the base of Last Chance Peak, at the head of the Methow Valley. It then climbs up the south face of this mountain and around its west face to the pass. Formations passed include the redbeds of the Ventura Formation near the Lost River crossing; massive arkosic sandstone of the Winthrop Formation on the southeast face of Last Chance Peak; intrusive dikes at Dead Horse Point high on the southwest face (not the place to stop for those afraid of heights!); and fossiliferous marine shales. From Harts Pass a 2-mile drive and a short but steep hike to the summit of Slate Peak (7,000 feet) provides one of the great views in the Northwest—north into British Columbia, west and southwest across the North Cascades, and east toward the rugged and lonely country of the Chewack River drainage. Here one really feels the impact of glaciation, for the great Cordilleran Ice Sheet passed right over the top of Harts Pass as it flowed south toward the Columbia Plateau. The valley of the Pasayten River at your feet to the north is a classic example of a glaciated U-shaped valley, and the steep cirque headwalls produced by alpine glaciation are in evidence everywhere, especially on the north and east sides of the higher peaks. The drive back to Mazama provides a good view of the glaciated upper Methow Valley, Washington's answer to Yosemite.

Mazama to Pateros (*57 miles*). From Mazama past Winthrop, the exposures along the road are poor. The valley walls, however, have good outcrops of Cretaceous sandstones and shales (west side) and the Midnight Peak Volcanics (east side). Highly folded and fractured slates of the Newby Formation are seen in road cuts between Winthrop and Twisp.

South of Twisp, the valley narrows, and the lower reaches of the Methow River meander in a moderately tight canyon. This reflects in part a change to more resistant crystalline bedrock, in part the fact that the mainstream flow of the large Methow glacier was straight toward Brewster, across the flank of Lookout Mountain east of Carlton. The bedrock geology of the lower Methow Valley is complex. Units seen in road cuts include schists and gneisses of the Leecher Metamorphics north of Carlton, the Methow Gneiss around the town of Methow, and migmatites just west of Pateros. Large gravel terraces, deposited between the stagnant glacial ice (as it wasted away) and the valley sides, are most prominent between Twisp and Pateros.

TRIP 6. U.S. HIGHWAY 2

everett–stevens pass–leavenworth–wenatchee–waterville–spokane

This is not the easiest route between the Seattle area and Spokane, but it does have some interesting geology en route. From Monroe east to the junction with U.S. Highway 97 near Leavenworth, the roadside geology is detailed in Information Circular 38 of the Washington Division of Mines and Geology.

Everett to Stevens Pass (*65 miles*). From Everett to Monroe, the route is across the broad floodplain of the Snohomish River, and no bedrock is seen. The road turns east up the Skykomish River but stays away from the rock walls until the vicinity of Index. A few miles west of there, a very large morainal and deltaic embankment of sand and gravel, deposited between the mountain front and the Puget Lobe of glacial ice (see Chapter 18), towers almost 1,000 feet above the highway and the river.

Mount Index, an Early Tertiary granodiorite intrusion, rises precipitously above the south valley wall southeast of Index (Figure 7-11). Exposures of the rock are seen in road cuts opposite it and in the river bed several miles to the east. The cement plant at Grotto, now dormant (or maybe extinct!), quarried Permian limestone of the Chilliwack Group from the south valley wall. East of here for a few miles are more exposures of granodiorite and of metamorphosed Swauk sandstone.

Phyllites of the Shuksan Suite are seen in a road cut 5 miles east of Skykomish. From here to the summit, most outcrops are of granodiorite of the Cretaceous Mount Stuart Batholith. This rock is particularly well exposed in road cuts between Tunnel Creek and the top of the pass.

Stevens Pass to Wenatchee (*63 miles*). The first several miles are in the batholith. From the contact (near the east portal of the Great Northern tunnel) to a mile or so west of the Lake Wenatchee turnoff, the occasional outcrops are of biotite schist and gneiss of the Cascade core (equivalent to the Skagit Suite). Near Winton, the Leavenworth Fault along the west side of the Chiwaukum Graben is crossed. Swauk sandstone and shale crop out from here to the bridge across the Wenatchee River at the head of Tumwater Canyon. There the highway crosses the fault again; and, in the 8-mile-long canyon, granodiorite, diorite, gabbro, and serpentine of the Mount Stuart complex are exposed.

The Leavenworth Fault is crossed for the third (and last) time at the mouth of Tumwater Canyon at Leavenworth. From here to Wenatchee, exposures are of the folded sandstones and shales of the Swauk Formation, well displayed in road cuts and along the hills that flank the Wenatchee River. The hill just south of Leavenworth is mantled by till deposited by alpine glaciers that flowed out of the Icicle Creek canyon to the west. Presumably, some of the prominent river terraces downstream relate to the glaciation, but this is unclear.

Wenatchee to Spokane, via Waterville (*159 miles*). The road runs up the east side of the Columbia River to Orondo, then east up Corbally Canyon to the Waterville Plateau. It extends east to Coulee City, and on across the Channeled Scablands to Spokane. The first exposures of rock are of Swakane Gneiss opposite Rocky Reach Dam. The Swauk Formation can be seen to wedge out between this unit and the Columbia River Basalt under the rim of the plateau southeast of the dam. The steep ascent up Corbally Canyon exposes numerous nearly vertical dikes of gabbroic rock that have intruded the gneiss. Near the plateau rim are found good outcrops of the Columbia River Basalt and some interbedded sandstone and shale. Basalt is seen again in the small coulee cut by Douglas Creek east of Waterville. Otherwise, no outcrops are passed to the edge of Moses Coulee, 20 miles distant. The coulee is magnificent. The road winds down past layers of basalt and over the upstream end of a gigantic gravel bar, built by the waters of one of the Spokane Floods (Chapter 17). The coulee itself is immense—the rival of the Grand Coulee to the east. Unlike the latter, its origin cannot be explained in part by a diverted Columbia River, for at its upstream end it dies out on top of the Waterville Plateau, far above the Columbia River valley. Good exposures of the basalt flows are seen along the road as it climbs the east wall.

The stretch of plateau crossed between Moses and Grand Coulees is made interesting by the road's crossing of the Withrow Moraine, a line of low hills running from Chelan to near Dry Falls formed along the position of southernmost advance of the Okanogan glacial lobe. South of the moraine the topography is smooth, while the moraine itself and the plateau surface to the north are lumpy and mantled with large basaltic (and in places, granitic) erratics. As the road descends to Banks Lake in Grand Coulee, it runs down the dip of the east-facing Coulee Monocline.

The highway crosses Grand Coulee on top of the low rock dam that confines Banks Lake in the upper coulee. One mile south is the spectacular lip of Dry Falls, and the lower Grand Coulee runs from there 7 miles along the monoclinal axis to Soap Lake. A side trip down State Highway 17 at least to the Sun Lakes is recommended strongly. There is an excellent, small museum at Dry Falls.

The route from Coulee City to Spokane crosses the northern part of the Columbia Plateau. Major floodways floored by basalt lie between Wilbur and Davenport and east of Reardan. Good exposures of basalt occur along the highway where it descends into Spokane.

TRIP 7. INTERSTATE HIGHWAY 90

seattle–snoqualmie pass–ellensburg–moses lake–spokane

This is the major east-west highway across the state. The geology from Seattle to the Teanaway Junction (U.S. Highway 97) is described in Information Circular 38 of the Washington Division of Mines and Geology.

Seattle to Ellensburg (*107 miles*). Only minor amounts of bedrock are exposed west of North Bend. They include:

1. Oligocene marine sandstone, with clam fossils, 2 miles east of Eastgate;
2. Eocene (Puget Group) sandstone and mudflow breccia, rich in volcanic clasts, several miles east of Issaquah and near the Fall City turnoff;
3. Volcanic andesite of Oligocene (?) age 2 miles west of North Bend.

All other outcrops are of glacial sediments of the Fraser Glaciation. Especially prominent are cross-bedded deltaic sands and gravels at Eastgate and the front of a gigantic delta that built into Glacial Lake Sammamish immediately north of Issaquah.

Mount Si dominates the North Bend area. Careful inspection of its south face, noting the change from smoothed lower to craggy upper slopes, reveals the elevation (about 2,000 feet) reached by the Puget Lobe of glacial ice. The highway travels up the South Fork of the Snoqualmie River. It climbs the steep morainal embankment built across the valley mouth east of North Bend, within sight of a similar deposit that dams the Middle Fork (see Chapter 18). From here to the summit, outcrops are quite varied—Permian Chilliwack Group sediments along the south wall at the top of the embankment; Lower Tertiary intrusive diorite in a quarry a few miles to the east; and a mixture of Eocene sedimentary rocks and Miocene granodiorite of the Snoqualmie Batholith for the last several miles up to the summit. Thick beds of clay were deposited in the lake upvalley from the ice dam of the Puget Lobe. These are prominent just east of the aforementioned quarry.

The descent from the pass to Lake Keechelus offers good exposures of glacial till. These form a series of arcuate moraines produced during the final recessional stage of alpine glaciers. The road along the north shore of Lake Keechelus has numerous outcrops of the Keechelus Volcanic Series (Chapter 12), a major sequence of Lower Tertiary lavas, breccias, and interbedded sediments in the Central Cascades of Washington. The lake is dammed downstream by a large recessional moraine, on which has been constructed a low rock dam. Outcrops of Eocene sedimentary and volcanic strata, the Naches Formation, occur from the east end of Lake Keechelus intermittently for 9 miles. The Easton Schist, correlative with the Shuksan Suite of the North Cascades (Chapter 7) crops out at the small bridge across an arm of Lake Easton and underlies the hills to the south. The layered cliff to the northeast consists of several thousand feet of basaltic flows of the Eocene Teanaway Basalt (Chapter 7).

No rock is exposed from Lake Easton to Cle Elum. The highway runs near the axis of the Cle Elum Syncline, but the bedrock, the Roslyn Formation, lies concealed beneath glacial deposits and sand and gravel laid down by the Yakima River. The river is crossed at Cle Elum, and the new route to Ellensburg climbs over the east end of the Cle Elum Ridge, with exposures of Yakima Basalt. White pumiceous sand and gravel of the Pliocene Ellensburg Formation are seen to the east, dipping south at a few degrees off the flank of the Wenatchee Mountains. To the south and west, the gentle dip of the Yakima Basalt off the east flank of the Cascade Arch is apparent. The last few miles of this stretch is across the alluvial cover of the Kittitas Valley.

Ellensburg to Spokane (175 miles). The route crosses the Kittitas Valley, slowly climbs the hills to the east, and then makes the long descent to the Columbia River at Vantage. Exposures are of the Yakima Basalt Formation, including the interbedded Vantage Sandstone Member (Chapter 17). The formation dips east toward the river at about the same angle as the road grade, so the Vantage Sandstone is exposed in many places.

The Vantage area is one of the most interesting places on the entire Columbia Plateau. The Ginkgo Museum is a required stop. It contains an outstanding collection of fossil wood and leaves collected from the Vantage Sandstone, as well as good displays on the origin of the Columbia Plateau. The Yakima Basalt is exposed magnificently in bluffs on both sides of the river, and gentle folding of the strata is apparent. The highway crosses the river and turns north, but 1 mile south on the road to Othello is an excellent exposure of pillow basalt of the Frenchman Springs Member. The pillows formed when a flow entered a lake. Another worthwhile sidetrip is down the old road that descends Frenchman Springs Coulee, 4 or 5 miles north of the bridge. (Turn left at a small house painted red, yellow, and blue!) The coulee was one of the escape channels by which Spokane Flood waters returned to the Columbia River, and its immense size gives some feeling for the drama of those floods. Also of interest is the Roza Basalt Member which caps the section here. The columns are immense—10 to 12 feet across—and erosion has produced a most unusual scene. Waste piles from the strip mining of diatomite (the Quincy Member) can be seen on the plateau surface to the north.

The highway runs east to Moses Lake across the Quincy Basin. This valley is a slight structural downwarp that accumulated some flood sediments from waters that issued from Grand Coulee and the various floodway channels to the east.

Some of the floodwaters spilled south around the east end of the Frenchman Hills, the low ridge south of this stretch of the highway, cutting the Drumheller Channels south of Moses Lake. The rest escaped west into the Columbia River via Frenchman Springs Coulee and The Potholes cataracts to the north. Evidence of the floods, such as large boulders, giant gravel bars, and sand waves are found in the Quincy Basin but are not too obvious from the road. Postglacial sand dunes are well developed south of Moses Lake (Figure 2-8).

From Moses Lake, the road runs east along a flood channel to Ritzville, then northeast along a major scabland tract to Spokane. Loess hills line the floodway, and scabby basaltic bedrock is exposed in its floor. The highway drops down off the plateau surface at Spokane and runs into Idaho on the alluvial floor of the Spokane Valley. East of Spokane, the surrounding hills consist of Precambrian strata that stood above the Miocene floods of Yakima Basalt.

TRIP 8. STATE HIGHWAY 410

tacoma-chinook pass-yakima (chinook pass closed in winter)

Tacoma to Yakima (*156 miles*). The highway runs east past Puyallup on the floodplain of the Puyallup River, then crosses the divide on glacial sediments to Buckley and the valley of the White River. The toe of the Osceola Mudflow (Chapter 13), a 5,000-year-old deposit derived from Mount Rainier, is crossed near Buckley; and the road runs on this unit past Enumclaw and into the Cascades. Intermittent exposures for the next 20 miles are of the lower Miocene Fifes Peak Andesite (Chapter 12), a major unit in the Central and Southern Cascades. Columnar structure is apparent in some outcrops, but many are quite massive. About 5 miles north of the Crystal Mountain road, the contact with the underlying Ohanapecosh Formation is crossed. The best exposures of this are seen in the first mile of the side road up to Crystal Mountain. Granodiorite of the Tatoosh Pluton is first encountered near the side road to Sunrise (a recommended side trip); but, above this Ohanapecosh strata crop out again over both Chinook and Cayuse Passes. Exposures are excellent, as is the view of nearby Mount Rainier.

The highway descends the glaciated American River valley from Chinook Pass to the junction with the Naches River. Outcrops are of Ohanapecosh and Fifes Peak strata, plus outliers of the Miocene Bumping Lake intrusion of granodiorite. Yakima Basalt forms cliffs along the road near the Naches River junction, and one road cut immediately north of the Sawmill Flat Campground contains gigantic basalt pillows up to 22 feet in diameter. Pre-Yakima Basalt volcanic strata crop out again south from here for 6 miles, then this Cascade-type bedrock is lost permanently under the basalt.

From Nile to Naches, the route lies under the southwest face of Cleman Mountain, a large, asymmetric fold in the Yakima Basalt. The moderately steep southwest dip of the lava is apparent along the mountain front. A thick section of boulder conglomerate and mudflow breccia, belonging to the Pliocene Ellensburg Formation, forms a hill west of the river at Nile. From Naches to Yakima, finer-grained Ellensburg strata (gravels, sands, and silts) are exposed locally. Especially noteworthy are the cuts just above Naches on the road to the Wenas Valley.

TRIP 9. U.S. HIGHWAY 12

aberdeen-centralia-white pass-yakima

Aberdeen to Chehalis (*57 miles*). This stretch follows the Chehalis River, with little bedrock exposed except for a few outcrops of marine Oligocene sandstone and shale. The river valley itself is of interest. Its large size and big meanders suggest a river very much larger than the present one. In fact, 13,000 to 15,000 years ago a river as large as the modern Columbia flowed down this valley to the ocean. This was when the normal drainage of the Puget Lowland to the Pacific via the Strait of Juan de Fuca was blocked by glacial ice, and all of the meltwater plus the normal runoff from the west flank of the Cascades and east flank of the Olympics exited via the Chehalis River. (See Chapter 18.)

Chehalis to Yakima (*149 miles*). The route turns east 7 miles beyond Chehalis and enters the Cascades along the valley of the Cowlitz River. Eocene basalt is exposed along the valley sides near Mayfield and Mossyrock Dams. Beyond Kosmos, the wallrock strata consist of andesitic flows, breccias, and sedimentary interbeds correlative with the Ohanapecosh and Fifes Peak Formations in the Mount Rainier area (Chapter 12), but the highway stays on the valley floor until a few miles past Packwood. There it turns east up Clear Creek toward White Pass. Some of the most spectacular columns in the Northwest are seen at The Palisades, a scarp revealing a Quaternary valley-filling lava flow. Somewhat older (but post-Miocene) andesitic lava caps the range around White Pass.

Outcrops around the Tieton Reservoir are particularly significant, for they lie within the only patch of pre-Tertiary strata exposed in the Cascade Range between the Snoqualmie Pass area and the Sierra Nevada Province. They consist of somewhat metamorphosed sedimentary and volcanic rocks that have been correlated recently with the Upper Paleozoic Chilliwack Group. From here to the Naches River, outcrops are of Cenozoic lava. The Yakima Basalt, dipping east off the Cascade Arch, is most prominent.

The last 19 miles to Yakima is discussed in Trip 8.

D oregon trip guides

The following guides presume some familiarity with Chapters 5, 10, 11, 12, 13, 14, 15, and 17.

TRIP 1. U.S. HIGHWAY 101

brookings–coos bay–newport–tillamook–astoria

California State Line to Coos Bay (120 miles). This is one of the more rugged stretches of the Oregon coast, providing many fine vistas of steep sea cliffs, uplifted wave-cut terraces, and numerous offshore rocks or stacks. From Langlois (near Cape Blanco) south, the bedrock consists of Mesozoic sedimentary and volcanic strata of the Klamath Mountains Province. These are highly folded and fractured, and cut locally by masses of soft, green serpentine. The volcanic rocks have been altered mostly to hard, dense greenstone; the sandstones and shales locally to slate or phyllite. (Refer to Chapter 10 for a more thorough discussion of the stratigraphy of the Klamath Mountains.) Casual investigation of outcrops provides an excellent feeling for the general lithology of the area (and, in fact, for the Mesozoic stratigraphy of the entire Cordilleran eugeosyncline) and for the structural complexities.

The physiography is most interesting. Wave action is the most obvious geologic agent at work. The tendency for waves to deflect toward headlands and to concentrate their energy there rather than in intervening bays is apparent. Currents moving along the shore transport sediment derived from wave erosion and supplied by rivers, and they deposit the sand and silt in the more protected bays between rocky headlands. In this way, the coastline becomes relatively smooth and straight. The processes are gradual, however, and the Coast Ranges and Klamath Mountains are geologically young features, so equilibrium has not yet been achieved. Raised wave-cut benches, found as high as 1,000 feet and more above present sea level, attest to the recency of the mountain building. Furthermore, some of these terraces are demonstrably warped, that is, they differ in elevation from point to point along the coast. This demonstrates that the uplift of the coastal ranges has involved more than a simple, uniform north-south arching.

North of Langlois, the bedrock is entirely of Tertiary age, and the Mesozoic-Tertiary contact is used arbitrarily to define the border between the Klamath Mountains and Coast Ranges. From Port Orford to Coos Bay, the highway runs generally inland away from the coast, and the exposures of bedrock are more limited (virtually nonexistent south of Bandon.) Occasional road cuts between Prosper and Coos Bay consist of dark gray mudstone and sandstone beds of the Eocene Umpqua and Coaledo Formations (Chapter 11). Excellent exposures of the Coaledo strata are found along the sea cliffs at Sunset Bay and Cape Arago southwest of Coos Bay (Figure 11-8).

Coos Bay to Newport (101 miles). This part of the highway runs, for the first 55 miles, just inland of the impressive sand dunes region of the Oregon coast. The sand has been deposited by longshore currents in the indentation between Cape Arago and Heceta Head (Figure 11-13). The prevailing onshore winds have piled the sand into large dunes that lie immediately behind and parallel to the beach, with smaller and more active fingers of sand projecting inland. The dunes have sealed off the mouths of many of the streams and small rivers that flow westward off the Central Coast Ranges, producing numerous lakes south and north of Reedsport.

Road cuts provide good exposures of the internal stratification of sand dunes but little of the underlying bedrock except near Reedsport. There, outcrops of evenly bedded sandstones and siltstones of the marine Eocene Tyee Formation are found. This unit underlies much of the Central Coast Ranges to the east (Chapter 11).

Hard volcanic bedrock plunges steeply into the sea from the Cape Mountain–Heceta Head region north to Yachats (Figure 11-7). These strata are massive flows, breccias, and near-surface intrusions of basaltic composition, of late Eocene age. They dip gently seaward and are in the lower part of the Coaledo Formation, resting here on the Tyee Formation. Wave erosion along a steeply dipping fault has excavated the popular Sea Lion Caves.

North of Yachats, the Coaledo outcrop belt moves inland and the volcanic members pinch out. Exposures along the highway from Yachats to Newport are not very numerous. They are of massive marine sandstones and some conglomerate beds. These strata are part of the Yaquina Formation of Oligocene age. Tuffaceous Miocene mud-

D-1 *Index map to trips in Oregon.*

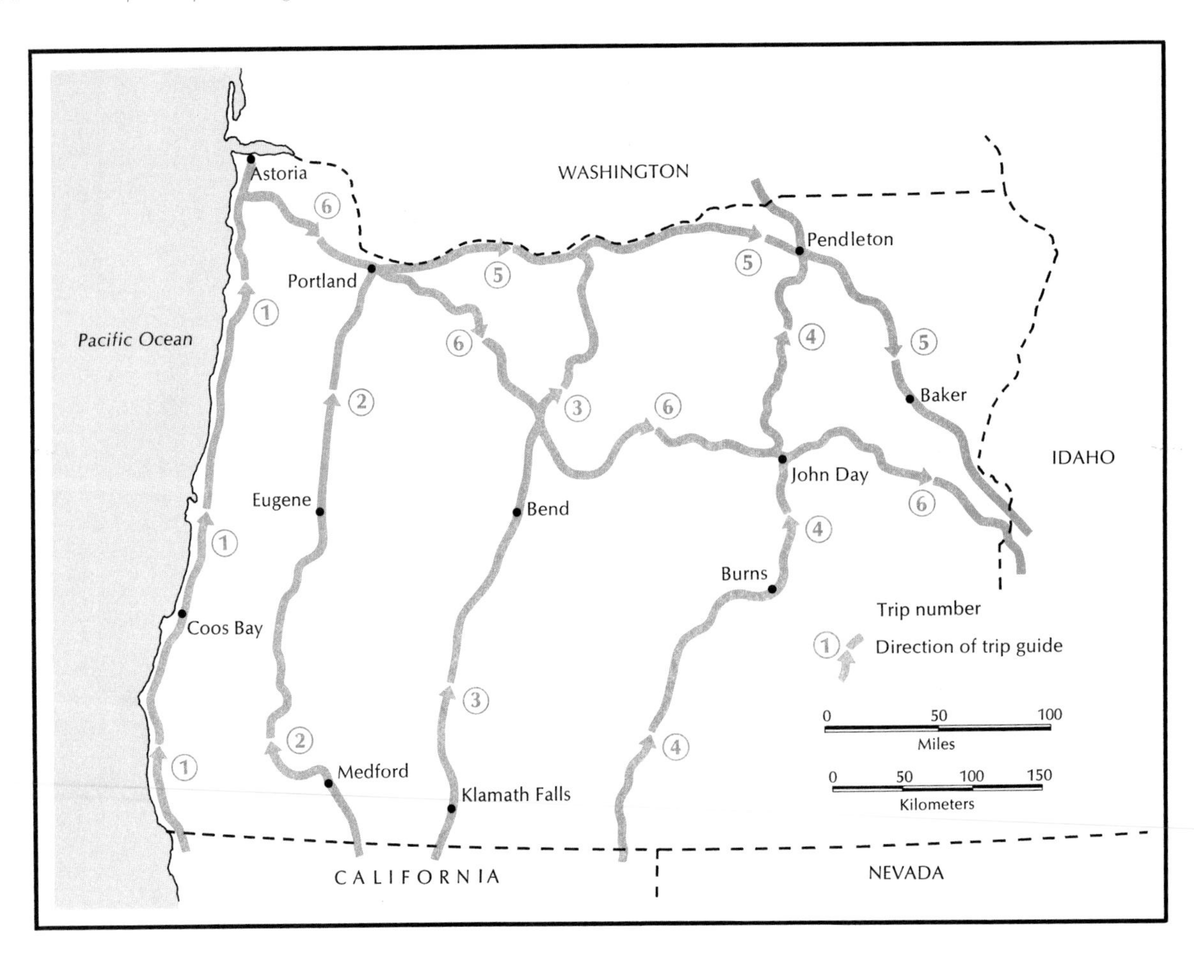

stones and sandstones crop out around Newport. Seal Rock north of Waldport is composed of dark, fine-grained gabbro, emplaced as an intrusive plug, probably during the Miocene Epoch.

Newport to Astoria (147 miles). This segment of the coast highway is quite varied, both topographically and geologically. Bold promontories such as Foulweather, Lookout, Meares, and Falcon Capes, and Cascade and Tillamook Heads are connected by long, low, gently curving beaches, some with well-developed dunes. The rocky capes, like those to the south, are made up for the most part of hard, volcanic rock. Cascade Head is basaltic greenstone of Eocene age. The rest are erosional remants of the nonmarine Columbia River Basalt of middle to late Miocene age. The basalt is notably less altered than are the older Eocene basalts, and columnar jointing is quite prominent in places.

Tertiary sedimentary strata are not well exposed along the highway, although Miocene sandstones and tuffaceous shales crop out around Tillamook Bay, inland from Tillamook Head, and near Astoria.

TRIP 2. INTERSTATE HIGHWAY 5

siskiyou pass–grants pass–eugene–portland

California State Line to Grants Pass (52 miles). From the state line to a point a few miles east of Gold Hill, the highway follows close to the boundary between the Klamath Mountains Province and the Cascade Range. Outcrops of light-colored sandstone and shale of the marine Upper Cretaceous Hornbrook Formation (Chapter 10) are prominent at the summit of Siskiyou Pass. From here to Ashland, the rocks exposed belong to the border of a Mesozoic granitic batholith in which Mount Ashland is centrally located. From Ashland north almost to Gold Hill, the highway follows close to the bottom of the valley of Bear Creek, and there are few bedrock exposures. To the east in the foothills of the Cascades, andesitic flows and breccias of Oligocene and early Miocene age form prominent cliffs locally.

East of Gold Hill, the highway turns west into the Klamath Mountains Province, and outcrops from there to Grants Pass are of metamorphosed volcanic rocks (now greenstone) and sedimentary strata (now slate and phyllitic sandstone) of the Upper Paleozoic (?) and Triassic Applegate Group (Chapter 10). Another Mesozoic granitic batholith extends westward from Grants Pass but is not exposed in the valley bottom along the highway.

Grants Pass to Eugene (129 miles). The road turns north away from the Rogue River at Grants Pass, crosses the Umpqua River drainage basin around Roseburg, and finally enters the headwaters region of the Willamette River near Cottage Grove. Rock exposures are numerous. From Grants Pass to Sexton Pass (about 12 miles), the route is through granitic rocks that are not well exposed. From there to Canyonville, most outcrops are of volcanic greenstone or sedimentary strata of the Rogue and Galice Formations–Jurassic marine units distributed widely throughout the northern Klamath Mountains.

From Canyonville to Myrtle Creek, Upper Jurassic and Cretaceous sediments of the Riddle and Days Creek Formations floor the valley of the Umpqua River but are not well exposed along the highway. A large northeast-trending band of serpentine and greenstone is crossed just north of Myrtle Creek, and beyond this almost to Roseburg outcrops are of the Jurassic Dothan Formation. These strata are in fault contact with Eocene basaltic rocks (Umpqua Formation) prominent in the hills around Roseburg, and the fault is taken as the dividing line between the Klamath Mountains and Coast Ranges Provinces.

Umpqua greenstone exposures are numerous along the highway as far as Wilbur. Dark sandstone and shale beds (also part of the Umpqua) underlie the route for the next 12 miles, followed by more basaltic greenstone north of Rice Hill. Eocene sandstone and shale of the Tyee Formation crop out around Curtin, while nonmarine pyroclastic rocks, also of Eocene age, are exposed from there north to Cottage Grove.

From Cottage Grove to Eugene, the exposed bedrock consists for the most part of dark gray, dense basalt. The rock is used locally for road construction and the best outcrops are in quarry walls. Most masses formed as intrusive sills, dikes, and plugs. They are of middle to late Miocene age and probably correlate with the Columbia River Basalt Group. Several large landslides, of some concern to the Highway Department, are prominent in the hillsides south of Eugene.

Eugene to Portland (*110 miles*). This is high-speed geology, for outcrops are few and far between. The highway runs mostly on the flat floor of the Willamette Valley, and exposures are of Quaternary alluvial sediments. (My prejudice is showing. To some, this is the good stuff.) An excellent exposure of another Miocene sill, with prominent columnar jointing, is close to the highway several miles south of Coburg. Miocene basalt is crossed just south of Salem and in the hills for the last 10-mile stretch into Portland, but the rest of this stretch offers little exposure of pre-Quaternary units. The bedrock geology of much of the Willamette Valley must be deduced from exposures in the neighboring Coast Ranges and Cascade Range, coupled with subsurface drilling and geophysical data.

TRIP 3. U.S. HIGHWAY 97

klamath falls–bend–biggs

California State Line to Bend (*156 miles*). This stretch of highway crosses the northwestern corner of the Basin and Range Province and the western edge of the High Lava Plains. The boundary between the two is not obvious. Outcrops along the road are not too numerous except for the segment east of Upper Klamath Lake, beneath the impressive Modoc Escarpment. The cliffs face west and have been produced by relatively recent movement on large, steeply dipping faults. Such faults, generally trending almost north-south, as here, are characteristic of the entire Basin and Range Province—the basins generally being formed by relative downdropping between large faults. A large fault is exposed beautifully near Algoma, 11 miles north of Klamath Falls. The surface of the fault has been eroded very little, and it preserves the grooves and scratches (known as slickensides) that are produced by the abrasion of one side moving against the other. Upper Cenozoic lavas make up most of the exposed cliff sections around the lake, but locally some tuffaceous sedimentary strata crop out.

The highway follows the bottom of the Klamath Lake basin for many miles north of the present lake shore. Crater Lake is near at hand in the Cascade Range to the west, but little suggestion of its proximity is gained along the route. Actually, close observation of some of the road cuts (and they are infrequent) should make one suspect proximity to a major volcano, for the ground surface is covered by several feet or more of light-colored volcanic ash. This is the 7,000-year-old Mazama Ash that spewed from the Crater Lake volcano (Mount Mazama) and drifted to the north and east as far as central British Columbia and central Montana (Chapter 13).

The Lava Butte area 10 miles south of Bend is a "must" stop. The butte itself marks the northernmost of eight eruptive vents along the Northwest Rift Zone of the Newberry Volcano. The butte is a cinder cone. A blocky basaltic lava flow issued from a vent on the south flank and flowed about 6 miles to the west and north into the valley of the Deschutes River. A good road leads to the summit of the butte, where there is a small museum and a trail around the crater rim. The view westward of the blocky flow and the nearby Cascades (especially Bachelor Butte, Broken Top, and the Three Sisters) is magnificent. Despite the extremely youthful appearance of Lava Butte and its flow, radiocarbon dates suggest that it was built about 6,000 years ago.

Just east of U.S. Highway 97 and 1½ miles south of Lava Butte is Lava River Cave. This lava tube is a mile long. It formed in one of the older fluid basalt flows of Newberry Volcano, and the area is a state park. The cave can be entered where the roof collapsed.

Bend to Biggs (*147 miles*). The highway crosses the western end of the Deschutes-Umatilla Plateau, the Oregon part of the Columbia Plateau Province. The route descends toward the Columbia River, paralleling two of its larger tributaries, the Deschutes River to the west and the John Day River to the east.

Between Bend and Redmond, the outcrops consist of Late Cenozoic basalt, and the slight relief on the plateau surface reflects pressure ridges or large wrinkles on the surface of this very extensive lava flow. The rock is quite vesicular (spongy) as the result of gas pockets contained in the congealing flow. A thin layer of Mazama Ash covers some of the outcrops. A few miles north of Redmond, the highway crosses the nearly vertical-sided canyon of the Crooked River. This is one of the best places in the region to see in cross-sectional view the flat-lying lava flows that underlie central Oregon. Several miles to the east, Smith Rocks (Figure 14-10) tower above the general level of the plateau. They consist of very massive, silicic lava and are the remains of the throat of an Early Cenozoic volcano—a probable source for one or more of the welded tuff flows in the Clarno or John Day Formations. From Madras north, the highway has few exposures until it starts to descend more steeply into the Columbia River Gorge near Wasco. Here one can see not only the many flows of the Miocene Columbia River Basalt but also in places the Pliocene sands and gravels of The Dalles-Deschutes nonmarine sequence.

TRIP 4. U.S. HIGHWAY 395

lakeview-burns-john day-pendleton-pasco (wash.)

Nevada State Line to Burns (*155 miles*). Here is a route that crosses the center of the Basin and Range Province of southeastern Oregon and provides a very good feeling for the geology of the region. The predominant lithologies are Upper Cenozoic nonmarine volcanic and sedimentary rocks. The characteristic block faulting has produced some impressive range front scarps in which the strata are clearly displayed. Late Pleistocene lakes filled many of the basins. Although most of these have all but disappeared, they have left behind beach deposits and wave-cut benches on the valley sides. Three remnant lakes lie to the west of the route—Goose Lake straddling the state line, Lake Abert north of Lakeview, and small Alkali Lake 20 miles to the north.

Abert Rim east of Lake Abert is the most impressive geologic feature seen. It is a west-facing eroded fault scarp that rises more than 2,000 feet from the valley floor. Layer upon layer of eastward-dipping basaltic lava are exposed. These flows are of Miocene age, correlative with the Steens Basalt in Steens Mountain to the east and regionally with the Columbia River Basalt north of the Blue Mountains. Were it not for the block faulting, these Miocene lavas would be concealed beneath the relatively thin veneer of younger sedimentary and volcanic strata. North of Lake Abert and scarp gradually diminishes; and, from Alkali Lake to Burns, the route is across a moderately low relief surface. Upper Cenozoic basalt is exposed in the rim of a few shallow draws.

Burns to John Day (*70 miles*). The highway climbs up the relatively gentle south flank of the Blue Mountains uplift and then drops rapidly down Canyon Creek between the flanks of the Aldrich Mountains (to the west) and Strawberry Mountains (to the east) into the town of John Day. For the first 25 miles, outcrops consist mostly of Miocene and Pliocene sedimentary rocks with some interbedded tuffs and silicic flows. The strata rise to the north at a greater angle than does the road, so progressively older rocks are encountered. South of Silvies, the first exposures of Mesozoic strata are seen in the valley bottom. These correlate with the Upper Triassic and Jurassic marine lavas and sedimentary formations that have been well studied in the Suplee-Izee district 20 miles to the west (Chapter 14).

Slightly metamorphosed Paleozoic formations, cut by gabbro and serpentine intrusions of probable Triassic age, are exposed in the valley walls south of Canyon City. The intrusions were responsible for the gold-mining activity in this area in the late nineteenth century. From Canyon City to John Day, the highway traverses the south flank of the John Day Syncline, cut by a large fault. Both structures trend east-west, essentially parallel to this stretch of the John Day River, but they are not very obvious from the highway.

John Day to Pendleton (128 miles). For 8 miles, the route runs west along the valley bottom, at the foot of the Aldrich Mountains. The highway is close to the axis of the John Day Syncline. This is accentuated most clearly by the dip toward the valley bottom from both sides of the resistant bench produced by the Rattlesnake Tuff, a Pliocene unit (Figure 14-13). Excellent exposures of this tuff are present at Mount Vernon, where the route turns north away from the valley. Less than a mile from the junction, the road crosses into volcanic sediments and flows of the Lower Cenozoic Clarno Formation. Two miles farther, it passes into serpentine and Paleozoic metasedimentary rock, and these are exposed intermittently for several miles. Then the Beech Creek Fault is crossed. This is a steep fracture that trends northwest, with the northeast side having been displaced relatively down. From here to the Blue Mountain front, the highway runs mostly in Columbia River Basalt, although some Mascall beds floor the valley around Fox, and Clarno beds are intersected in several short stretches.

The structure of the northern part of the Blue Mountain uplift is relatively simple. A distinct northerly dip to the basalt layers is apparent south of Pilot Rock as the north flank of the arch is approached. The lava appears almost to cascade off the mountain front and flatten out again under the Columbia Basin to the north. The tilting of the layers occurred, of course, long after the flows had spread and hardened in horizontal sheets. Quite likely it continues today.

Pendleton to Pasco (66 miles). This stretch runs northwest for 32 miles to the Columbia River just above McNary Dam and then runs upstream along the east bank of the Columbia. Particularly noteworthy are the many excellent exposures of Columbia River Basalt in the canyon walls close to the river. Wallula Gateway, the eastern end of the Columbia River Gorge, is passed close to the state line. This constriction was largely responsible for checking the great floods of water that swept across the Columbia Plateau in the Late Pleistocene (Chapter 17), thereby producing large temporary lakes in the Pasco Basin. Some of the sediment deposited in these lakes is apparent in places near the highway south of Pasco.

TRIP 5. INTERSTATE HIGHWAY 80N AND U.S. HIGHWAY 30

portland–the dalles–pendleton–baker–ontario

The first part of this trip provides a rare cross-sectional view of the internal architecture of the Cascade Range. The last part sweeps over the northeastern end of the Blue Mountains uplift, but only near the Idaho border does the traveler get a glimpse of the pre-Cenozoic core of the range.

Portland to The Dalles (85 miles). The highway follows the south bank of the Columbia River from about 20 miles east of Portland past The Dalles. Near Portland, outcrops are scarce, as most of the lowland surface is composed of alluvial sand and silt. Around Troutdale, 14 miles east of the city limits, sands and gravels of the Pliocene Troutdale Formation are exposed in some of the hillsides. The first outcrops of more solid rock are encountered as the valley of the Columbia (the mouth of the gorge) is reached a few miles to the east. Exposures are mostly basaltic lava flows of the Columbia River Basalt Group; they dip westward at a low angle off the Cascade Arch. Progressively older (deeper) flows are seen at road level as the Cascades are entered, until finally the axis of the arch is crossed near Bonneville. Here the gorge has been eroded completely through the basalt so that outcrops of Lower Cenozoic lithified mudflow deposits, a part of the Eagle Creek Formation (Chapter 12), occur close to the river. These exposures are not nearly so dramatic as the many cliffs and road cuts carved out of the basalt, such as where the columnar structure is arranged in fanning patterns (Figure 17-5).

The Bridge of the Gods area at Bonneville Dam is the toe of a gigantic landslide, perhaps no more than a few hundred years old. It originated high on the southeast face of Table Mountain, north of the river, and moved 7 miles downslope to dam the Columbia River very near the site of Bonneville Dam. The slide is apparent not only because of the scar on Table Mountain but also because of the extremely hummocky, irregular surface of this gigantic unstable mass.

Deeply eroded volcanoes that emitted some of the Upper Cenozoic lavas that constitute the Cascade Andesite Series (Chapter 12) are preserved on both sides of the river. Some of the more prominent volcanic necks on the Oregon side are Pepper Mountain, Larch Mountain, and Mount Talapus west of Bonneville, and Mount Defiance east of there. Similar volcano remnants across the river are Mount Pleasant and Mount Zion opposite Crown Point and Underwood Mountain near Hood River.

Shellrock Mountain and Wind Mountain are opposite one another and very close to the river (Figure 12-12). These are shallow igneous intrusions, quite silicic in composition and of probable Pliocene age.

From Cascade Locks eastward, the Columbia River Basalt, here about 2,000 feet thick, dips generally to the east, so that the road is going up the section. This eastern flank of the arch is not uniform, however. Some impressive northeast-trending folds cross the gorge in the stretch from Hood River to The Dalles. In order, these are the Bingen Anticline, the Mosier Syncline (Figure 12-9), the Ortley Anticline, and The Dalles Syncline whose axis passes about under the center of the city.

The Dalles to Pendleton (*130 miles*). This stretch continues along the Columbia River down the east flank of the Cascade Arch, then climbs up onto the plateau surface near Hermiston. The gorge walls provide more excellent exposures of the Columbia River Basalt and the sediments of the overlying Dalles Formation. Another noteworthy geologic feature is the scabland rock surface produced by the scouring of the torrents of the Spokane Floods (Chapter 17). Some are 100 feet or more above the present river level. A particularly impressive surface is on the inside of the river bend opposite The Dalles. The strong winds that sweep through the gorge have produced some magnificent sand dunes, whose sparkling whiteness contrasts markedly with the dark and somber basaltic cliffs. Unfortunately, some of the most impressive dune fields, developed on islands in the river bed, have now been inundated by the reservoir behind John Day Dam.

The canyon walls on the Washington side rise far above the plateau rim south of the Columbia along much of the upper gorge (Figure 17-15). This reflects the Columbia Hills structure—a somewhat complex folded and faulted structure north of the river.

The last 46 miles into Pendleton crosses the plateau surface and provides little rock exposure.

Pendleton to Ontario (*168 miles*). A few miles east of Pendleton, the highway climbs up the steep Blue Mountain front, with excellent road-cut exposures of the Columbia River Basalt, including some sedimentary interbeds. On top of the range, the rolling upland surface provides relatively few outcrops. The large valley around La Grande is a downdropped fault block (a graben). The highway turns south through some low basaltic hills and into the Baker Valley. From here on, most of the Blue Mountain Province is underlain by Mesozoic and Paleozoic marine sedimentary and volcanic rocks, all at least slightly metamorphosed and some quite so. A large Cretaceous batholithic intrusion cores the Elkhorn Mountains to the west, and there is a similar one in the Wallowa Mountains 25 miles to the northeast.

At Pleasant Valley, the route picks up the Burnt River and follows it down to Huntington. Outcrops are numerous. Some consist of Miocene and Pliocene lake and stream deposits, in part deposited in grabens between active block faults. Most outcrops are of Paleozoic and Mesozoic bedrock. Especially prominent are Triassic lavas (greenstone), sandstone, and shale near Huntington, and limestone (not surprisingly) around the nearby town of Lime. From Huntington to Ontario, the road sweeps through low hills of Upper Tertiary sediments.

TRIP 6. U.S. HIGHWAY 26

seaside–portland–madras–prineville–john day–nyssa

This route provides an exceptionally interesting crossing of the state. The trip requires more time than highways to the north or south (but then, high-speed travel and geologic investigation have never been compatible).

Seaside to Portland (*71 miles*). The road runs south for several miles, then turns east into the Coast Ranges. Columbia River Basalt outcrops are prominent in the hillside just southeast of the junction. Just east of Necanium, the highway crosses into Oligocene marine siltstones and shales of the Keasey Formation. Because these are relatively soft, they are not very well exposed; and they have given rise to numerous landslides. A quarry north of the road and 5 miles east of Elsie exposes a basaltic intrusion with well-developed columnar jointing. It probably correlates with the Eocene Tillamook Volcanic Series. Outcrops around Sunset Tunnel, 5 miles west of Buxton, are again of the Keasey sediments. Several miles farther on, grey sandstone beds of the middle Oligocene Pittsburg Bluff Formation crop out.

Outcrops are scarce between Buxton and the Portland Hills. The road is underlain by gently folded Columbia River Basalt capped by younger sediment. Mount Sylvania, just north of the highway near Portland, is a Pliocene or early Pleistocene low basaltic shield volcano on top of the Columbia River lavas. The Portland Hills immediately west of the city are a northwest-trending anticline in the Columbia River Basalt, which is exposed in road cuts as the route descends to the Willamette River.

Portland to Madras (*111 miles*). This segment runs southeast over the Cascade crest just south of Mount Hood. The geology en route is not too varied. From Portland past Gresham and all the way to Wemme, the road is on Quaternary sediments and alluvium, although toward the end of this stretch the surrounding hillsides consist of mostly andesitic lava of the Sardine and Cascade Andesite Series (Chapter 12). Numerous exposures of the latter are seen in the climb up to Wapinitia Pass from Government Camp, and on a clear day the glimpses seen of majestic Mount Hood close by to the north are spectacular.

The descent down the relatively even east flank of the Cascades is not noteworthy until the highway drops into the canyon of the Deschutes River at Warm Springs. This provides an excellent look at the units that underlie the plateau country east of the Cascades—a cap of Pliocene or Pleistocene basalt over light-colored tuffaceous sands and gravels of the Dalles-Deschutes sequence. Regionally, these are underlain by the Columbia River Basalt, but this is not exposed very well along Highway 26.

Madras to John Day (*142 miles*). The road runs southeast across the plateau surface to Prineville, where the basalt cap makes a most impressive rimrock above the town. The cap persists for a few miles as the road climbs into the Ochoco Mountains to the northeast, but the 25-mile ascent to the pass is more scenic than geologically instructive. The descent to Mitchell is something else. Near the Ochoco crest, road cuts are in sediments, flows, and breccias of the Eocene Clarno Formation—a widespread unit of considerable significance in north-central Oregon (Chapter 14). Below these are outcrops of Cretaceous marine sandstones and shales, locally severely folded and fractured, and in one road cut sliced by a horizontal thrust fault. Black Butte and White Butte south of Mitchell are remnants of Clarno (or John Day?) volcanoes or shallow silicic intrusions. The last few miles into Mitchell are past outcrops of both Clarno and Cretaceous strata. A short side trip to Painted Hills State Park (Figure 14-8) is very worth while. There you get a good look at the variegated tuffaceous sediments of the John Day Formation.

The road climbs steeply east from Mitchell, through mostly volcanic strata of the Clarno Formation, to a plateau underlain by Columbia River Basalt. After about 15 miles, it starts to descend to the John Day River via the canyon cut by Rock Creek. The principal faults and folds in this area run almost east-west, parallel to the creek; and outcrops passed as the highway swings back and forth are in the Columbia River Basalt, John Day, Clarno, and Mascall Formations. A caprock of the Pliocene Rattlesnake Tuff forms several prominent mesas south and west of the junction of Highways 26 and 19 in Picture Gorge.

The geology immediately around Picture Gorge is well exposed and extremely interesting. The reader is referred to Chapter 14 for maps, diagrams, photographs, and text. A short side excursion north on Oregon Highway 19 is recommended for examination of the John Day Formation and, if time permits, fossil hunting.

From Picture Gorge to John Day the highway follows the John Day River, which in turn lies close to the axis of the John Day Syncline. The Rattlesnake Tuff layer dips gently toward the axis. One outcrop between Dayville and Mount Vernon shows the Mascall-Columbia River Basalt contact, tipped up on end and slightly overturned. The Aldrich Mountains to the south consist mostly of Triassic strata; the hills north of the valley are composed of basalt.

John Day to Nyssa (142 miles). This segment runs east out the end of the John Day Valley via Dixie Pass and then turns southeast to Vale via the Unity Plateau and Lost Valley. The Strawberry Mountains southeast of John Day are capped by several thousand feet of Miocene andesitic to rhyolitic lava and ash erupted from local volcanoes. The highway is in the Rattlesnake Formation around Prairie City, then turns northeast into Clarno strata. A Cretaceous granitic intrusion underlies Dixie Butte north of the pass. From Dixie Pass to Ironside, the bedrock is chiefly Upper Cenozoic river and lake beds and interbedded lava flows. Two narrow tracts of Mesozoic sedimentary and volcanic rock are crossed east of Ironside. From Brogan to Vale, the road follows Willow Creek and no beds are exposed, although lake beds compose the surrounding hills. No rock is exposed on the surface of the Snake River Plain, traversed between Vale and Nyssa.

E idaho trip guides

The following guides presume some familiarity with Chapters 5, 6, 14, 16, and 17.

TRIP 1. U.S. HIGHWAY 95

jordan valley (ore.)-weiser-lewiston-coeur d'alene-eastport

Jordan Valley to Weiser (107 miles). The first 22 miles runs north, just west of the Oregon-Idaho border, across low hills that bound the Snake River Plain to the northeast. The highway then turns into Idaho and descends to the floodplain of the Snake River, crossed near Homedale. The bedrock consists of Miocene basalt, the Owyhee Formation, largely concealed beneath alluvial and lake beds of the Payette Formation and Idaho Group and flows of the silicic Idavada Volcanics. Exposures are generally poor. From Homedale to Weiser the route lies a few miles east of the Snake River, crossing the Boise River near Parma and the Payette River near Payette. The Payette beds are well exposed in bluffs along both sides of the valley for the last 30 miles.

Weiser to Lewiston (236 miles). Much of this stretch runs near the eastern edge of preservation of the Columbia River Basalt. The large Cretaceous intrusive complex known as the Idaho Batholith underlies the Sawtooth and Clearwater Ranges to the east, but these crystalline rocks are crossed only briefly south of Riggins. Marine sedimentary and volcanic strata of primarily Permian and Triassic age underlie the Columbia River Basalt at the northeastern end of the Blue Mountains Province and in the Seven Devils Mountains. These are best exposed in deep canyons, such as Hells Canyon on the Snake River, and along the Salmon River near Riggins. The Columbia River Basalt, as the result of Late Cenozoic uplift, dips west and northwest off the flanks of the Idaho Batholith and the Blue Mountains.

Columbia River Basalt crops out in the hills flanking the road north of Weiser and intermittently along the road from Cambridge to New Meadows. Eight miles north of New Meadows are excellent exposures of crystalline granitic rocks. They belong to the border phase of the Idaho Batholith, and some are gneissose. They continue for about 3 miles, and are followed by another 5 miles of basalt. Triassic sandstones and shales are entered just south of Pollock and exposed intermittently into Riggins. From Riggins to White Bird, the road follows the canyon of the Salmon River, which offers excellent exposures of the Seven Devils Volcanics. These are somewhat altered marine andesites and basalts of Permian and Triassic age and have a wide distribution in the eastern Blue Mountains. The Columbia River Basalt drops down again near White Bird, concealing the older rocks underneath; and, except for a small granitic outlier south of Ferdinand, only the basalt is seen from White Bird to Lewiston. Exposures are particularly good in the valley of the Clearwater and Snake Rivers around Lewiston. The basalt has been warped by a gigantic fold—a monocline that trends east-west with the south side dropped down several thousand feet relative to the north.

Hells Canyon Side Trip. A trip to the bottom of Hells Canyon is strongly recommended for those not subject to claustrophobia. The best access is

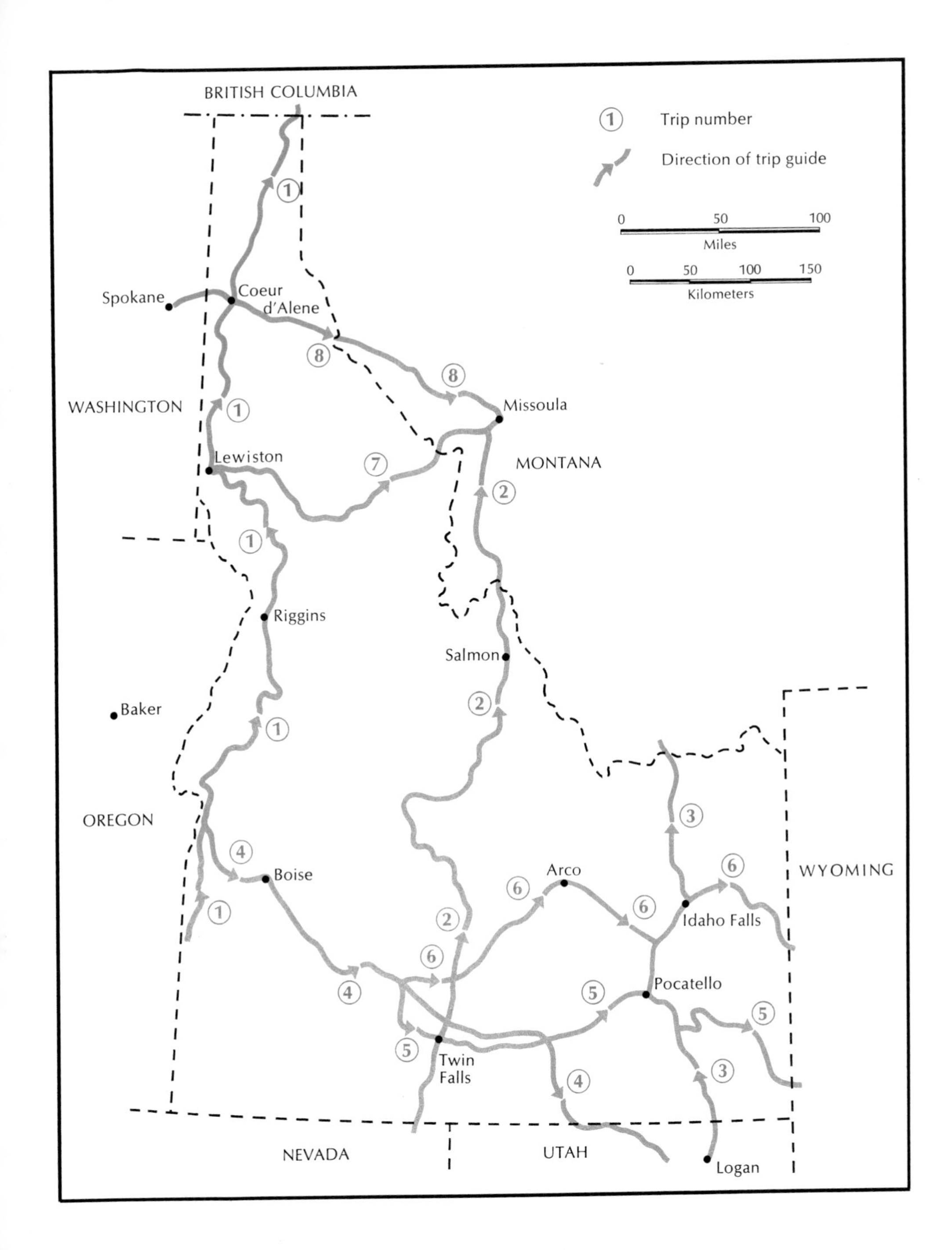

E-1 *Index map to trips in Idaho.*

northwest from Cambridge via Brownlee Summit on Idaho Highway 71 or east from Baker (Oregon) via Oregon Highway 86. The canyon rim is formed by nearly flat-lying Columbia River Basalt flows; but, as the route descends, Triassic marine sandstones and shales are crossed and are well exposed. A reasonably good road runs north (downstream) along the river as far as the Hells Canyon Dam; and some other Permian and Lower Mesozoic strata crop out en route, most notably limestone north of Oxbow Dam and the Seven Devils Volcanics around Hells Canyon Dam (Figure 14-3). The deepest part of the canyon lies north of the Hells Canyon Dam (Figure 14-2) but is not accessible by car.

The adventuresome traveler, not prone to attacks of vertigo, can exit the canyon via Kleinschmidt Grade. The road (dirt) runs from near Oxbow Dam southeast to Council, and the first few miles climb steeply up the east wall of the canyon. Stop your car before admiring the view!

Lewiston to Coeur d'Alene (*118 miles*). The route runs north, first climbing updip across the Lewiston Monocline in the Columbia River Basalt and then across the plateau surface to Moscow. The basalt is buried for the most part on the plateau by the Palouse soil. Outcrops around Moscow include the southernmost outliers of the Precambrian Belt Supergroup, Cretaceous granitic rocks, and the basalt cover. From Viola to Coeur d'Alene, all three of these units are exposed—Belt quartzites and argillites for 10 miles south of Desmet, granitic rocks for 10 miles south of Coeur d'Alene, and basalt around Potlatch and Worley. The Palouse Formation is also quite extensive.

Coeur d'Alene to Yahk (*B.C.*) (*117 miles*). The highway follows the Purcell Trench, a major north-south valley occupied by the Kootenai River near the border, and by Kootenay Lake farther north. This trench was the site of a significant lobe of glacial ice in the Pleistocene. The glacier flowed south into the Idaho Panhandle and dammed the Clark Fork River at Sandpoint, producing Glacial Lake Missoula (Chapter 17). The trench is floored by a mixture of alluvium, glacial outwash deposits, and sediment deposited by the Spokane Floods. Bedrock is not well exposed along the route.

From Coeur d'Alene to Sandpoint the highway runs close to the eastern edge of the Cretaceous Kaniksu Batholith, intruded here into the Belt Supergroup. From Sandpoint to Copeland, near the border, the bedrock beneath the valley consists of Kaniksu granitic rocks, although a major fault follows the trench and some Belt strata crop out in the hills west of Bonners Ferry. From Copeland, the route runs northeast into the Purcell Range, which consists largely of Belt (Purcell) rocks, some of which are exposed in road cuts.

At Yahk the highway joins Provincial Route 3 (Appendix B, Trip 2).

TRIP 2. U.S. HIGHWAY 93

jackpot (nev.)–twin falls–salmon–missoula (mont.)

Note: The geology along this highway is the subject of an excellent publication of the Idaho Bureau of Mines and Geology.

Jackpot to Twin Falls (*55 miles*). The first 15 miles passes through exposures of the nearly flat-lying silicic volcanics and associated sediments of the Idavada Formation. (Idavada is the name of the former "town" on the Idaho side of the state line.) These Pliocene strata are overlapped by Pleistocene basalt of the Snake River Group at the edge of the plain, several miles south of Rogerson. Exposures are poor from there to Twin Falls.

Twin Falls to Salmon (*249 miles*). The highway crosses the gorge of the Snake River just north of Twin Falls. This stretch of the river offers excellent cross-sectional views of the many thin basalt flows that underlie the Snake River Plain. (See Figure 16-8, which shows the north canyon wall 7 miles to the east of Highway 93.) The road runs north across the plain and into the mountains via the Wood River valley. A distinctly recent flow is crossed for several miles north of Shoshone. The hills at the edge of the plain east of Magic Reservoir expose Lower Tertiary volcanic and sedimentary strata of the extensive Challis Volcanics. Within the mountains, the highway follows the alluvial floodplain of the Wood River. Paleozoic sediments (somewhat metamorphosed) crop out in the Pioneer Mountains to the east, the same units plus some outliers of the Cretaceous Idaho Batholith in the Sawtooth Mountains to the west. Patches of Challis Volcanics are preserved, and this unit lies on both sides of the route from Ketchum (Sun Valley) northwest to a few miles south of Obsidian. Here the contact between the Idaho Batholith and Paleozoic strata is crossed, but along the road it is concealed beneath the alluvium at the head of the valley of the Salmon River. The granitic rocks of the batholith underlie the highway north to Stanley, then east almost to Clayton, where the eastern margin of the intrusive complex is again crossed. From here to the vicinity of Challis, various Paleozoic formations are exposed in the surrounding hills and occasional road cuts. These are overlain by volcanic strata of the Challis Volcanics, including prominent light-colored tuffs and tuffaceous sediments of the Germer Tuff Member. Excellent exposures of the Challis Formation are seen in the Upper Salmon River Gorge, north of Challis. Steeply dipping and massive Precambrian beds of the Belt Series crop out along the valley 10 miles to the north, just before the highway runs out onto the floodplain south of Salmon. The Germer Tuff Member forms very prominent bluffs within the valley basin, whereas the Bitterroot Range to the east and Salmon River Mountains to the west consist largely of Belt strata.

Salmon to Missoula (141 miles). The route runs north, crossing over the Bitterroot Range into Montana at Lost Trail Pass and descending to the broad, alluvial valley of the Bitterroot River. The Germer Tuff crops out for 12 miles; and the base of the unit, resting on upturned and truncated Belt strata, is displayed beautifully in the canyon wall at the end of this stretch (Figure 16-6). The Salmon River turns west at North Fork to start its run to Riggins through the high country of the Idaho Batholith. The highway follows up the North Fork to the pass, with road cuts in Belt rocks. North of Lost Trail Pass, the route traverses a salient of the Idaho Batholith. Some long-range panoramic views into the crystalline terrain are had looking west from the Bitterroot Valley floor up into the heavily glaciated range. No rock is exposed in the valley bottom. Shoreline terraces and beach deposits from Glacial Lake Missoula (Chapter 17) are preserved locally along the valley walls, most strikingly just east of Missoula.

TRIP 3. U.S. HIGHWAY 91 AND INTERSTATE HIGHWAY 15

logan (utah)–pocatello–idaho falls–monida pass

Logan to Pocatello (93 miles). This stretch lies in the Basin and Range Province, characterized by miogeosynclinal sedimentary strata of Paleozoic and Mesozoic age, disrupted by Cretaceous and Early Tertiary thrust faults and younger high-angle normal faults. The highway stays along the valley bottom of Bear River, Marsh Creek, and Portneuf River; and little bedrock is seen. These valleys are floored by alluvium and the Salt Lake Formation, a series of gravels, sands, and silts deposited along river channels and in lakes during Late Cenozoic time. Quaternary basalt of the Snake River Group is preserved in patches within the valley from McCammon north to Pocatello.

Pocatello to Monida Pass (129 miles). The highway runs north diagonally across the eastern end of the Snake River Plain, then crosses the Centennial Range into Montana via Monida Pass. Snake River Basalt is seen in places along the Snake River (as at Idaho Falls), but much of the route is across a relatively featureless surface. From Spencer to Humphrey, Upper Tertiary volcanics, equivalent to the Idavada Volcanics south of the plain, are traversed. These form much of the south flank of the Centennial Range, although north of Humphrey, Cretaceous sedimentary rocks that underlie the volcanics are exposed.

TRIP 4. INTERSTATE HIGHWAY 80N AND U.S. HIGHWAY 30S

ontario (ore.)–mountain home–boise–rupert–tremonton (utah)

Ontario to Jerome (*175 miles*). The highway runs southeast across the surface of the Snake River Plain. Roadside exposures are few, although Upper Tertiary sediments and lava flows of the Payette Formation and the Idaho and Snake River Groups are seen in bluffs along the valley side. These units are exposed best in hillsides along the Snake River near Bliss. (See, for example, Figure 16-7.)

Jerome to Tremonton (*163 miles*). For 56 miles, the route is east across the Snake River Plain north of the river. Beyond Rupert, it turns south, and you pass (via U.S. Highway 30S) up the valley of the Raft River into the Basin and Range Province and then southeast into Utah. The highway runs across alluvial basin fill, but metamorphosed Paleozoic strata and Cretaceous granitic rock are exposed in the Albion and Raft River Ranges to the west, and Paleozoic sedimentary formations crop out in the mountains to the east.

TRIP 5. U.S. HIGHWAYS 30 AND 30N

bliss–twin falls–pocatello–soda springs–wyoming border

Bliss to Pocatello (*166 miles*). The highway runs south from Bliss, crossing the Snake River a few miles south of Hagerman. The area of the crossing is quite dramatic. Immediately upstream on the east bank is the famous Thousand Springs area, with great torrents of water issuing from within and between the exposed flows of the Snake River Basalt. The bluffs west of the river consist of flat-lying sediments of the Idaho Group (Figure 16-7), whose light color contrasts markedly with the somber shade of the basalt. The road climbs up out of the valley through Upper Tertiary sediments and flows and turns east into Twin Falls. From here it runs east, then northeast, to Pocatello, remaining south of the river on the margin of the plain. Lake silts underlie the route south of the American Falls Reservoir.

Pocatello to the Wyoming Border (*120 miles*). This stretch crosses the northeastern corner of the Basin and Range Province. The first 22 miles is along Interstate 15 (see Trip 3). The highway turns east at McCammon and runs to the southeast corner of the state, across the structural grain (north-south) of the folds and faults. The first few miles follow the Portneuf River across Paleozoic sedimentary rocks. The valley floor around Bancroft is underlain by an outlier of Snake River Basalt. From Soda Springs past Montpelier, the highway stays on the floor of the Bear River, finally turning east into the mountains southeast of Dingle. From here into Wyoming, the bedrock consists of Triassic and Jurassic marine sedimentary formations, strongly tilted by Late Cretaceous and Early Tertiary folding and faulting.

TRIP 6. U.S. HIGHWAY 26

bliss–arco–idaho falls–alpine (wyo.)

Bliss to Arco (*111 miles*). This stretch follows the northern edge of the Snake River Plain, up the Little Wood River to Carey and then along the foot of the Pioneer Mountains past Craters of the Moon National Monument. The latter is the only feature of real note en route, but it is geologically of such interest as to strongly suggest use of this highway for east-west passage of the plain. Most of the trip from Bliss to Craters of the Moon is across a moderately mature lava surface, although recently erupted basalt crops out between Gooding and Richfield. The young lavas of the Craters of the Moon area are encountered a few miles east of Carey. Some of the flow surface is moderately blocky. Elsewhere, smooth lava has been rippled by large pressure ridges, ten feet or more high and hundreds of yards long. The many cones along the

Craters of the Moon rift system are clearly seen, extending southeast many miles toward the center of the plain (see Chapter 16). The monument itself represents a unique opportunity to view fresh volcanic structures at close hand (Figures 2-2, 16-11).

The Pioneer Mountains, immediately north of the road, consist primarily of partly metamorphosed Paleozoic formations, capped in places by lava of the Lower Tertiary Challis Volcanics.

Arco to Alpine (155 miles). The highway runs southeast across the plain to Blackfoot, then turns northeast (with Interstate 15) to Idaho Falls. Arco lies along the Lost River, which runs southeast into the plain from between the Lost River Range and Pioneer Mountains. The river loops around to the northeast and disappears into the basaltic bedrock surface of the Snake River Plain. It is stated often that the Thousand Springs area along the Snake River northwest of Twin Falls is the outlet for the Lost River, but this is a bit misleading. The plain does not conceal long underground rivers. Instead, the plain surface and subsurface, consisting of alternate layers of fractured basalt and sediments (including stream gravels), is quite porous. Much of the runoff from the surrounding ranges sinks into the plain surface, so that the strata underground are saturated. The Snake River has cut deeply into the plain in places (such as Twin Falls), and the many springs along the canyon walls are fed by groundwater that has traveled *slowly* downdip along permeable beds, mostly from the north and northeast. Thus the waters of the Lost River make a contribution to the saturation of the rocks and hence the springs, but they do not flow in some open, underground cavern. The highway crosses the Lost River 15 miles southeast of Arco and about 10 miles southwest of the place where the last of the water sinks underground.

East of the junction with the cutoff to Idaho Falls (U.S. Highway 20), two hills rise prominently above the general level of the plain (Figure 16-9). Known as Twin Buttes, these hills consist of ancient lava cones that preceded all of the recent basaltic eruptions of the Snake River Plain. The younger basaltic flows have partly buried these older cones. An even larger example of an old, partly buried cone is Big Southern Butte, 6 miles to the southwest. Pressure ridges in the Snake River Basalt are well developed in places along the highway.

From Blackfoot to Idaho Falls, the route parallels the Snake River. It runs east from Idaho Falls, crossing the river and cutting off a big bend in the Snake, which is rejoined again east of Ririe. From here, the highway is located along the Snake River valley. The mountains on either side consist of folded Upper Paleozoic and Triassic sedimentary strata. Lower in the valley are remnants of Upper Tertiary silicic lava, tuff, and sedimentary beds (Idavada Volcanics and Salt Lake Formation) and flows of Snake River Basalt.

TRIP 7. U.S. HIGHWAY 12

lewiston–orofino–kooskia–missoula (mont.)

Lewiston to Kooskia (74 miles). The route is upstream along the valley of the Clearwater River. The first few miles follows the Lewiston Monocline, a large east-west fold in the Columbia River Basalt (see Trip 1). Basalt crops out in hillside exposures along this entire stretch. The flows rest unconformably on gneissose crystalline rocks belonging to the border phase of the Cretaceous Idaho Batholith, and these are exposed (poorly) low in the valley around Orofino. Granitic rock of the batholith proper crops out below the basalt from a few miles southeast of Orofino to Kamiah. Thereafter, the basalt again reaches to the valley bottom.

Kooskia to Missoula (139 miles). At Kooskia, the highway turns east up the Middle Fork of the Clearwater, then up the valley of the Lochsa River to Lolo Pass. From there, it descends the east flank of the Bitteroots to Missoula. For 10 miles east from Kooskia, the Columbia River Basalt persists, but then its eastern edge (an erosional scarp) is passed and all the outcrops are of the Idaho Batholith—border phase gneisses past Lowell, then granitic rocks nearly to the pass. A small patch of metamorphosed Belt strata, preserved near the top of the batholith, is crossed 7 miles southwest of the summit. The main contact between the batholith and the Belt Supergroup is crossed a few miles northeast of the summit, and Belt strata are exposed partway down to the Bitterroot Valley floor. Remnants of shoreline deposits from Glacial Lake Missoula (Chapter 17) are found along the main valley sides.

TRIP 8. INTERSTATE HIGHWAY 90 AND U.S. HIGHWAY 10

spokane (wash.)-coeur d'alene-kellogg-missoula (mont.)

Spokane to Kellogg (72 miles). From Spokane to Coeur d'Alene, the route lies on glacial sediments in the Spokane River valley. Metamorphic rocks crop out in the surrounding hillsides except near the west end of Coeur d'Alene Lake. Here the bedrock consists of a southern arm of the Kaniksu Batholith (Cretaceous), capped by a remnant of Columbia River Basalt. Outcrops of slates, argillites, and quartzites of the Belt Supergroup are seen first along the north shore of the lake. This unit is well exposed along the route for the next hundred miles. Many of the exposures are quite spectacular—great cliffs of steeply dipping, massive beds. The structural grain of the region is not so apparent. It consists of many large, nearly vertical faults that trend west-northwest, essentially parallel to the route. A second set of faults runs north-northwest; and, although this is not so important here, it is the dominant trend farther north. The major west-northwest fracture belt is known as the Lewis and Clark Line. At least some of the faults have seen major horizontal movement along them in the past. For instance, the Osburn Fault, which extends from Coeur d'Alene to Missoula, is thought to have moved at least 16 miles, the north side shifting east relative to the south side. The origin of the Lewis and Clark Line is not clearly understood, but it seems likely that this zone represents a major flaw in the earth's crust.

Kellogg to Missoula (129 miles). The highway climbs to the summit of the Bitterroot Range at Lookout Pass, with many outcrops along the route. It then follows the valley of the Saint Regis River all the way to Missoula. Belt strata are exposed in hillsides and some road cuts to approximately Alberton. From near here to the east, most roadside exposures are of Quaternary deposits, including some excellent cuts in lake beds deposited in Glacial Lake Missoula (Chapter 17). Shoreline deposits of this glacial lake are seen locally on hillsides around Missoula.

additional reading and references

(See also Reading and Reference list for Chapter 5.)

CHAPTERS 1, 2, AND 3

some general books on geology

Albritton, Claude C., Jr. (ed.): "Fabric of Geology," W. H. Freeman and Company, San Francisco, 1963. [Essays that illustrate the philosophy and some basic tenets of geology.]

American Geological Institute: "Dictionary of Geological Terms," Dolphin Books, Doubleday & Co., Inc., Garden City, N. Y., 1962 (paperback). [The best geologic dictionary available, containing definitions of approximately 7,000 terms.]

———: "Geology and Earth Sciences Sourcebook," Holt, Rinehart and Winston, Inc., New York, 1969. [A comprehensive listing of a variety of source materials.]

Dyson, James L.: "The World of Ice," Alfred A. Knopf, Inc., New York, 1962. [A very readable account of glaciers and glaciation.]

Fenton, C. L., and M. A. Fenton: "The Fossil Book," Doubleday & Co., Inc., Garden City, N. Y., 1958. [A good, nontechnical introduction to fossils and paleontology.]

Matthews, William H.: "Geology Made Simple," Doubleday & Co., Inc., Garden City, N. Y., 1967.

Rittman, A.: "Volcanoes and Their Activity," John Wiley & Sons, Inc., New York, 1962. [The most comprehensive book available on this subject.]

Shelton, John S.: "Geology Illustrated," W. H. Freeman and Company, San Francisco, 1966. [Perhaps the best book on geology for the public published in many years. Principles are elucidated by magnificent photographs and diagrams. Numerous regions are studied, including parts of the Northwest.]

Shimer, John A.: "This Sculptured Earth: The Landscape of America," Columbia University Press, New York, 1959. [Treats particular types of landscapes (volcanic regions, glaciated landscapes, etc.) with many examples discussed briefly.]

———: "This Changing Earth: An Introduction to Geology," Harper & Row, Publishers, Incorporated, New York, 1968.

some books on rocks, minerals, and gems

Fritzen, D. K.: "The Rock-hunter's Field Manual—a Guide to Identification of Rocks and Minerals," Harper & Row, Publishers, Incorporated, New York, 1959.

Pearl, Richard M.: "Gems, Minerals, Crystals and Ores," The Odyssey Press, Inc., New York, 1964.

———: "How to Know the Minerals and Rocks," New American Library of World Literature, Inc., New York, 1965.

Pough, Frederick H.: "A Field Guide to Rocks and Minerals," Houghton Mifflin Company, Boston, 1953.

Sinkankas, John: "Gemstones of North America," Van Nostrand Reinhold Company, New York, 1959.

———: "Gemstones and Minerals—How and Where to Find Them," Van Nostrand Reinhold Company, New York, 1961.

———: "Mineralogy for Amateurs," Van Nostrand Reinhold Company, New York, 1964.

Vanders, Iris, and Paul F. Kerr: "Mineral Recognition," John Wiley & Sons, Inc., New York, 1967.

Zim, Herbert S., and Paul R. Shaffer: "Rocks and Minerals," Golden Press, New York, 1957 (paperback).

some standard college texts and references

Berry, L. G., and Brian Mason: "Mineralogy," W. H. Freeman and Company, San Francisco, 1959.

Berry, William B. N.: "Growth of a Prehistoric Time Scale," W. H. Freeman and Company, San Francisco, 1968.

Clark, Thomas H., and C. W. Stearn: "Geological Evolution of North America," 2d ed., The Ronald Press Company, New York, 1968.

Compton, Robert R.: "Manual of Field Geology," John Wiley & Sons, Inc., New York, 1962.

Dunbar, Carl O.: "Historical Geology," 2d ed., John Wiley & Sons, Inc., New York, 1960.

Easterbrook, D. J.: "Principles of Geomorphology," McGraw-Hill Book Company, New York, 1969.

Gilluly, James, A. C. Waters, and A. O. Woodford: "Principles of Geology," 3d ed., W. H. Freeman and Company, San Francisco, 1968.

"Glossary of Geology and Related Sciences," 2d ed., American Geological Institute, Washington, D.C., 1960.

Hamilton, E. I.: "Applied Geochronology," Academic Press, Inc., New York, 1965.

Hurlbut, C. S., Jr.: "Dana's Manual of Mineralogy," 17th ed., John Wiley & Sons, Inc., New York, 1959.

Leet, L. Don, and Sheldon Judson: "Physical Geology," 4th ed., Prentice-Hall, Inc., Englewood Cliffs, N.J., 1971.

Longwell, C. R., R. F. Flint, and J. E. Sanders: "Physical Geology," John Wiley & Sons, Inc., New York, 1969.

Moore, R. C.: "Introduction to Historical Geology," 2d ed., McGraw-Hill Book Company, New York, 1958.

Pearl, Richard M.: "Geology," 3d. ed., Barnes & Noble, Inc., New York, 1963 (paperback).

Strahler, A. N.: "The Earth Sciences," Harper & Row, Publishers, Incorporated, New York, 1963.

———: "Introduction to Physical Geography," John Wiley & Sons, Inc., New York, 1965.

Thornbury, William D.: "Principles of Geomorphology," 2d ed., John Wiley & Sons, Inc., New York, 1969.

Verhoogen, John, F. J. Turner, L. E. Weiss, Clyde Wahrhaftig, and W. S. Fyfe: "The Earth: An Introduction to Physical Geology," Holt, Rinehart and Winston, Inc., New York, 1970.

Woodford, A. O.: "Historical Geology," W. H. Freeman and Company, San Francisco, 1965.

CHAPTER 4

Bullard, Sir Edward: The origin of the oceans, *Scientific American,* vol. 221, no. 3, pp. 66–75, September, 1969. [This issue of *Scientific American* is devoted to the oceans, and contains highly readable and authoritative articles.]

Emery, K. O.: The continental shelves, *Scientific American,* vol. 221, no. 3, pp. 106–122, September, 1969.

Hurley, Patrick M.: The confirmation of continental drift, *Scientific American,* vol. 218, no. 4, pp. 52–64, April, 1968. [An excellent summary of the geologic evidence, much of it newly found, that supports the concept of continental drift.]

LePichon, Xavier: Sea-floor spreading and continental drift, *Jour. Geophys. Res.,* vol. 73, pp. 3661-3697, 1968. [A technical discussion of spreading and its implications.]

Menard, H. W.: "Marine Geology of the Pacific," McGraw-Hill Book Company, New York, 1964. [A good general summary that predates much of the "New Global Tectonics."]

———: The deep-ocean floor, *Scientific American,* vol. 221, no. 3, pp. 126-142, September, 1969.

Phinney, Robert A. (ed.): "The History of the Earth's Crust," Princeton University Press, Princeton, N.J., 1968. [A moderately technical but quite readable symposium emphasizing plate tectonic theory.]

Raff, A. D., and R. G. Mason: A magnetic survey off the west coast of North America, 40°N. latitude to 52°N. latitude, *Geol. Soc. America Bull.,* vol. 72, pp. 1267-1270, 1961.

Takeuchi, H., S. Uyeda, and H. Kanamori: "Debate about the Earth: Approach to Geophysics through Analysis of Continental Drift," Freeman, Cooper & Co., San Francisco, 1967. [A layman's guide to solid-earth geophysics and continental drift. The history of the theory is traced carefully. Excellent.]

Vine, F. J.: Spreading of the ocean floor: new evidence, *Science,* vol. 154, no. 3755, pp. 1405-1415, 1966. [One of the cornerstones of the "New Global Tectonics."]

Wilson, J. Tuzo: Transform faults, oceanic ridges, and magnetic anomalies southwest of Vancouver Island, *Science,* vol. 150, no. 3695, pp. 482-485, 1965. [Plate theory applied to the Northwest.]

CHAPTER 5

on the northwest or the north american continent

Eardley, A. J.: "Structural Geology of North America," 2d ed., Harper & Row, Publishers, Incorporated, New York, 1962. [A standard technical reference on this subject; includes much stratigraphic and structural information and many references.]

Ekman, Leonard C.: "Scenic Geology of the Pacific Northwest," Binfords & Mort, Portland, Ore., 1962. [A nontechnical treatment with emphasis on types of geologic features. Brief mention of numerous areas.]

Fenneman, Nevin: "Physiography of Western United States," McGraw-Hill Book Company, New York, 1931. [A classic. The physiography of the West has not changed much in forty years, although we know much more about the geology of the region.]

Freeman, Otis W., and Howard H. Martin: "The Pacific Northwest," 2d ed., John Wiley & Sons, Inc., New York, 1954. [A standard reference on the geography of the Northwest.]

Geological Survey of Canada: Geological map of Canada, *Geol. Survey Canada Map* 1250A, 1:5,000,000, 1969.

Griggs, G. B., and L. D. Kulm: Sedimentation in Cascadia deep-sea channel, *Geol. Soc. America Bull.,* vol. 81, pp. 1361-1384, 1970. [Discusses the configuration and sedimentation history of the Cascadia Basin off the coast of Washington and Oregon. Technical.]

King, P. B.: "The Geological Evolution of North America," Princeton University Press, Princeton, N.J., 1959. [A technical treatment of the subject.]

Matthews, William H.: "Guide to National Parks, Their Landscape and Geology, Vol. 1: Western Parks," Natural History Press, New York, 1968. [A well-written and authoritative book for the nongeologist. Separate chapters on Crater Lake, Mount Rainier, and Olympic National Parks.]

Morgan, Neil: "The Pacific States," Time-Life Books, A Division of Time Inc., New York, 1967. [A popular geography for Washington, Oregon, and California.]

Raisz, Erwin: Landforms of the northwestern states (maps), Inst. of Geographical Exploration, Harvard University, Cambridge, Mass., 1941. [The best physiographic maps available.]

Thornbury, William D.: "Regional Geomorphology of the United States," John Wiley & Sons, Inc., New York, 1965. [Intended primarily for professional geologists and students. Authoritative, with many references. Chapters on the Northern Rockies, Columbia Intermontane Province, Sierra-Cascade Province, and Pacific Border Province are included.]

U.S. Geological Survey: Geologic map of the United States, *U.S. Geol. Survey Map,* 1:2,5000,000, 1932.

———: Geologic map of North America, *U.S. Geol. Survey Map,* 1:5,000,000, 1965.

———: National atlas: geology, *U.S. Geol. Survey Sheet,* 1:7,5000,000, 1966.

———: National atlas: tectonic features, *U.S. Geol. Survey Sheet* 70, 1:7,500,000, 1967.

on british columbia

Geological Survey of Canada: Geological map of British Columbia, *Geol. Surv. Canada Map* 932A, 1:1,267,200, 1962. [Almost unique for its honesty, it leaves blank regions that have not been studied.]

Gunning, H. C., and W. H. White (eds.): Tectonic history and mineral deposits of the western Cordillera, *Canadian Inst. of Mining and Metallurgy Sp. Vol.* 8, 1966. [Technical articles on the geology and mineral deposits of British Columbia and adjacent parts of Washington, Idaho, Montana, Alaska, and the Yukon. Many excellent maps. The best geologic summary of the region available.]

Holland, S. S.: Landforms of British Columbia, a physiographic outline, *B.C. Dept. Mines Petrol. Resources. Bull.* 48, 1964. [An excellent treatment, well-illustrated with oblique air photographs. Includes some general comments on the bedrock geology.]

Roddick, J. A., J. O. Wheeler, H. Gabrielse, and J. G. Souther: Age and nature of the Canadian part of the circum-Pacific orogenic belt, *Tectonophysics,* vol. 4, pp. 319–337, 1967. [A brief summary of the geologic evolution of western Canada. Technical.]

Wheeler, J. O. (ed.): Structure of the southern Canadian Cordillera, *Geol. Assoc. Canada Sp. Paper 6,* February, 1970. [Emphasizes particular study areas. Up to date and authoritative. For geologists.]

White, W. H.: Cordilleran tectonics in British Columbia, *Am. Assoc. Petroleum Geologists Bull.,* vol. 43, pp. 60–100, 1959. [A technical summary.]

on washington

Campbell, C. D.: Introduction to Washington geology and resources, *Washington Div. Mines and Geology Inf. Circ.* 22R, 1962. [A good introduction, now superseded somewhat by Livingston (1969).]

Easterbrook, Don J., and David A. Rahm: "Landforms of Washington," Dept. of Geology, Western Washington State College, Bellingham, 1970. [An up-to-date treatment of the present landscape, with particular emphasis on the Pleistocene and Holocene history of the state. Numerous aerial photographs and references. For the nonspecialist.]

Huntting, M. T., W. A. G. Bennett, V. E. Livingston, Jr., and W. S. Moen: Geologic map of Washington, *Washington Div. Mines and Geology Map,* 1:500,000, 1961. [A "must" for serious study of the state's geology.]

Livingston, Vaughn E., Jr.: Fossils in Washington, *Washington Div. Mines and Geology Inf. Circ.* 33, 1959. [An excellent summary for the nonspecialist.]

———: Geologic history and rocks and minerals of Washington, *Washington Div. Mines and Geology Inf. Circ.* 45, 1969. [Current and readable, with good maps and photographs.]

U.S. Geological Survey: Mineral and water resources of Washington, *Washington Div. Mines and Geology Reprint* 9, 1966. [An excellent inventory, with many short articles (some quite detailed) on the state's resources, and on the general geology of each province. Many references.]

Weissenborn, A. E.: Geologic map of Washington, *U.S. Geol. Survey Misc. Geol. Inv. Map* I-583, 1:2,000,000, 1969. [Reproduced at front of book.]

on oregon

Baldwin, E. M.: "Geology of Oregon," 2d ed., J. W. Edwards, Publisher, Incorporated, Ann Arbor, Mich., 1964 (distributed by Univ. of Oregon Cooperative Book Store, Eugene, Ore.). [The standard reference on the subject. A province-by-province approach with particular emphasis on the stratigraphy. Very well illustrated. For geologists and laymen.]

Beaulieu, J. D.: Geologic formations of western Oregon, *Oregon Dept. Geology and Mineral Industries Bull.* 70, 1971. [A brief technical discussion of 72 formations in western Oregon. Includes correlation charts. (Published too late for the data to be incorporated in this book.)]

Peterson, N. V., and E. A. Groh (eds.): Lunar geological field conference guide book, *Oregon Dept. Geology and Mineral Industries Bull.* 57, 1965. [Excellent field trip guides to some volcanic areas of the Oregon Cascades and High Lava Plains.]

U.S. Geological Survey: Mineral and water resources of Oregon, *Oregon Dept. Geology and Mineral Industries Bull.* 64, 1969. [Detailed and authoritative. Includes general descriptions of the geology of each of the state's provinces. Many references.]

Walker, G. W., and P. B. King: Geologic map of Oregon, *U.S. Geol. Survey Misc. Geol. Inv. Map* I-595, 1:2,000,000, 1969. [Reproduced at front of book.]

Wells, F. G., and D. L. Peck: Geologic map of Oregon west of the 121st meridian, *U.S. Geol. Survey Misc. Geol. Inv. Map* I-325, 1:500,000, 1961. [Includes the Cascades, Willamette Valley, Coast Ranges, and Klamath Mountains.]

Wilkinson, W. D. (ed.): Field guidebook; geologic trips along Oregon highways, *Oregon Dept. Geology and Mineral Industries Bull.* 50, 1959. [A mile-by-mile description of the geology along some of the state's major highways. Unfortunately currently out of print.]

on idaho

Rhodenbaugh, Edward F.: "Sketches of Idaho Geology," 2d ed., The Caxton Printers, Ltd., Caldwell, Idaho, 1961. [A popular treatment.]

Ross, Clyde P.: Geology along U.S. Highway 93 in Idaho, *Idaho Bur. Mines and Geology Pamphlet* 130, 1963. [A mile-by-mile road log with strip maps for much of east-central and south-central Idaho. Includes sections on the general stratigraphy and geologic history of the region.]

——— and J. Donald Forrester: Geologic map of the state of Idaho, *Idaho Bur. Mines and Geology Map,* 1:500,000, 1947. [Somewhat dated, but the best available.]

——— and J. Donald Forrester: Outline of the geology of Idaho, *Idaho Bur. Mines and Geology Bull.* 15, 1958. [A good generalized summary, intended to accompany state map (Ross and Forrester, 1947). Includes sections on the state's mining history. Somewhat technical but includes a glossary.]

U.S. Geological Survey: Mineral and water resources of Idaho, *Idaho Bur. Mines and Geology Sp. Rept.* 1, 1964. [A detailed discussion of Idaho's resources, with a generalized treatment of the state's geology, and a black-and-white geologic map. Many references.]

primary sources of information and materials on the geology of the northwest

U.S. Geological Survey
Denver Federal Center
Denver, Colorado 80225

British Columbia Department of Mines and Petroleum Resources
Victoria, British Columbia, Canada

Washington Division of Mines and Geology
Department of Natural Resources
Olympia, Washington 98501

Geological Survey of Canada
Department of Energy, Mines and Resources
Ottawa, Ontario, Canada

Idaho Bureau of Mines and Geology
Moscow, Idaho 83843

Oregon Department of Geology and Mineral Industries
1069 State Office Building
Portland, Oregon 97201

CHAPTER 6

Crosby, Percy: Tectonic, plutonic, and metamorphic history of the central Kootenay Arc, British Columbia, Canada, *Geol. Soc. America Sp. Paper* 99, 1968. [A technical discussion of a part of this complex arc.]

Holland, S. S.: Landforms of British Columbia, a physiographic outline, *B.C. Dept. Mines Petrol. Resources Bull.* 48, 1964. [Includes an excellent discussion of the glaciation of British Columbia.]

Price, R. A., and E. W. Montjoy: Geologic structure of the Canadian Rocky Mountains between Bow and Athabasca Rivers—a progress report, in J. O. Wheeler (ed.), Structure of the southern Canadian Cordillera, *Geol. Assoc. Canada Sp. Paper* 6, pp. 7–26, 1970. [A good summary of the thrust faulting that characterizes the eastern flank of the Rockies in Canada and Montana. Technical.]

Ross, C. P.: The Belt Series in Montana, *U.S. Geol. Survey Prof. Paper* 346, 1963. [The most detailed discussion of Belt stratigraphy. Technical.]

Ross, John V.: Structural evolution of the Kootenay Arc, southeastern British Columbia, in J. O. Wheeler (ed.), Structure of the southern Canadian Cordillera, *Geol. Assoc. Canada Sp. Paper* 6, pp. 53–65, 1970. [Contains a brief summary of the regional stratigraphy. Mostly a technical discussion of the complex metamorphic history of the Kootenay Arc.]

Wheeler, J. O.: Eastern tectonic belt of western Cordillera in British Columbia, in H. C. Gunning and W. H. White (eds.), Tectonic history and mineral deposits of the western Cordillera, *Canadian Inst. Mining and Metallurgy, Sp. Vol.* 8, pp. 27–46, 1966. [A good summary of the geologic evolution of the eastern Cordillera in British Columbia, includes a generalized stratigraphic correlation chart. Technical.]

Yates, R. G.: Northeastern Washington, in U.S. Geological Survey, Mineral and water resources of Washington, *Washington Div. Mines and Geology Reprint* 9, pp. 15–22, 1966. [A concise summary of the geologic evolution of the region.]

———, G. E. Becraft, A. B. Campbell, and R. C. Pearson: Tectonic framework of northeastern Washington, northern Idaho, and northwestern Montana, in H. C. Gunning and W. H. White (eds.), Tectonic history and mineral deposits of the western Cordillera, *Canadian Inst. Mining and Metallurgy Sp. Vol.* 8, pp. 47–59, 1966. [Stresses the structure of the region. Excellent maps.]

CHAPTER 7

Coates, J. A.: Stratigraphy and structure of Manning Park area, Cascade Mountains, British Columbia, in J. O. Wheeler, (ed.), Structure of the southern Canadian Cordillera, *Geol. Assoc. Canada Sp. Paper* 6, pp. 149–154, 1970. [A concise summary of the Late Mesozoic stratigraphic and structural history of the south end of the Pasayten Graben, with maps and cross sections. Technical.]

Crowder, D. F., and R. W. Tabor: "Routes and Rocks: Hikers Guide to the North Cascades from Glacier Peak to Lake Chelan," *The Mountaineers,* Seattle, 1965. [A fine combination of geology and information for hikers. Nontechnical.]

Danner, W. R.: Limestone resources of western Washington, *Washington Div. Mines and Geology Bull.* 52, 1966. [An exhaustive, technical inventory of the many limestone bodies in the North Cascades and elsewhere in western Washington. Contains much valuable stratigraphic information, especially for the Chilliwack Group.]

Grant, A. R.: Chemical and physical controls for base metal deposition in the Cascade Range of Washington, *Washington Div. Mines and Geology Bull.* 58, 1969. [A formidable title for a volume that contains a fine summary of Cascade geology. Moderately technical.]

Huntting, M. T., W. A. G. Bennett, V. E. Livingston, Jr., and W. S. Moen: Geologic map of Washington, *Washington Div. Mines and Geology Map,* 1:500,000, 1961.

McTaggart, K. C.: Tectonic history of the northern Cascades Mountains, in J. O. Wheeler (ed.), Structure of the southern Canadian Cordillera, *Geol. Assoc. Canad Sp. Paper* 6, pp. 137–148, 1970. [A concise summary that ties in well with the work of Misch (1966) south of the border. Technical.]

McTaggart, K. C., and R. M. Thompson, Geology of part of the northern Cascades in southern British Columbia, *Can. Jour. Earth Sci.,* vol. 4, pp. 1191–1228, 1967.

Misch, Peter: Tectonic evolution of the northern Cascades of Washington state, H. C. Gunning and W. H. White (eds.), Tectonic history and mineral deposits of the western Cordillera, *Canadian Inst. Mining and Metallurgy Sp. Vol.* 8, pp. 101–148, 1966. [The standard reference on the geology of the North Cascades. Technical.]

Tabor, R. W., and D. F. Crowder: "Hikers Map of the North Cascades; Routes and Rocks in the Mount Challenger Quadrangle," *The Mountaineers,* Seattle, 1968. [Excellent. Nontechnical.]

University of Washington Geology Department Staff: A geologic trip along Snoqualmie, Swauk, and Stevens Pass highways, *Washington Div. Mines and Geology Inf. Circ.* 38, 1963. [An outcrop-by-outcrop description and road log. Nontechnical.]

Weissenborn, A. E., and F. W. Cater: The Cascade Mountains, in U.S. Geological Survey, Mineral and water resources of Washington, *Washington Div. Mines and Geology Reprint* 9, pp. 27–37, 1966. [A summary of the entire range, with many references. Necessarily too concise. Moderately technical.]

CHAPTER 8

Campbell, R. B.: Tectonics of the south central Cordillera of British Columbia, in H. C. Gunning and W. H. White (eds.), Tectonic history and mineral deposits of the western Cordillera, *Canadian Inst. Mining and Metallurgy Sp. Vol.* 8, pp. 61–72, 1966. [A good summary of the overall stratigraphy of the southern part of the Interior Plateau Province.]

Roddick, J. A.: Coast Crystalline Belt of British Columbia, in H. C. Gunning and W. H. White (eds.), Tectonic history and mineral deposits of the western Cordillera, *Canadian Inst. Mining and Metallurgy Sp. Vol.* 8, pp. 73–82, 1966. [Gives a good feel for the variety and complexity of the Coast Mountains Province. Excellent maps.]

Souther, J. G.: Volcanism and its relationship to recent crustal movements in the Canadian Cordillera, *Can. Jour. Earth Sci.,* Vol. 7, pp. 553–568, 1970. [A thoughtful discussion of the possible relationship between plate tectonics and volcanism in British Columbia.]

——— and J. E. Armstrong: North Central Belt of the Cordillera of British Columbia, in H. C. Gunning and W. H. White (eds.), Tectonic history and mineral deposits of the western Cordillera, *Canadian Inst. Mining and Metallurgy Sp. Vol.* 8, pp. 171–184, 1966. [A companion paper to Campbell (1966) for the interior of British Columbia north of latitude 54°.]

Sutherland-Brown, A.: Tectonic history of the Insular Belt of British Columbia, in H. C. Gunning and W. H. White (eds.), Tectonic history and mineral deposits of the western Cordillera, *Canadian Inst. Mining and Metallurgy Sp. Vol.* 8, pp. 83–100, 1966. [A precise summary of the geologic history of Vancouver Island and the Queen Charlotte Islands. Mostly maps. Technical.]

———: Geology of the Queen Charlotte Islands, British Columbia, *B.C. Dept. Mines Petrol. Resources Bull.* 54, 1968. [The standard reference on this region.]

CHAPTER 9

Danner, W. R.: Limestone resources of western Washington, *Washington Div. Mines and Geology Bull.* 52, 1966. [Contains the only readily available modern description of the stratigraphic units of the San Juan Islands, though emphasis is on the occurrence of limestone. Technical.]

Huntting, M. T., W. A. G. Bennett, V. E. Livingstone, Jr., and W. S. Moen: Geologic map of Washington, *Washington Div. Mines and Geology Map,* 1:500,000, 1961. [The scale is large, but this is the best map of the San Juan Islands available.]

McLellan, R. D.: The geology of the San Juan Islands, *Univ. of Washington Publ. in Geol.,* vol. 2, 1927. [Now out of print and somewhat out of date, it remains the most complete treatment of the subject. Moderately technical.]

CHAPTER 10

Baldwin, Ewart M.: "Geology of Oregon," 2d ed., J. W. Edwards, Publisher, Incorporated, Ann Arbor, Mich., pp. 77–86 (Klamath Mountains), 1964. [Includes a geologic map of the southeastern part of the Klamath Mountains Province of Oregon.]

Dott, R. H., Jr.: Mesozoic-Cenozoic tectonic history of the southwestern Oregon coast in relation to Cordilleran orogenesis, *Jour. Geophys. Research,* vol. 70, no. 18, pp. 4687–4707, 1965. [Technical summary of some of the major stratigraphic puzzles in the Klamaths.]

Raisz, Erwin: Landforms of the northwestern states (maps), Inst. of Geographical Exploration, Harvard University, Cambridge, Mass., 1941. [The best physiographic maps available.]

Ramp, Len: Geology of the Klamath Mountains Province in U.S. Geological Survey, Mineral and water resources of Oregon, *Oregon Dept. Geology and Mineral Industries Bull.* 64, pp. 47–52, 1969. [A brief, up-to-date résumé, with references.]

Wells, F. G., and D. L. Peck: Geologic map of Oregon west of the 121st meridian, *U.S. Geol. Survey Misc. Geol. Inv. Map* I-325, 1:500,000, 1961. [Must be consulted for any comprehensive discussion of the Klamath Mountains.]

CHAPTER 11

Baldwin, Ewart M.: "Geology of Oregon," 2d ed., J. W. Edwards, Publisher, Incorporated, Ann Arbor, Mich., pp. 1–42 (Coast Range), 1964. [A comprehensive summary of Oregon Coast Range stratigraphy; includes a detailed correlation chart of western Oregon.]

Brown, R. D., Jr., H. D. Gower, and P. D. Snavely, Jr.: Geology of the Port Angeles–Lake Crescent Area, Clallam County, Washington, *U.S. Geol. Survey Oil and Gas Inv. Map* OM-203, 1960.

Cooper, W. S.: Coastal sand dunes of Oregon and Washington, *Geol. Soc. America Memoir* 72, 1958. [A detailed discussion; for geologists but quite readable.]

Crandell, D. R.: Glacial history of western Washington and Oregon, in H. E. Wright, Jr., and D. G. Frey (eds.), "Quaternary of the United States," Princeton University Press, Princeton, N.J., pp. 341–354, 1965. [Brief discussion of glaciation of Olympic Mountains is included.]

Danner, W. R.: "Geology of Olympic National Park," University of Washington Press, Seattle, 1955. [For the general reader.]

Easterbrook, Don J., and David A. Rahm: "Landforms of Washington," Dept. of Geology, Western Washington State College, Bellingham, 1970.

Huntting, M. T., W. A. G. Bennett, V. E. Livingston, Jr., and W. S. Moen: Geologic map of Washington, *Washington Div. Mines and Geology Map,* 1:500,000, 1961. [A "must" for serious study of the state's geology.]

Raisz, Erwin: Landforms of the northwestern states (maps), Inst. of Geographical Exploration, Harvard University, Cambridge, Mass., 1941. [The best physiographic maps available.]

Snavely, P. D., Jr., and H. C. Wagner: Tertiary geologic history of western Oregon and Washington, *Washington Div. Mines and Geology Rept. Inv.* 22, 1963. [An excellent summary, with maps that emphasize the changing environment through time.]

——— and H. C. Wagner: Geologic sketch of northwestern Oregon, *U.S. Geol. Survey Bull.* 1181-M, 1964.

———, H. C. Wagner, and N. S. MacLeod: Geology of western Oregon north of the Klamath Mountains in U.S. Geological Survey, Mineral and water resources of Oregon, *Oregon Dept. Geology and Mineral Industries Bull.* 64, pp. 32–46, 1969. [Similar to Snavely and Wagner (1963), although somewhat updated.]

Tabor, Rowland W.: "Geologic guide to the Deer Park area, Olympic National Park," Olympic Natural History Association, Inc., Port Angeles, Wash., 1965. [For the general reader.]

———: "Geologic guide to the Hurricane Ridge area," Olympic Natural History Association, Inc., Port Angeles, Wash., 1969. [Excellent treatment of this much-visited area. For nongeologists.]

Wagner, H. C., and P. D. Snavely Jr.: Western Washington in U.S. Geological Survey, Mineral and water resources of Washington, *Washington Div. Mines and Geology Reprint* 9, pp. 37–46, 1966. [Follows Snavely and Wagner (1963).]

Wells, F. G., and D. L. Peck: Geologic map of Oregon west of the 121st meridian, *U.S. Geol. Survey Misc. Geol. Inv. Map* I-325, 1:500,000, 1961. [Includes the Cascades, Willamette Valley, Coast Ranges, and Klamath Mountains.]

CHAPTER 12

Fiske, R. S., C. A. Hopson, and A. C. Waters: Geology of Mount Rainier National Park, Washington, *U.S. Geol. Survey Prof. Paper* 444, 1963. [A thorough summary of the Park, with considerable emphasis on the pre-Quaternary history of the Cascade Range. Technical.]

Grant, A. R.: Chemical and physical controls for base metal deposition in the Cascade Range of Washington, *Washington Div. Mines and Geology Bull.* 58, 1969. [A formidable title for a volume that contains a fine summary of Cascade geology. Moderately technical.]

Griggs, A. B.: Geology of the Cascade Range in U.S. Geological Survey, Mineral and water resources of Oregon, *Oregon Dept. Geology and Mineral Industries Bull.* 64, pp. 53–59, 1969. [A good, concise summary.]

Hammond, Paul, "Structure and Stratigraphy of the Keechelus Volcanic Group and Associated Tertiary Rocks in the West-central Cascade Range, Washington," unpublished doctoral dissertation, University of Washington, 1961.

Mackin, J. H., and A. S. Cary: Origin of Cascade landscapes, *Washington Div. Mines and Geology Inf. Circ.* 41, 1965. [A popular treatment, well written and well illustrated, although somewhat dated.]

Peck, D. L., A. B. Griggs, H. G. Schlicker, F. G. Wells, and H. M. Dole: Geology of the central and northern parts of the western Cascade Range in Oregon, *U.S. Geol. Survey Prof. Paper* 449, 1964. [A good compilation of the stratigraphy of this complex region. Technical.]

Raisz, Erwin: Landforms of the northwestern states (maps), Inst. of Geographical Exploration, Harvard University, Cambridge, Mass., 1941. [The best physiographic maps available.]

University of Washington Geology Department Staff: A geologic trip along Snoqualmie, Swauk, and Stevens Pass highways, *Washington Div. Mines and Geology Inf. Circ.* 38, 1963. [An outcrop-by-outcrop description and road log. Nontechnical.]

Waters, A. C.: The Keechelus problem, Cascade Mountains, Washington, *Northwest Sci.*, vol. 35, no. 2, pp. 39–57, 1961. [A detailed analysis of the state of confusion in our knowledge of the stratigraphy of the southern and central Cascades in Washington. A good summary of earlier work.]

Weissenborn, A. E., and F. W. Cater: The Cascade Mountains, in U.S. Geological Survey, Mineral and water resources of Washington, *Washington Div. Mines and Geology Reprint* 9, pp. 27–37, 1966. [A summary of the entire range, with many references. Necessarily too concise. Moderately technical.]

CHAPTER 13

Coombs, Howard A.: Mount Baker, a Cascade volcano, *Geol. Soc. America Bull.*, Vol. 50, pp. 1493–1510, 1939. [A brief, moderately technical account—the standard reference for Mount Baker.]

Crandell, D. R.: Surficial geology of Mount Rainier National Park, Washington, *U.S. Geol. Survey Bull.* 1288, 1969. [This and the following two bulletins are models of scientific writing for the public. They are well written, well illustrated, and authoritative.]

———: The geologic story of Mount Rainier, *U.S. Geol. Survey Bull.* 1292, 1969.

——— and D. R. Mullineaux: Volcanic hazards at Mount Rainier, Washington, *U.S. Geol. Survey Bull.* 1238, 1967.

Dole, H. M. (ed.): Andesite conference guidebook, *Oregon Dept. Geology and Mineral Industries Bull.* 62, 1968. [An excellent volume with articles on the McKenzie Pass area, Crater Lake, Newberry Crater, Mount Hood, and the composition and origin of the Cascade lavas. Fine photographs and maps, some in color.]

Fiske, R. S., C. A. Hopson, and A. C. Waters: Geology of Mount Rainier National Park, Washington, *U.S. Geol. Survey Prof. Paper* 444, 1963. [Traces the evolution of the volcano as well as that of the underlying range. Technical.]

McBirney, A. R.: Petrochemistry of the Cascade andesite volcanoes, in H. M. Dole (ed.), Andesite conference guidebook, *Oregon Dept. Geology and Mineral Industries Bull.* 62, pp. 101–107, 1968.

Peterson, N. V., and E. A. Groh (eds.): Lunar geological field conference guide book, *Oregon Dept. Geology and Mineral Industries Bull.* 57, 1965. [Contains discussions and road logs for various volcanic regions in the Oregon Cascades and High Lava Plains. Excellent photos and maps. A "must" for touring these areas.]

Tabor, R. W., and D. F. Crowder: On batholiths and volcanoes—intrusion and eruption of Late Cenozoic Magmas in the Glacier Peak area, North Cascades, Washington, *U.S. Geol. Survey Prof. Paper* 604, 1969. [Excellent technical discussion of the history of Glacier Peak volcano.]

Taylor, Edward M.: Roadside geology, Santiam and McKenzie Pass highways, Oregon, in H. M. Dole (ed.), Andesite conference guidebook, *Oregon Dept. Geology and Mineral Industries Bull.* 62, pp. 3–33, 1968.

Verhoogen, John: Mount Saint Helens, a Recent Cascade volcano, *Univ. Calif. Publ. in Geol. Sci. Bull.* vol. 24, pp. 263–302, 1937. [Somewhat dated, but the only reference available on the bedrock of this volcano.]

Walker, G. W., R. C. Greene, and E. C. Pattee: Mineral resources of the Mount Jefferson Primitive Area, Oregon, *U.S. Geol. Survey Bull.* 1230-D, pp. D1–D32, 1966. [Contains information on Mount Jefferson volcano as well as on the underlying range.]

Williams, Howel: The geology of Crater Lake National Park, Oregon, with a reconnaissance of the Cascade Range southward to Mount Shasta, *Carnegie Inst. Washington Pub.* 540, 1942. [The standard reference on Mount Mazama and Crater Lake. A fine example of volcanological analysis and reasoning.]

———: Volcanoes of the Three Sisters region, Oregon Cascades, *Univ. Calif. Dept. Geol. Sci. Bull.*, vol. 27, pp. 37–84, 1944.

——— and Gordon Goles: Volume of the Mazama Ash-fall and the origin of Crater Lake Caldera, in H. M. Dole (ed.), Andesite conference guidebook, *Oregon Dept. Geology and Mineral Industries Bull.* 62, pp. 37–41, 1968. [Modifies somewhat calculations made by Williams (1942) on the volume of Mount Mazama not accounted for.]

Wise, William S.: Geology of the Mount Hood volcano, in H. M. Dole (ed.), Andesite conference guidebook, *Oregon Dept. Geology and Mineral Industries Bull.* 62, pp. 81–98, 1968. [The best general summary available for this interesting cone.]

CHAPTER 14

Baldwin, E. M.: "Geology of Oregon," 2d ed., J. W. Edwards, Publisher, Incorporated, Ann Arbor, Mich., pp. 94–116 (Blue Mountains), 1964. [Treats western and eastern parts of the range separately.]

Brown, C. E., and T. P. Thayer: Geologic map of the Canyon City quadrangle, northeastern Oregon, *U.S. Geol. Survey Misc. Geol. Inv. Map* I-447, 1:250,000, 1966. [Covers the central part of the Blue Mountains Province. Includes short text and references.]

Crandell, D. R.: Glacial history of western Washington and Oregon, in H. E. Wright, Jr., and D. G. Frey (eds.), "Quaternary of the United States," Princeton University Press, Princeton, N.J., pp. 341–354, 1965. [Includes a discussion of glaciation in the Wallowas.]

Dickinson, W. R., and L. W. Vigrass: Geology of the Suplee-Izee area, Crook, Grant and Harney Counties, Oregon, *Oregon Dept. Geology and Mineral Industries Bull.* 58, 1965. [A detailed description of this region of primarily Mesozoic strata. Moderately technical but contains a glossary.]

Raisz, Erwin: Landforms of the northwestern states (maps), Inst. of Geographical Exploration, Harvard University, Cambridge, Mass., 1941. [The best physiographic maps available.]

______ "The Geologic setting of the John Day Country, Grant County, Oregon," *U.S. Geological Survey*, Washington, D.C., 1969. [A very well-illustrated booklet for the general reader.]

Swanson, Donald A.: Reconnaissance geologic map of the east half of the Bend quadrangle, Crook, Wheeler, Jefferson, Wasco, and Deschutes Counties, Oregon, *U.S. Geol. Survey Misc. Geol. Inv. Map* 1-568, 1:250,000, 1969. [An up-to-date map of the western Blue Mountains, with detailed stratigraphic descriptions and many references.]

Thayer, T. P., and N. S. Wagner: Geology of the Blue Mountain region, in U.S. Geological Survey, Mineral and Water Resources of Oregon, *Oregon Dept. Geology and Mineral Industries Bull.* 64, pp. 68–74, 1969.

Wilkinson, W. D. (ed.): Field guidebook: geologic trips along Oregon highways, *Oregon Dept. Geology and Mineral Industries Bull.* 50, 1959. [Includes a road log from Prineville to Picture Gorge and north through Fossil to the Columbia River.]

Raisz, Erwin: Landforms of the northwestern states (maps), Inst. of Geographical Exploration, Harvard University, Cambridge, Mass., 1941. [The best physiographic maps available.]

Staples, L. W.: Origin and history of the thunder egg, *Ore Bin*, vol. 27, no. 10, pp. 195–204, 1965. [A popular account, well done. The *Ore Bin* is a magazine published monthly by the Oregon Department of Geology and Mineral Industries. It contains short articles of interest to the public and the profession and is highly recommended to anyone who wishes to keep abreast of Oregon geology.]

Walker, G. W., and N. V. Peterson: Geology of the Basin and Range Province, *in* U.S. Geological Survey, Mineral and water resources of Oregon, *Oregon Dept. Geology and Mineral Industries Bull.* 64, pp. 83–88, 1969. [A concise summary by two authorities on the area.]

CHAPTER 15

Allison, Ira S.: Fossil Lake, Oregon: its geology and fossil faunas, *Oregon State Univ. Press Studies in Geology* 9, 1966. [Describes the history of one of the Pleistocene lakes southeast of Newberry Volcano.]

Baldwin, Ewart M.: High Lava Plains (and) Basin and Range and Owyhee Upland areas, in E. M. Baldwin, "Geology of Oregon," 2d ed., J. W. Edwards, Publisher, Incorporated, Ann Arbor, Mich., pp. 117–139, 1964. [Good stratigraphic resume.]

Higgins, M. W., and A. C. Waters: Newberry Caldera field trip, in H. M. Dole (ed.), Andesite conference guidebook, *Oregon Dept. Geology and Mineral Industries Bull.* 62, pp. 59–77, 1968. [Beautifully illustrated, in color.]

Peterson, N. V., and E. A. Groh (eds.): Lunar geological field conference guidebook, *Oregon Dept. Geology and Mineral Industries Bull.* 57, 51 pp., 1965. [Contains a short summary and road log for a trip to Newberry Crater (pp. 10–18) and the Hole-in-the-Ground-Fort Rock-Devils Garden area (pp. 19–28). Mostly maps and oblique aerial photographs.]

CHAPTER 16

Armstrong, R. C., and S. S. Oriel: Tectonic development of Idaho-Wyoming Thrust Belt, *Am. Assoc. Petroleum Geologists Bull.*, vol. 49, pp. 1847–1866, 1965. [A technical account of Late Mesozoic-Early Cenozoic thrust faulting in southeastern Idaho and western Wyoming.]

Idaho Bureau of Mines and Geology: Idaho's mineral industry—the first hundred years: *Idaho Bur. Mines and Geology Bull.* 18, 1961. [An excellent summary of the history of mining in the state.]

Malde, H. E., and H. A. Powers: Upper Cenozoic stratigraphy of western Snake River Plain, Idaho, *Geol. Soc. America Bull.*, vol. 73, no. 10, pp. 1197–1219, 1962. [A detailed, member-by-member description of the units. Technical.]

Raisz, Erwin: Landforms of the northwestern states (maps), Inst. of Geographical Exploration, Harvard University, Cambridge, Mass., 1941. [The best physiographic maps available.]

Ross, C. P., R. R. Reid, and A. E. Weissenborn: Geology, in U.S. Geological Survey, Mineral and water resources of Idaho, *Idaho Bur. Mines and Geology Sp. Rept.* 1, pp. 23–40, 1964. [A good overview of the state's geologic history. Many references.]

Stearns, Harold T.: "Geology of the Craters of the Moon National Monument, Idaho," Craters of the Moon Natural History Association, Arco, Ida., 1963. [Well written and illustrated. For general readers.]

Travis, W. I., H. A. Waite, and J. F. Santos: Water resources, in U.S. Geological Survey, Mineral and water resources of Idaho, *Idaho Bur. Mines and Geology Sp. Rept.* 1, pp. 255–308, 1964.

Yates, R. G., G. E. Becraft, A. B. Campbell, and R. C. Pearson: Tectonic framework of northeastern Washington, northern Idaho, and northwestern Montana, in H. C. Gunning and W. H. White (eds.), Tectonic history and mineral deposits of the western Cordillera, *Canadian Inst. Mining and Metallurgy Sp. Vol. 8,* pp. 47–59, 1966. [Stresses the structure of the region. Excellent maps.]

CHAPTER 17

Baldwin, Ewart M.: "Geology of Oregon," 2d. ed., J. W. Edwards, Publisher, Incorporated, Ann Arbor, Mich., pp. 87–93 (Deschutes-Umatilla Plateau), 1964. [A stratigraphic summary of the Oregon part of the Columbia Plateau Province.]

Bretz, J. H.: Washington's Channeled Scabland, *Washington Div. Mines and Geology Bull.* 45, 1959. [The culmination of nearly a half century of research in the area by the author. Detailed but quite readable. Stresses the flood hypothesis of origin.]

———, H. T. U. Smith, and G. E. Neff: Channeled Scabland of Washington: new data and interpretations, *Geol. Soc. America Bull.,* vol. 67, pp. 957–1040, 1956. [Somewhat more technical than Bretz (1959).]

Easterbrook, D. J., and D. A. Rahm: "Landforms of Washington," Dept. of Geology, Western Washington State College, Bellingham, pp. 102–150 (Columbia Basin), 1970. [A good description, well illustrated with photographs.]

Gilmour, Ernest H., and Dale Stradling (eds.): "Proceedings of the Second Columbia River Basalt Symposium," Eastern Washington State College Press, Cheney, 333 pp., 1970. [Technical papers from a comprehensive symposium held in March, 1969.]

Griggs, A. B.: Columbia Basin, in U.S. Geological Survey, Mineral and Water Resources of Washington, *Washington Div. Mines and Geology Reprint* 9, pp. 22–27, 1966. [A very brief summary.]

Jones, Fred O.: "Grand Coulee from Hell to Breakfast" Binfords & Mort, Portland, Ore., 1947. [A very readable nontechnical account of the geologic evolution of the Grand Coulee area, plus information on the area's history and the building of Grand Coulee Dam.]

Mackin, J. H.: A stratigraphic section in the Yakima Basalt and the Ellensburg Formation in south-central Washington, *Washington Div. Mines and Geology Rept. Inv.* 19, 1961. [The best description of individual flows of the Columbia River Basalt.]

Newcomb, R. C.: Geology of the Deschutes-Umatilla Plateau, in U.S. Geological Survey, Mineral and water resources of Oregon, *Oregon Dept. Geology and Mineral Industries Bull.* 64, pp. 60–66, 1969. [Includes a generalized map of the structure of the region.]

Pardee, Joseph T.: Unusual currents in Glacial Lake Missoula, Montana, *Geol. Soc. America Bull.,* vol. 53, pp. 1569–1600, 1942. [Historically important as it confirmed Glacial Lake Missoula as the source of the water for Bretz's Floods.]

Raisz, Erwin: Landforms of the northwestern states (maps), Inst. of Geographical Exploration, Harvard University, Cambridge, Mass., 1941. [The best physiographic maps available.]

Richmond, G. M., R. Fryxell, G. E. Neff, and P. W. Weis: The Cordilleran ice sheet of the Northern Rocky Mountains, and related Quaternary history of the Columbia Plateau, in H. E. Wright, Jr., and D. G. Frey (eds.), "Quaternary of the United States," Princeton University Press, Princeton, N.J., pp. 231–242, 1965. [An excellent technical summary.]

Shelton, John S.: "Geology Illustrated," W. H. Freeman and Company, San Francisco, 1966. [Contains an excellent section on the geologic evolution of the Columbia Plateau.]

Thornbury, William D.: "Principles of Geomorphology," 2d ed., John Wiley & Sons, Inc., New York, 1969.

Waters, A. C.: Stratigraphic and lithologic variations in the Columbia River Basalt, *Am. Jour. Sci.*, vol. 259, no. 8, pp. 583–611, 1961. [Contains a proposed subdivision of the Columbia River Basalt based in part on chemical and mineralogical criteria.]

CHAPTER 18

Armstrong, J. E., D. R. Crandell, D. J. Easterbrook, and J. B. Noble: Late Pleistocene stratigraphy and chronology in southwestern British Columbia and northwestern Washington, *Geol. Soc. America Bull.*, vol. 76, no. 3, pp. 321–330, 1965. [Correlates glacial sequences between northern Washington and the Fraser River lowland area. Technical.]

Baldwin, Ewart M.: "Geology of Oregon," 2d ed., J. W. Edwards, Publisher, Incorporated, Ann Arbor, Mich., pp. 43–58 (Willamette Valley), 1964. [Emphasizes the Cenozoic stratigraphy. Includes a geologic map of the Portland area.]

Crandell, D. R.: Glacial history of western Washington and Oregon, in H. E. Wright, Jr., and D. G. Frey (eds.), "Quaternary of the United States," Princeton University Press, Princeton, N.J., pp. 341–354, 1965. [An excellent summary, with references, maps.]

Easterbrook, Don J.: Pleistocene chronology of the Puget Lowland and San Juan Islands, Washington, *Geol. Soc. America Bull.*, vol. 80, pp. 2273–2286, 1969.

——— and D. A. Rahm: "Landforms of Washington," Dept. of Geology, Western Washington State College, Bellingham, pp. 42–72 (Puget Lowland), 1970. [Excellent general treatment of the glaciation of the Puget Lowland; good maps and photographs. References.]

Raisz, Erwin: Landforms of the northwestern states (maps), Inst. of Geographical Exploration, Harvard University, Cambridge, Mass., 1941. [The best physiographic maps available.]

Schultz, C. B., and H. T. U. Smith (eds.): Pacific Northwest, *INQUA Guidebook for Field Conference* J, 1965. [A good sourcebook, with information on the glaciation of the Puget Lowland and Washington Cascades not available elsewhere. Unfortunately this volume is not easily obtained.]

Trimble, D. E.: Geology of Portland, Oregon, and adjacent areas: *U.S. Geol. Survey Bull.* 1119, 1963. [Detailed and moderately technical.]

University of Washington Geology Department Staff: A geologic trip along Snoqualmie, Swauk, and Stevens Pass highways, *Washington Div. Mines and Geology Inf. Circ.* 38, 1963. [Discusses glacial features between Seattle and the Cascade front east of North Bend.]

glossary

abyssal plain flat, nearly level area that occupies the deepest portion of an ocean basin.

agate a variety of silica consisting mainly of the mineral chalcedony in varigated bands or other patterns.

alluvium sediment deposited by running water.

anatexis the natural melting of rock, producing a magma.

angle of repose the maximum angle at which loose rock remains at rest.

angular unconformity an unconformity in which the older strata dip at a different angle (generally steeper) than the younger strata.

antecedent river a river that has maintained its course across a rising mountain mass.

anticline a fold that is convex upward.

aquifer stratum or zone below the earth's surface capable of producing water, as from a well.

arc islands or mountains arranged in a great curve.

arkose a sandstone rich in feldspar minerals. Adjective: *arkosic*.

ash, volcanic uncemented volcanic ejecta consisting of fragments mostly under 4 millimeters in diameter.

ash flow an avalanche of hot volcanic ash and gases.

asthenosphere a soft, plastic layer near the top of the earth's mantle.

badland topography an intricate maze of narrow ravines and sharp crests and pinnacles.

bar an embankment built by waves and currents in sand, gravel, or alluvium.

basement a complex of igneous and metamorphic rocks at the base of a stratigraphic section.

basin and range structure regional structure dominated by fault-block mountains separated by sediment-filled basins.

batholith a body of intrusive igneous rock with a surface area greater than 40 square miles.

bed a distinct stratigraphic layer, bounded by bedding planes.

bed load soil, rock particles, and debris rolled along the bottom of a stream.

bedrock any solid rock exposed at the surface of the earth or overlain by unconsolidated material.

bombs, volcanic pyroclastic ejecta consisting of fragments of lava that were liquid or plastic at the time of ejection and having a rounded or streamlined form.

breccia a rock made of highly angular, coarse fragments.

caldera a large basin-shaped volcanic depression, the diameter of which is many times greater than that of the included vent or vents.

Cascadian orogeny Late Cenozoic mountain building.

cast a mold that has been filled naturally with some mineral substance.

catastrophe a sudden violent change in the physical conditions of a part of the earth's surface.

cinder cone a conical elevation formed by the accumulation of volcanic ash or cinders around a vent.

cirque a deep, steep-walled recess in a mountain, caused by glacial erosion.

clay 1) very fine-grained sediment, smaller than silt. 2) a mineral group.

clean sediment well-sorted sediment.

cleaver a rock spur separating two valley glaciers.

columnar jointing cracks that break rock into polygonal columns. Usually forms in volcanic rocks due to cooling and shrinkage.

composite cone a volcanic cone built of alternating layers of lava and pyroclastic material. Synonym: *stratovolcano.*

contact the surface where two different kinds of rocks come together.

continental drift the concept that the continents or large pieces of continents can drift on the surface of the earth.

continental margin zone separating continents from the deep sea bottom; includes the continental shelf, slope, and rise.

continental plate thick crust underlying a continent.

continental shelf gently sloping portion of the continental margin extending from the shore to the top of the continental slope.

continental slope portion of the continental margin between the outer edge of the continental shelf and the abyssal plain or, if present, an oceanic trench.

convection current movement of material due to differences in density, generally brought about by heating.

Cordillera the mountainous western part of North America, between the Central Plains and the Pacific Ocean.

correlation the determination of the equivalence in geologic age of stratigraphic units in separated areas.

country rock rock surrounding and penetrated by igneous intrusions.

crater, volcanic a steep-walled depression directly above a volcanic vent.

creep an imperceptibly slow downslope movement of soil or loose rock.

cross-bedding inclined bedding between main bedding planes in sediments or sedimentary rock.

crust outer layer of the earth, above the mantle.

crystalline rock a general term for intrusive igneous and metamorphic rocks.

daughter element an element formed from another by radioactive decay.

debris avalanche very rapid downslope movement of loose rock material.

decomposition the breakdown of minerals in rocks through chemical processes during weathering.

delta an alluvial deposit at the mouth of a river.

deposition the laying down of sediments. Synonym: *sedimentation.*

differential erosion more rapid erosion of one portion of the earth's surface, compared to another.

dike a tabular body of igneous rock that cuts across the structure of adjacent rocks or cuts massive rocks.

dip the angle at which any planar feature is inclined from the horizontal.

dirty sediment a poorly sorted, dark sediment.

disconformity unconformity between parallel strata.

disintegration the natural mechanical breaking down of rock during weathering.

dome cone a roughly dome-shaped volcano with steep sides, formed by eruption of very viscous lava.

drowned coast results of subsidence of a coast that transforms lower portions of old river valleys into tidal estuaries.

epoch geologic time unit that is a subdivision of a period.

equilibrium a condition of balance in a system.

era a geologic time unit of the highest order, comprising more than one period.

erosion processes whereby rock material is loosened or dissolved and removed from any part of the earth's surface.

erratic a transported rock fragment different from the bedrock on which it lies; applied generally to those rocks transported by glacial or floating ice.

estuary drainage channel adjacent to the sea in which the tide ebbs and flows.

eugeosyncline that part of the geosyncline in which volcanic rocks are abundant.

fault a fracture or fracture zone along which there has been displacement of the sides relative to one another parallel to the fracture.

fault zone a fault that, instead of being a single clean fracture, may be a zone of fracture thousands of feet wide.

fauna the animals of any place or time that lived in association with each other.

fiord, fjord a long deep arm of the sea, occupying a channel having high steep walls.

fissure an extensive fracture in rocks.

flora the plants of any place or time that lived in association with each other.

fold a bend in strata or any planar structure.

formation a mappable rock unit.

fossils the remains or traces of animals or plants that have been preserved by natural causes in the earth's crust.

fractionation separation of a substance from a mixture.

geochemistry the study of the distribution and changes in elements within the earth's crust.

geochronology the study of time in relationship to the earth's history or a system of dating developed for that purpose.

geode a hollow, globular body filled with an interior lining of inward projecting crystals.

geomagnetic pertaining to the magnetic field of the earth.

geomorphology the study of the form of the earth's surface, including the evolution of landforms.

geophysics the study of the structure, composition, and development of the earth.

geosyncline large, generally linear trough that subsided throughout a long period of time in which a thick succession of stratified sediments and possibly extrusive volcanic rocks accumulated. The trough is eventually deformed and uplifted to form a mountain range.

glacial groove a large furrow cut by the abrading action of rock fragments contained in a glacier.

glacial lobe a large, tonguelike projection from the main mass of a continental glacier.

glacial maximum the position or time of greatest advance of a glacier.

glacial recession reduction in area and thickness of a glacier.

glacial retreat occurs when the front of a glacier recedes, although ice may be still flowing toward the front of the glacier.

glacial striation fine-cut lines on the surface of bedrock inscribed by fragments carried in an overriding glacier.

glaciation alteration of the earth's surface through erosion and deposition by glacial ice or by processes beyond the margin of the ice.

glacier a flowing mass of ice.

glowing avalanche a volcanic eruption in which a hot cloud of gases and pyroclastic material is propelled by gravity downslope.

graben a block that has been downthrown along faults relative to the rocks on either side.

granitic a general term for coarse-grained intrusive igneous rocks with moderate to high silica content.

granitization the formation of a rock with granitic texture and composition by metamorphism rather than by igneous crystallization.

greenstone a dense, green rock formed by the alteration of silica-poor igneous rocks, typically basalt or gabbro.

groundwater subsurface water below the water table.

hydraulic damming temporary ponding of water produced by a channel constriction.

igneous formed by solidification from a molten state.

inclusion a fragment of older rock encased in an igneous rock.

index fossil fossil characteristic of a specific restricted time period.

intermittent stream a stream that flows only part of the time.

intrusion a body of igneous rock that invades older rock and crystallizes underground.

island arc curved chain of islands generally convex toward the open ocean, margined by a deep submarine trench.

isotope elements having an identical number of protons in their nuclei, but differing in the number of neutrons.

joint fracture in rock along which no appreciable movement has occurred.

kame terrace a body of stratified drift deposited between a glacier and an adjacent valley wall.

kettle a depression in drift, made by the wasting away of a detached mass of glacial ice that has been buried in drift.

lava fluid rock that issues from a volcano or a fissure in the earth's surface.

lava tube a tubular opening caused by the flow out of liquid material from a crusted-over section of lava.

law of superposition the concept that underlying strata must be older than overlying strata as long as neither inversion nor overthrusting has occurred.

levee a bank confining a stream channel or limiting areas subject to flooding.

lode a tabular deposit of valuable minerals.

magma a naturally occurring rock melt.

mantle the layer of the earth's interior between the crust and the core.

mass-wasting the downslope movement of loose rock material under the influence of gravity.

metamorphism a process by which solid rocks are altered in composition, texture, or internal structure by pressure, heat, and the introduction of new chemical substances. Metamorphism does not involve significant melting of the rock.

mid-ocean ridge great median arch of the sea bottom extending the length of an ocean basin and roughly paralleling the continental margins.

migmatite rock consisting of a mixture of granitic rock and schist or gneiss.

mineral a homogeneous naturally occurring, inorganic crystalline substance.

miogeosyncline that part of a geosyncline characterized by nonvolcanic rocks.

moraine glacial drift, deposited chiefly by direct glacial action.

mudflow a flow of debris lubricated with a large amount of water.

Nevadan orogeny Late Jurassic mountain building.

nonconformity an unconformity where the older rocks are of plutonic origin, that is, are granitic or metamorphic rocks.

orogeny the process of forming mountains, particularly by folding and thrusting.

outcrop exposure of bedrock.

outwash drift deposited by meltwater streams beyond active glacier ice.

overthrust a thrust fault with a low dip and large net slip, generally measured in miles.

paleomagnetism faint magnetic polarization preserved in some rocks that reflects the earth's magnetic field as it existed at the time of the accumulation of the sediment or the solidification of the magma that formed the rock.

paleontology the study of life in past geological ages based on the fossil remains of organisms.

pegmatite igneous rock of relatively coarser grain than the large mass of plutonic rock it cuts as a dike or a vein.

period a major, worldwide standard geologic time unit, a subdivision of an era.

permeability the capacity that a rock has for transmitting a fluid.

petrology the study of the natural history of rocks, including their origins, present conditions, alterations, and decay.

pillow structure the peculiar structure exhibited by some lavas which consists of a group of rounded masses that resemble pillows. This is usually the result of underwater crystallization of the lava.

placer deposit a sediment in which gold, tin, or other heavy metals have been concentrated.

plastic substance a solid that yields by flowage when subjected to a directed stress in excess of a critical level.

plateau basalt those basaltic lavas that occur as vast accumulations of horizontal flows on a regional scale.

playa the shallow central basin of a desert plain, in which water gathers after a rain and is evaporated.

porosity the ratio of the volume of void space in a rock or soil to its total volume.

pressure ridge a large wave-like form on the surface of a congealed lava flow.

pyroclastic material detrital material ejected explosively from a volcano.

radioactive age determination the determination of the time that has elapsed since crystallization of a rock or mineral by study of the ratio or ratios between stable (or radioactive) daughter products and their parent elements.

recessional moraine a moraine formed during a temporary decrease in the rate of glacial retreat.

regional metamorphism large-scale metamorphism unrelated to obvious igneous intrusive bodies.

rock any consolidated, naturally formed mass of mineral matter.

root zone place where a low-angle thrust fault becomes steeper and disappears beneath the surface of the earth.

runoff the discharge of water through surface streams.

sea-floor spreading a process that involves the generation of oceanic crust along the axis of a mid-ocean ridge and its gradual migration away from the axis.

seamount a submarine mountain rising above the ocean floor; usually a volcanic cone.

sedimentary rock rock formed from the accumulation of sediment, which may be rock particles, the remains of plants or animals, or the product of chemical processes.

seismic sea wave (*tsunami*) commonly misnamed "tidal wave"; a long period wave caused by submarine seismic activity or volcanic eruption.

seismology a geophysical science concerned with the study of earthquakes and measurement of the elastic properties of the earth.

shield a continental block of the earth's crust that has been relatively stable over a long period of time.

shield volcano a broad, gently sloping volcanic cone.

sill a flat, relatively thin body of intrusive igneous rock emplaced parallel to the bedding of the intruded rocks.

slickenside polished and scratched rock surface resulting from friction along a fault plane.

sorting as applied to sediment, a measure of the distribution of sizes of the various particles. A well-sorted sediment consists primarily of particles of approximately the same size.

spit a small sandy or pebbly point projecting from the shore into a body of water.

stack a remnant of a headland; a small island near the shore, which was once part of the mainland but has been isolated by wave erosion.

stade a period of increased glacial activity during a longer epoch of glaciation.

stock an igneous intrusion that extends to an unknown depth and has an exposed surface area of less than 40 square miles.

stoping a process whereby blocks of older rock fall into a rising magma and are assimilated into it at depth.

stratigraphy the study of surface-accumulated (that is, sedimentary and volcanic) rocks.

stratovolcano see *composite cone*.

stratum a section of a formation that consists throughout of approximately the same kind of rock material. Plural: *strata*.

structural geology the study of the structural, rather than the compositional, features of rocks, and the geographic distribution of the features and their causes.

subduction a process involving the thrusting of oceanic crust beneath the margin of a continent.

submarine canyon steep valleys in the continental margin.

superposition the order in which rocks are placed above one another.

surficial related to the earth's surface; especially alluvial or glacial deposits lying on bedrock.

syncline a downfold.

talus the collection of fallen material that has formed a slope at the foot of a cliff.

tectonism crustal instability.

terminal moraine a moraine formed by a glacier at its farthest advance.

terminus the outer margin of a glacier.

thin-skinned tectonics deformation of strata at or near the earth's surface without involvement of deeper rocks.

thrust fault a fault with a low angle of inclination in which the upper block moves upward over the underlying block.

thunder egg geodelike body containing opal, agate, or chalcedony weathered out of welded tuff or lava.

till unsorted, unstratified sediment carried or deposited by a glacier. Typically a pebbly clay.

topography the physical features of a region, especially the relief and contour of the land.

trench, oceanic a long, narrow, steep-sided depression of the deep sea floor adjacent to a continental margin.

tuff a rock formed of compacted volcanic ash.

turbidity current a highly turbid and relatively dense current that moves down the bottom slope of a body of standing water.

type locality (*type section*) the place at which a formation is typically displayed and from which it is named.

unconformity a surface of erosion that separates younger strata from older. A time gap in the stratigraphic record.

uniformitarianism the concept that the present is the key to the past.

vein an occurrence of ore disseminated through country rock having regular development in length, width, and depth.

vent an opening at the earth's surface from which volcanic materials are erupted.

vesicle a small cavity in a volcanic rock formed by a gas bubble during solidification. Adjective: *vesicular.*

viscosity a measure of the resistance of a fluid to flow.

volcano a vent in the earth's crust from which volcanic products issue, or a mountain that has been built up by the eruptive products from a vent.

warp a gentle bend in the earth's crust.

water table the upper surface of the zone of underground saturation.

wave-cut bench a level of gently sloping erosional plane formed by wave action.

weathering the disintegration and decomposition of rock exposed to the atmosphere.

welded tuff a tuff that has been consolidated by the combined action of heat retained by the particles and the enveloping hot gases.

Index

A

B

C

D

E

F

G

H

I

J

K

L

M

N

O

P

Q

R

S

T

U

V

W

Y

Z

DATE DUE
